Recent Advances in Polymeric Membranes—Preparation and Applications

Recent Advances in Polymeric Membranes—Preparation and Applications

Guest Editors

María Ortencia González-Díaz
Manuel Aguilar Vega

MDPI

Basel • Beijing • Wuhan • Barcelona • Belgrade • Novi Sad • Cluj • Manchester

Guest Editors
María Ortencia González-Díaz
Materials Unit
Centro de Investigación
Científica de Yucatán A.C.
Merida
Mexico

Manuel Aguilar Vega
Materials Unit
Centro de Investigación
Científica de Yucatán A.C.
Merida
Mexico

Editorial Office
MDPI AG
Grosspeteranlage 5
4052 Basel, Switzerland

This is a reprint of the Special Issue, published open access by the journal *Membranes* (ISSN 2077-0375), freely accessible at: https://www.mdpi.com/journal/membranes/special_issues/0VXE047BK8.

For citation purposes, cite each article independently as indicated on the article page online and as indicated below:

Lastname, A.A.; Lastname, B.B. Article Title. *Journal Name* **Year**, *Volume Number*, Page Range.

ISBN 978-3-7258-6626-7 (Hbk)
ISBN 978-3-7258-6627-4 (PDF)
https://doi.org/10.3390/books978-3-7258-6627-4

Contents

About the Editors

María Ortencia González-Díaz

María Ortencia González-Díaz is a researcher at the Centro de Investigación Científica de Yucatan, specializing in polymeric membranes and their applications in water treatment, gas separation, and biomedical technologies. Her work focuses on membrane design, functionalization strategies, and sustainable fabrication methods. She has contributed to multiple peer-reviewed publications and collaborative projects aimed at advancing membrane science for industrial and biomedical applications.

Manuel Aguilar Vega

Manuel Aguilar Vega is a senior researcher at the Centro de Investigación Científica de Yucatán (CICY), where he is also involved in postgraduate programs in polymeric materials and renewable energy. His work focuses on transport processes in polymeric materials and the design of membranes for gas, vapor, and liquid separations, with applications in industrial gas separation, water treatment, and clean energy technologies. He leads projects on sour gas sweetening, valorization of gas streams, and industrial membrane separations, and has published extensively in peer-reviewed journals on polymer science and membrane technology.

Preface

This Reprint brings together recent contributions on polymeric membranes, highlighting their preparation, functionalization, and applications in water treatment, energy conversion and storage, environmental remediation and biomedical applications. It is aimed at researchers, engineers, industry professionals, and students interested in membrane science. These articles provide an overview of current developments and practical approaches in this versatile field.

María Ortencia González-Díaz and Manuel Aguilar Vega
Guest Editors

membranes

MDPI

Editorial

Recent Advances in Polymeric Membranes—Preparation and Applications

Maria Ortencia González-Díaz [1,2,*] and Manuel Aguilar-Vega [1,*]

1 Centro de Investigación Científica de Yucatán, A.C., Calle 43 No. 130, Chuburná de Hidalgo, Mérida C.P. 97200, Yucatán, Mexico
2 Secretaría de Ciencia, Humanidades, Tecnología e Innovación (SECIHTI), Av. Insurgentes Sur 1582, Col. Crédito Constructor, Demarcación Territorial, Benito Juárez C.P. 03940, Ciudad de Mexico, Mexico
* Correspondence: maria.gonzalez@cicy.mx (M.O.G.-D.); mjav57@gmail.com (M.A.-V.)

1. Introduction

Polymeric membranes have gained increasing importance due to their low energy consumption, ease of operation, and favorable chemical, mechanical, and thermal stability [1]. Their versatility and relatively low cost have positioned them as essential materials in modern separation and purification technologies, motivating continuous innovation to meet increasingly demanding environmental, industrial, and biomedical applications, among others.

Across these diverse fields, recent research has converged on improving membrane robustness, selectivity, and long-term stability, together with advancing more sustainable and energy-efficient fabrication practices. Current efforts include enhancing resistance to fouling, extreme pH, oxidizing agents, and high-pressure operation, as well as incorporating greener solvents and environmentally conscious processing routes [2]. Moreover, there is growing interest in membranes capable of maintaining stable, energy-efficient performance during extended operation, particularly under variable feed conditions and frequent cleaning cycles [3]. Advances in polymer architecture, surface functionalization, and controlled polymerization have expanded the ability to tune membrane porosity, transport behavior, and biocompatibility. These strategies support current progress in water treatment and energy conversion systems, where optimized ionic transport and electrochemical stability are increasingly required, and biomedical technologies [4,5]. In particular, the capacity to engineer biomimetic morphologies allows polymeric membranes to approximate key characteristics of the extracellular matrix (ECM), enabling controlled drug delivery, improved cell–material interactions, and enhanced tissue regeneration.

The development of new polymeric materials and their growing number of applications necessitates the development of better-suited membranes with different characteristics or new composite membranes and methods of preparation, capable of meeting increasingly complex industrial, environmental, and medical demands. Their preparation can be achieved through various techniques and configurations, and ongoing research is continuously broadening the scope of materials and functionalities available. The following Special Issue, *Recent Advances in Polymeric Membranes—Preparation and Applications*, presents eight contributions, including both original research and reviews, that highlight recent progress across multiple fields.

2. Overview of Special Issue Contributions

In the field of water treatment and environmental remediation, the contributions addressed both sustainability and performance improvements in polymeric membranes.

The authors of one study explored the revalorization of recycled high-impact polystyrene (HIPS) into asymmetric membranes for wastewater purification (Contribution 1). Through chemical sulfonation, sulfonic acid groups were successfully incorporated into the polymer structure, significantly improving surface hydrophilicity (with the contact angle decreasing from 83.8° to 66.1°) and enhancing antifouling resistance. As a result, the sulfonated membranes exhibited a marked increase in hydrophilicity and antifouling resistance, leading to a higher flux recovery ratio (FRR) compared with their non-sulfonated counterparts. The R-HIPS-5 membrane also achieved high rejection rates for Reactive Black 5 dye (94%) and divalent salts such as $MgSO_4$ (72%) and Na_2SO_4 (67%), demonstrating that recycled plastics can be transformed into high-performance functional materials for sustainable water treatment. In another study (Contribution 2), the authors evaluated the long-term durability of poly(vinylidene fluoride) (PVDF) ultrafiltration membranes under repeated alkaline cleaning over a period of two years in a pilot-scale setup. Despite progressive pore enlargement and an over 50% increase in permeate flux, the quality of the treated water was preserved due to the formation of a thin fouling-induced separation layer that maintained high rejection of chemical oxygen demand (COD), turbidity, and surfactants. This finding confirms that PVDF membranes possess strong chemical resilience and can sustain efficient operation under industrial cleaning regimes.

Environmental challenges were also addressed in two comprehensive review papers. The authors of the first study (Contribution 3) focused on recent developments in polymer inclusion membranes (PIMs), highlighting advances in selectivity through the use of ionic liquids, task-specific carriers, and polymer functionalization techniques. They also addressed improvements in structural stability achieved with high-performance polymers and cross-linking strategies. Particular attention was paid to the shift toward sustainable fabrication, including the adoption of biodegradable polymers, green solvents, and scalable techniques such as phase inversion and electrospinning. These efforts collectively point to a new generation of PIMs that combine high separation efficiency, mechanical durability, and reduced environmental impact. In the second review (Contribution 4), the authors analyzed the performance of pressure-driven membrane processes, microfiltration (MF), ultrafiltration (UF), nanofiltration (NF), and reverse osmosis (RO), for removing micro- and nanoplastics from diverse aqueous matrices, including wastewater, surface water, and landfill leachate. Reported removal efficiencies reached up to 100%, depending on membrane type, pore size, and hydrodynamic conditions. The authors also discussed how membrane fouling, plastic particle aggregation, and the lack of standardized testing protocols remain major barriers to large-scale implementation. The authors emphasized that future progress will rely on improving fouling control strategies, optimizing membrane surface properties, and integrating membrane systems with complementary technologies such as coagulation or adsorption to enhance process robustness and scalability. Both reviews highlight how polymeric membranes are evolving to meet environmental demands while aligning with principles of sustainability and the circular economy.

In the field of energy applications, the authors of another included article (Contribution 5) focused on quaternized polysulfone membranes as anion-exchange solid polymer electrolytes for zinc–air batteries. This work contributes to the growing interest in polymer-based electrolytes as safer and more stable alternatives to conventional aqueous systems. By optimizing the degree of functionalization, the membranes achieved a hydroxide ionic conductivity of 22.19 $mS \cdot cm^{-1}$ and a peak power density of 70 $mW \cdot cm^{-2}$, both of which surpassed those of the commercial benchmark membrane (Fumapem FAA-3-50). In addition, postmortem analysis revealed reduced zinc oxide formation on the electrode surface, indicating enhanced rechargeability and lower passivation during cycling. These results highlight the critical role of polymer chemistry in advancing energy-conversion and storage

https://doi.org/10.3390/membranes16010003

technologies and demonstrate the potential of solid-state membranes to enable efficient, durable, and environmentally friendly electrochemical devices.

Regarding biomedical applications, the contributions reflected the increasing importance of polymeric membranes in health-related technologies. Electrospun polyacrylonitrile (PAN) membranes incorporating melanin derived from pecan nutshell residues exhibited strong antioxidant, antimicrobial, and enzyme-inhibiting properties, confirming the effectiveness of natural pigments as active fillers in polymer matrices (Contribution 6). The membranes containing up to 5 wt% melanin showed radical scavenging rates exceeding 80% for ABTS and 60% for DPPH assays, demonstrating a marked improvement compared to unmodified PAN fibers. In addition, they inhibited key skin-aging enzymes such as collagenase, tyrosinase, and elastase by up to 37%, 36%, and 33%, respectively, and displayed antibacterial activity against common pathogenic strains. These multifunctional features suggest potential use in skin-care and regenerative applications, illustrating how natural bioactive compounds can be successfully integrated into synthetic membranes to produce sustainable, value-added biomaterials. In another study (Contribution 7), the authors optimized field-emission scanning electron microscopy (FE-SEM) conditions to visualize the delicate porous structure of polypropylene membranes used in extracorporeal membrane oxygenation (ECMO) devices, enabling high-resolution imaging while minimizing morphological damage. The authors of another contribution reviewed the development pathway of a new hydrophilic polysulfone dialyzer, including its design, in vitro characterization, and clinical evaluation (Contribution 8). The modified membrane demonstrated reduced protein adsorption, improved hemocompatibility, and stable performance, illustrating how innovations in surface chemistry can translate into safer and more effective dialysis treatments.

3. Conclusions

Together, the contributions included in this Special Issue demonstrate how polymeric membranes are being advanced to address pressing global challenges. From water purification and environmental remediation to next-generation batteries and biomedical applications, these studies highlight the versatility of polymeric membranes and the continuing innovation in their preparation and application. Despite significant progress, challenges remain in scaling up sustainable fabrication methods, improving long-term stability, and translating laboratory developments into industrial and clinical contexts. We would like to thank all of the authors and reviewers for their valuable contributions, and we hope that this Special Issue will continue to inspire further research and innovation in the field of polymeric membranes.

Author Contributions: The authors equally contributed to the Editorial. All authors have read and agreed to the published version of the manuscript.

Acknowledgments: We thank the editorial board of *Membranes* for their invaluable support.

Conflicts of Interest: The authors declare no conflicts of interest.

List of Contributions:

1. Yam-Cervantes, M.; Sulub-Sulub, R.; Hunh-Ibarra, M.; Duarte, S.; Uc-Fernandez, E.; Pérez-Canales, D.; Aguilar-Vega, M.; González-Díaz, M.O. Asymmetric Membranes Obtained from Sulfonated HIPS Waste with Potential Application in Wastewater Treatment. *Membranes* **2024**, *14*, 247. https://doi.org/10.3390/membranes14120247.
2. Gryta M.; Woźniak, P.; Impact of Long-Term Alkaline Cleaning on Ultrafiltration Tubular PVDF Membrane Performances. *Membranes* **2025**, *15*, 192. https://doi.org/10.3390/membranes15070192.

 https://doi.org/10.3390/membranes16010003

3. Nowik-Zając, A.; Sabadash, V. Recent Developments in Polymer Inclusion Membranes: Advances in Selectivity, Structural Integrity, Environmental Applications and Sustainable Fabrication. *Membranes* **2025**, *15*, 249. https://doi.org/10.3390/membranes15080249.
4. Pinto, P.E.; Giacobbo, A.; Almeida, G.M.d.; Rodrigues, M.A.S.; Bernardes, A.M. Pressure-Driven Membrane Processes for Removing Microplastics. *Membranes* **2025**, *15*, 81. https://doi.org/10.3390/membranes15030081.
5. Salazar-Gastélum, L.J.; Arredondo-Espínola, A.; Pérez-Sicairos, S.; Álvarez-Contreras, L.; Arjona, N.; Guerra-Balcázar, M. Quaternized Polysulfone as a Solid Polymer Electrolyte Membrane with High Ionic Conductivity for All-Solid-State Zn-Air Batteries. *Membranes.* **2025**, *15*, 102. https://doi.org/10.3390/membranes15040102.
6. García-García, M.; Jaime-Ferrer, J.S.; Medrano-Lango, F.N.; Quintana-Rodríguez, E.; Campos-García, T.; Rodríguez-Sevilla, E.; Orona-Tamayo, D. Electrospun Membranes Loaded with Melanin Derived from Pecan Nutshell (*Carya illinoinensis*) Residues for Skin-Care Applications. *Membranes* **2025**, *15*, 44. https://doi.org/10.3390/membranes15020044.
7. Fukuda, M.; Nishite, Y.; Murata, E.; Namekawa, K.; Mori, T.; Tanaka, T.; Sakai, K. Determining Accurate Pore Structures of Polypropylene Membrane for ECMO Using FE-SEM Under Optimized Conditions. *Membranes* **2025**, *15*, 174. https://doi.org/10.3390/membranes15060174.
8. Zawada, A.M.; Griesshaber, B.; Ottillinger, B.; Erlenkötter, A.; Crook, N.; Boyington, S.; Stauss-Grabo, M.; Kennedy, J.P.; Lang, T. Development and Investigation of a New Polysulfone Dialyzer with Increased Membrane Hydrophilicity. *Membranes* **2025**, *15*, 132. https://doi.org/10.3390/membranes15050132.

References

1. Hadi, M.K.; Wang, X.; Peng, Y.; Sangaraju, S.; Ran, F. Functional Polymeric Membrane Materials: A Perspective from Versatile Method and Modification to Potential Applications. *Polym. Sci. Technol.* **2025**, *1*, 366–412. [CrossRef]
2. Sean, M.Q.; Chua, S.F.; Ang, W.L.; Lau, W.J.; Mansourizadeh, M.; Thamaraiselvan, C. Advancements in polymeric membranes for challenging water filtration environments: A comprehensive review. *J. Environ. Chem. Eng.* **2024**, *12*, 112628. [CrossRef]
3. Ezugbe, E.O.; Rathilal, S. Membrane Technologies in wastewater Treatment: A review. *Membranes* **2020**, *10*, 89. [CrossRef] [PubMed]
4. Eniola, J.O.; Kujawa, J.; Nwokoye, A.; Al-Gharabli, S.; Avornyo, A.K.; Giwa, A.; Amusa, H.K.; Yusuf, A.O.; Okolie, J.A. Advances in electrochemical membranes for water treatment: A comphensive review. *Desalin. Water Treat.* **2024**, *319*, 100450. [CrossRef]
5. Hazarika, G.; Jadhav, S.V.; Ingole, P.G. Exploring the potential of polymeric membranes in cutting-edge chemical and biomedical applications: A review. *Mater. Today Commun.* **2024**, *39*, 109022. [CrossRef]

membranes

MDPI

Review

Recent Developments in Polymer Inclusion Membranes: Advances in Selectivity, Structural Integrity, Environmental Applications and Sustainable Fabrication

Anna Nowik-Zając [1,*] and Vira Sabadash [2]

1 Faculty of Science and Technology, Jan Dlugosz University in Czestochowa, Armii Krajowej 13/15, PL 42200 Czestochowa, Poland
2 Department of Ecology and Sustainable Environmental Management, Lviv Polytechnic National University, UA 79000 Lviv, Ukraine; vira.v.sabadash@lpnu.ua
* Correspondence: a.zajac@ujd.edu.pl; Tel.: +48-343614918

Abstract: Polymer inclusion membranes (PIMs) have undergone substantial advancements in their selectivity and efficiency, driven by their increasing deployment in separation processes, environmental remediation, and sensing applications. This review presents recent progress in the development of PIMs, focusing on strategies to enhance ion and molecule selectivity through the incorporation of novel carriers, including ionic liquids and task-specific extractants, as well as through polymer functionalization techniques. Improvements in mechanical and chemical stability, achieved via the utilization of high-performance polymers such as polyvinylidene fluoride (PVDF) and polyether ether ketone (PEEK), as well as cross-linking approaches, are critically analyzed. The expanded application of PIMs in the removal of heavy metals, organic micropollutants, and gas separation, particularly for carbon dioxide capture, is discussed with an emphasis on efficiency and operational robustness. The integration of PIMs with electrochemical and optical transduction platforms for sensor development is also reviewed, highlighting enhancements in sensitivity, selectivity, and response time. Furthermore, emerging trends towards the fabrication of sustainable PIMs using biodegradable polymers and green solvents are evaluated. Advances in scalable manufacturing techniques, including phase inversion and electrospinning, are addressed, outlining pathways for the industrial translation of PIM technologies. The review concludes by identifying current limitations and proposing future research directions necessary to fully exploit the potential of PIMs in industrial and environmental sectors.

Keywords: polymer inclusion membranes; carrier-mediated transport; green membrane technologies; deep eutectic solvents; ionic liquids; heavy metal removal; gas separation; membrane stability; analytical sensing membranes; sustainable separation processes

1. Introduction

1.1. Overview of Polymer Inclusion Membranes (PIMs)

Polymer inclusion membranes (PIMs) are a unique type of liquid membrane system. They involve the trapping of a selective carrier and often a plasticizer within a solid polymer matrix. This setup creates a hybrid structure that combines the high selectivity and transport methods typical of supported liquid membranes (SLMs) with the strength and versatility of polymer films. Since they were first developed in the early 1990s, PIMs have become useful materials for various separation and sensing tasks, especially in water treatment, hydrometallurgy, gas purification, and analytical chemistry.

In a typical PIM, the active transport agent (carrier) is evenly spread throughout a polymer matrix. This matrix may include substances such as cellulose triacetate (CTA), polyvinyl chloride (PVC), or polyvinylidene fluoride (PVDF). A plasticizer is sometimes added to improve flexibility and diffusion. The carrier interacts selectively and reversibly with the target species at the membrane surface. This interaction allows transport through cycles of complexation and decomplexation. This carrier-mediated transport method allows for precise adjustments in selectivity by changing the carrier chemistry, the membrane's composition, and interactions between polymer and carrier. The structure of a PIM is illustrated in Figure 1.

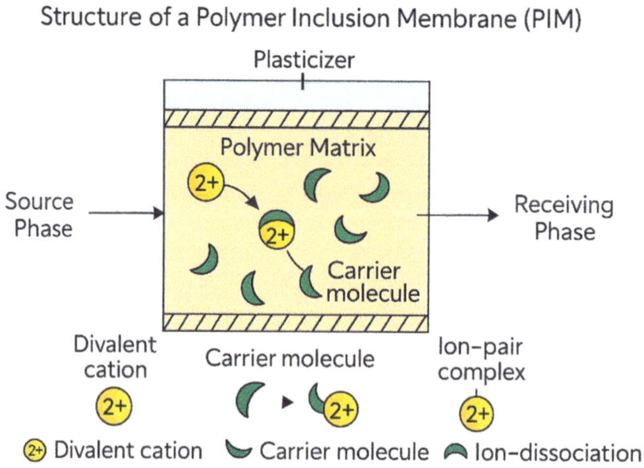

Figure 1. Schematic diagram of the structure of PIMs.

While polymer inclusion membranes have emerged as a promising class of selective separation materials, it is essential to evaluate their performance in comparison with conventional membrane technologies such as reverse osmosis (RO), nanofiltration (NF), and ceramic membranes. These mature techniques have long dominated the industrial separation landscape due to their high throughput, mechanical robustness, and scalability. Reverse osmosis and nanofiltration are pressure-driven processes commonly used in water purification and desalination. While effective for the bulk separation of ions and molecules, they often face challenges such as high energy consumption, fouling, and limited selectivity for specific target species. RO membranes exhibit nearly complete ion rejection but lack the tunable specificity typical of PIMs. Additionally, their operation usually requires high pressures of up to 60 bar, which can be a limiting factor in resource-constrained environments. PIMs function at ambient pressure. They can be customized for highly selective separations by using specific carriers designed for targeted ions. For example, Aliquat 336 is effective for separating chromium (VI), while D2EHPA is used for rare-earth elements [1,2]. This selectivity enables the efficient extraction of trace contaminants or valuable metals from complex matrices where RO and NF would indiscriminately remove both desired and interfering species. For instance, Uğur et al. [3] demonstrated the selective separation of Zn^{2+} and Cd^{2+} using a PIM that contains calix[4]resorcinarene. They achieved separation factors that cannot be obtained with conventional membranes. Furthermore, PIMs are less prone to biofouling due to their dense structures and hydrophobic nature. Their regeneration can be accomplished simply through pH manipulation instead of backflushing or chemical cleaning, thereby reducing operational costs and complexity. While PIMs generally exhibit lower flux than RO or NF membranes, their advantages in selectivity and ease of operation make them particularly well suited for analytical applications, resource recovery, and selective remediation [4]. Ceramic membranes are well known for their outstanding resistance to heat and chemicals, making them highly effective in challenging environments. They are commonly used in applications such as hot gas filtration, chemical wastewater treatment, and food processing [5]. These membranes can function at temperatures exceeding 300 °C and can withstand highly acidic or basic conditions, surpassing the durability of many polymeric materials. However, they exhibit low flexibility and high brittleness and often require significant capital investment for fabrication and operation due to the sintering process and use of metal oxide precursors (e.g., alumina and zirconia). PIMs generally exhibit lower stability under extreme conditions; however, they provide high selectivity for target analytes due to the inclusion of task-specific extractants and functional carriers. For example, Aliquat 336-based PIMs have demonstrated excellent selectivity for Cr(VI) and As(V) from real wastewater samples under ambient conditions, a property that ceramic membranes cannot exhibit without pre-treatment or functionalization steps [6–8]. Moreover, ceramic membranes typically perform size-based (sieving) separations (micro-, ultra-, and nanofiltration). At the same time, PIMs operate through facilitated transport mechanisms, allowing for the separation of ions with similar sizes and charges—a significant advantage in rare-earth element recovery and analytical preconcentration [9]. Ceramic membranes are not naturally compatible with hydrophobic carriers, such as calixarenes or organophosphorus compounds, because of their inorganic surface chemistry. In contrast, polymer inclusion membranes (PIMs) allow for the uniform dispersion of these extractants within the polymer matrix. This capability enables the recognition and transport of specific ions [10–13].

Polymer inclusion membranes offer several advantages over traditional separation methods [14,15]. They facilitate transport through carrier-mediated diffusion and can operate under ambient pressure and temperature conditions. Additionally, they can be easily customized by chemically modifying the polymer matrix or carrier molecule. PIMs also have low energy requirements and possess intrinsic selectivity, making them well suited for tasks that involve trace-level separation [16]. This includes applications like extracting heavy metals, removing pharmaceuticals, and concentrating analytes for analysis. Furthermore, PIMs can be regenerated simply by changing the phase conditions, eliminating the need for harsh cleaning protocols that are often required in reverse osmosis (RO) or nanofiltration (NF) systems [17]. While PIMs have unique advantages, they also face several challenges, including mechanical strength, long-term stability, and scaling up for industrial use. These issues need to be addressed before PIMs can compete effectively with established membrane technologies. To enhance their performance, hybrid approaches are being explored, such as incorporating nanofillers or utilizing high-performance polymers like PVDF and PEEK [18]. This comparative evaluation emphasizes the unique role of PIMs within the spectrum of membrane technology. Although PIMs are not intended to replace high-throughput systems like reverse osmosis in bulk applications, they provide exceptional selectivity and sustainability for fine separations where traditional membranes are less effective.

Despite their unique advantages, PIMs face several real limitations that must be addressed for wider industrial use. Although there has been remarkable progress in the design and application of PIMs, various practical challenges hinder their large-scale and long-term deployment. One of the main concerns is long-term fouling, particularly when PIMs are used with complex feed matrices such as industrial effluents, wastewater, or biological fluids [19,20]. The accumulation of organic matter, microbial growth, and particulate deposition can block membrane pores, reduce transport efficiency, and ultimately shorten the membrane's lifespan [21]. While surface modification techniques like PEGylation and fluorination have been explored to reduce fouling, their long-term effectiveness under real-world conditions remains uncertain [21–23]. A second critical issue is carrier leaching, particularly in systems where the carrier is physically trapped rather than chemically bonded to the polymer matrix. This problem worsens under dynamic flow conditions or prolonged operation in aggressive media, such as acidic or basic solutions. Carrier leaching not only undermines the selectivity and durability of the membrane, but it also raises environmental concerns due to the potential release of toxic extractants into the treated medium. Operational stability continues to be a significant challenge. Many polymer inclusion membranes perform exceptionally well in controlled laboratory conditions but tend to suffer a decline in performance when exposed to real-world stressors, such as varying temperatures, ionic strengths, pH levels, and continuous pressure. Biodegradable polymer matrices, like polyvinyl alcohol (PVA) or polylactic acid (PLA), are environmentally friendly, but they may experience hydrolytic degradation or a loss of mechanical integrity over long-term use [24]. Additionally, the economic viability of advanced PIM designs is hindered by the high cost of ionic liquids (ILs), particularly task-specific or highly pure types. Although ILs provide enhanced stability, tunable selectivity, and negligible volatility, their synthesis and purification processes are both expensive and energy-intensive. Prices for aprotic ILs commonly used in membrane systems range from approximately USD 2.50 to 50 per kg, significantly higher than those for conventional organic solvents [25]. This financial burden restricts their application in large-volume or low-margin scenarios unless they can be efficiently recovered and reused. Moreover, the incorporation of nanofillers or hybrid components (e.g., MOFs, COFs and carbon nanotubes) also increases the material costs and complexity of fabrication [26]. Addressing these limitations requires a multidisciplinary effort involving material innovation, scalable green fabrication strategies, and the development of standardized testing protocols under realistic operational scenarios. As the field matures, future research must strike a balance between performance, sustainability, and cost to enable the commercial viability of PIM-based systems in diverse separation processes.

In recent years, significant research has focused on enhancing the selectivity, chemical stability, and sustainability of PIMs. A key advancement has been the addition of task-specific ionic liquids (ILs) as carriers. These salts, which are liquid at room temperature, have low volatility, high thermal stability, and adjustable physical and chemical properties. For instance, Hernández-Fernández et al. (2023) showed that using imidazolium-based ILs in PIMs significantly improved the selective separation of heavy metal ions, such as Pb^{2+} and Zn^{2+}. This is due to strong coordination between the IL cations and anions [27]. Similarly, in a 2024 review titled "Ionic liquid-based extraction of metal ions via polymer inclusion membranes: a critical review", Adigun et al. discussed the strong attraction of IL-functionalized PIMs for metal ions and the potential for adjusting selectivity via IL functionalization [28].

In addition to ionic liquids, organophosphorus extractants such as di-(2-ethylhexyl) phosphoric acid (D2EHPA), Aliquat 336, and Cyanex 301 are widely used as carriers. They are effective in transporting metal ions. A recent study by Kazemi and Yaftian (2024) reported the development of a PVDF-HFP-based PIM that includes D2EHPA. This PIM showed excellent selectivity and stability for the extraction of Bi^{3+} ions, maintaining high transport efficiency over multiple reuse cycles [29].

Concurrently, progress in polymer science has enabled the design of PIMs with improved structural integrity and enhanced resistance to degradation. High-performance polymers, such as polyether ether ketone (PEEK), PVDF, and CTA, have shown superior chemical and thermal resistance compared to traditional matrices like PVC. Their use has increased the operational lifespan of PIMs, especially in harsh industrial or environmental conditions. Additionally, polymer cross-linking and blending with hydrophilic additives have strengthened the membranes, decreased brittleness, and reduced carrier leaching.

Beyond performance improvements, recent research has focused on the sustainable production of PIMs. An increasing amount of literature explores the use of biodegradable polymers (e.g., polylactic acid, chitosan), along with green solvents like ethanol, ethyl acetate, and Cyrene™, as ecofriendly alternatives to traditional toxic solvents, such as tetrahydrofuran (THF) and dichloromethane. In a 2025 study titled "Polymer Inclusion Membranes (PIMs) for Metal Separation-Toward Environmentally Friendly Production and Applications", Senila et al. reviewed the integration of green chemistry principles into PIM production, including solvent replacement, bio-based polymers, and recyclable components [30].

The application scope of PIMs has expanded considerably in recent years. In environmental remediation, PIMs have shown effectiveness in the selective removal of heavy metals (e.g., Cd^{2+}, Cu^{2+}, Zn^{2+} and Ni^{2+}) from industrial wastewater, acid mine drainage, and leachates from electronic waste recycling. Their high separation factors, low energy requirements, and reusability make them attractive options to traditional methods. For example, polymer inclusion membranes based on Aliquat 336 embedded in CTA have shown over 95% removal efficiency for Zn^{2+} from complex aqueous matrices, while maintaining structural stability across ten extraction cycles [31].

Analytical platforms have also been integrated with PIMs to preconcentrate and speciate metal ions, pharmaceuticals, and organic pollutants. The immobilized carrier environment enables the enrichment of trace analytes, improving detection sensitivity in optical and electrochemical sensors. Recent developments include optode-based PIM sensors with colorimetric responses to mercury(II) and silver(I), demonstrating rapid and selective detection in field conditions [32].

The potential of PIMs in gas separation, particularly for CO_2 capture, is gaining attention. Ionic liquid-modified membranes have shown good results in selectively transporting CO_2 over N_2 or CH_4, offering new possibilities for their application in flue gas treatment and carbon capture technologies. Although gas-selective PIMs are still in the early stages of commercialization, initial studies highlight their scalability and compatibility with low-pressure operations.

Scalable fabrication methods such as phase inversion, solution casting, and electrospinning are currently being explored. These methods aim to shift PIMs from laboratory prototypes to industrial modules. They allow for control over the thickness, porosity, and morphology of the membranes, which affects their transport kinetics and durability. Combining modular units and hybrid membrane systems represents an exciting area in PIM development, particularly for continuous-flow processes.

In summary, polymer inclusion membranes are in a transformative stage marked by the careful design of carriers, sustainable fabrication approaches, and interdisciplinary applications. Their adjustable selectivity, operational stability, and potential for green engineering make them a flexible technology for separation science in the 21st century. The following sections of this review will provide more details on these advancements, focusing on the specific strategies, materials, and applications that are reshaping the scope and potential of PIMs.

1.2. Historical Development and Current Research Trends

The concept of polymer inclusion membranes (PIMs) emerged from the broader field of liquid membrane technologies, which began in the mid-20th century. In 1963, Bloch proposed immobilizing liquid extractants in solid polymer supports, enabling selective separations without the phase instability associated with traditional solvent extraction systems [33]. This basic concept set the stage for the development of PIMs. In the 1990s, researchers such as Almeida, Cattrall, and Kolev defined the PIM structure by adding carriers like Aliquat 336 or D2EHPA into polymers such as cellulose triacetate (CTA) or polyvinyl chloride (PVC) [2,9]. They used a carrier-mediated transport mechanism. The analyte would form a reversible complex with the carrier at the source phase, diffuse through the membrane, and decomplex at the receiving phase. This process allowed for the targeted transport of specific species.

During the 2000s, the research focus shifted to mechanistic understanding and optimization. Studies characterized how polymer–carrier ratios, plasticizer concentration, and matrix hydrophobicity influenced membrane permeability, selectivity, and reusability [34]. Researchers also began exploring alternative carriers, such as calixarenes and phosphonic acids, enabling the separation of rare-earth elements and toxic heavy metals under environmentally relevant conditions.

From 2010 onward, PIM applications broadened substantially, particularly into environmental remediation and analytical chemistry. This period saw the emergence of PIMs for removing dyes, pharmaceuticals, and heavy metals from real wastewater matrices. Mechanistic models based on Nernst-Planck equations and Fickian diffusion helped quantify multi-ion systems and predict transport behavior [8]. At the same time, chemically resistant polymers such as PVDF, PEEK, and modified CTA improved durability under harsh chemical conditions.

A breakthrough in PIM technology occurred in the early 2020s: the introduction of ionic liquids (ILs) and deep eutectic solvents (DESs) as task-specific carriers. These carriers offered better chemical adjustment, thermal stability, and selectivity for both inorganic and organic targets. In a detailed 2024 review, Adigun et al. noted that IL-based PIMs are superior in terms of extraction capacity, chemical resistance, and long-term reusability, especially when handling complex effluents [28]. A similar wave of innovation focused on using DES-based carriers, leveraging their biodegradability and lower toxicity. Simultaneously, sustainability in PIM fabrication became a central research priority. A 2025 review by Senila et al. summarized advances in green PIM production, including the use of biodegradable polymers (e.g., PLA and chitosan), biofillers (e.g., acetylated lignin), and green solvents (e.g., Cyrene™, 2-MeTHF, and ethyl acetate) [30]. Studies have demonstrated that these materials could maintain high transport efficiencies while reducing the environmental footprint of membrane production. For instance, Kaczorowska et al. (2022) fabricated CTA-based PIMs incorporating lignin for Ni(II) recovery with industrial wastewater relevance [35].

At the same time, structural and functional improvements to PIMs gained traction. Kazemi and Yaftian (2024) developed PVDF-HFP membranes incorporating D2EHPA, which exhibited a transport efficiency of >90% Bi(III) and strong operational stability over multiple reuse cycles [29]. Other innovations included cross-linked systems, nanocomposites, and biofiller-enhanced matrices that combined performance with material resilience [36]. A particularly active research trend is the integration of PIMs with sensing technologies. Colorimetric and electrochemical PIM-based sensors have been successfully developed for detecting mercury(II), silver(I), and phosphate ions, showing fast response times and high sensitivity [35,37,38]. These sensors are being adapted for field-deployable monitoring systems with environmental and biomedical applications.

In parallel, advances in scalable fabrication techniques such as phase inversion, electrospinning, and solvent casting enable industrial translation. These methods enable precise control over membrane thickness, porosity, and mechanical strength, which are crucial for continuous-flow operation and long-term use in real-world systems. Figure 2 shows a historical timeline of PIM development, giving a visual overview of this progression. Table 1 provides a detailed breakdown of important periods, innovations, and representative publications.

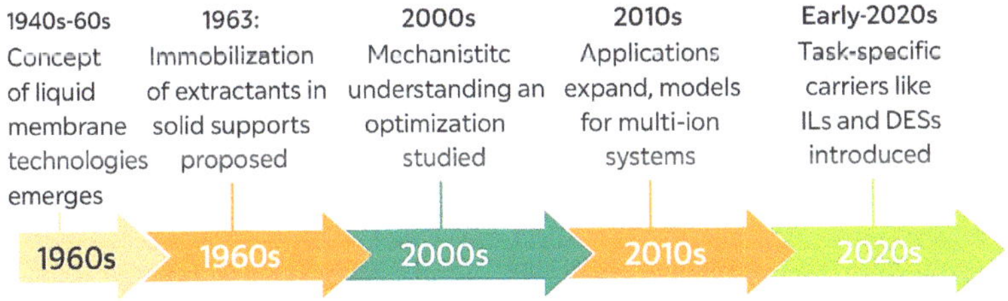

Figure 2. Historical Timeline of PIM Development.

Table 1. Key Milestones in PIM Development and Representative Publications.

Period	Key Developments	Representative Publications
1960s–1980s	Conceptual foundation; immobilized extractants; early SLMs	Bloch et al. [33]
1990s	Formalization of PIMs; use of CTA/PVC; Aliquat 336 and D2EHPA	Almeida et al. [9], Hakami et al. [11], Zielińska et al. [12]
2000s	Mechanistic studies; polymer-carrier optimization; initial modeling	Nghiem et al. [39], Cattrall et al. [40]
2010–2015	Broader applications in analytical/environmental science	Gherasim et al. [41], Kolev et al. [42,43]
2016–2020	ILs and DESs introduced as carriers; modeling transport of complex ions	Hernández-Fernández et al. [27], Matsumoto et al. [44], Kończyk et al. [45]
2021–2022	Green solvents and biodegradable polymers; sensor integration	Senila et al. [30], Sánchez-Ponce et al. [46]
2023–2025	Scalable production; biofillers; gas separation and industrial wastewater use	Adigun et al. [28], Kazemi and Yaftian [29], Kaczorowska et al. [35]

This development shows that PIMs have changed from basic solvent-immobilization systems to customizable, sustainable, and increasingly scalable materials. The combination of green chemistry, new materials, and process engineering ensures that PIMs will remain important in separation science and sustainable technologies in the future.

2. Advances in the Selectivity and Functionalization of PIMs

Selectivity is the defining characteristic of polymer inclusion membranes (PIMs), distinguishing them from other separation technologies. Selectivity can be dramatically enhanced by two main strategies: (1) designing advanced carriers such as ionic liquids (ILs), deep eutectic solvents (DESs), and task-specific extractants, and (2) modifying the chemical and physical structure of the polymer matrix. These complementary methods provide synergistic control over transport behavior, enabling PIMs to selectively extract target species from increasingly complex mixtures, such as industrial leachates, environmental waters, and biological fluids.

2.1. Novel Carriers and Extractants: Ionic Liquids, Task-Specific Molecules

The transport carrier is essential to a polymer inclusion membrane (PIM). It plays a direct role in recognizing, complexing, and moving target species. In recent years, there has been significant progress in this field. This is especially true for the use of ionic liquids

(ILs), deep eutectic solvents (DESs), and task-specific organic extractants. These offer customizable chemical environments that dramatically enhance selectivity.

Ionic liquids are room-temperature molten salts composed of organic cations—such as imidazolium, pyridinium, or phosphonium—and large, non-coordinating anions like $[PF_6]^-$ or $[NTf_2]^-$. Their structural diversity and negligible volatility make them attractive alternatives to traditional liquid extractants. When embedded in PIMs, ILs can provide tailored ion–pair interactions and controlled diffusion environments. For example, membranes formulated with 1-octyl-3-methylimidazolium bis(trifluoromethanesulfonyl)imide ($[C_8mim][NTf_2]$) and Cyanex 301 in a PVDF-HFP matrix demonstrated high selectivity for Co^{2+} and Ni^{2+}, enabling over 90% recovery of Co^{2+} from lithium-ion battery leachate. The separation factor (α) for Co^{2+} over Li^+ exceeded 30, making this system highly relevant for the recovery of critical metals from electronic waste [27].

In addition to imidazolium-based ILs, phosphonium ionic liquids, such as trihexyl(tetradecyl)phosphonium chloride ($[P_{66614}][Cl]$), have emerged as powerful carriers for soft acid metals like Ag^+ and Hg^{2+}. Their large hydrophobic cations provide a highly non-polar environment that facilitates selective complexation through soft—soft Lewis interactions. Importantly, phosphonium ILs exhibit high thermal and chemical stability, making them suitable for challenging aqueous matrices and industrial effluents [28].

A growing class of carrier media is deep eutectic solvents (DESs). These are formed by the eutectic mixture of a hydrogen bond donor and acceptor (e.g., choline chloride with glycerol, urea, or ethylene glycol). DESs have many useful properties similar to ILs but are usually easier to synthesize, biodegradable, and more cost-effective. DES-based PIMs have shown potential in selectively separating of Cd^{2+}, Zn^{2+}, and Pb^{2+}, performing comparably to IL-based systems. These systems are particularly beneficial in low-resource settings or when the sustainability of membranes is important [47]. DESs can also enhance membrane antifouling properties, consistent with trends described in Kandel et al. (2025) in natural antifouling agents in membrane systems [48]. The use of these solvents not only facilitates selective ion separation but also reduces biofouling through various mechanisms, including the modification of hydrophilic surfaces, disruption of bacterial adhesion, and generation of osmotic stress in microbial cells. In addition, DESs composed of naturally derived constituents such as betaine, glycerol, or lactic acid exhibit antimicrobial activity, further extending their potential role in sustainable membrane engineering. This dual functionality makes DES-based carriers highly promising for developing multifunctional PIMs suitable for complex wastewater matrices.

Another group of extractants that has attracted significant interest includes task-specific molecules, such as calixarenes, crown ethers, and phosphorylated ligands. These molecules provide size-selective and highly directional complexation mechanisms, resembling processes of biological recognition. For example, calix[4]arene derivatives bearing phosphine oxide groups have been incorporated into CTA-based membranes for the separation of rare-earth elements, achieving separation factors > 100 for La^{3+} over Eu^{3+} [49,50]. Their rigid cavity structure and ability to form stable inclusion complexes make them ideal for high-precision separation. On the other hand, crown ethers are effective for alkali metal cations, particularly in PIMs where ion size and hydration shell properties matter. A membrane containing 18-crown-6 and Aliquat 336 showed high selectivity between Na^+ and K^+. This provides precise control in systems such as nutrient recovery or water softening [50]. Importantly, greener carriers such as DESs, task-specific ILs, and bio-based extractants are not merely auxiliary components but are equally vital as membrane-support materials for achieving comprehensive sustainability. While polymer matrices dictate the mechanical properties and stability of PIMs, it is the carrier or extractant that governs transport selectivity, efficiency, and reusability. A poorly selected extractant, even when combined with a green polymer, may jeopardize environmental safety due to its toxicity, potential for bioaccumulation, or leaching. Thus, carriers must be optimized for both their functionality and ecological compatibility.

The effectiveness of these carriers depends not only on their chemical attraction but also on their compatibility with the polymer matrix, their diffusion coefficients, and their tendency to leach into nearby phases. Carriers with high molecular weights or lipophilicity tend to remain immobilized within the membrane, maintaining long-term performance. Meanwhile, functional groups capable of hydrogen bonding or ion pairing ensure the reversibility of complexation, which is essential for facilitated transport. Natural or semi-synthetic carriers derived from renewable resources present promising benefits as safer alternatives. For instance, cyclodextrins can selectively encapsulate hydrophobic organic pollutants through host–guest interactions. In the context of PIMs, CDs are being studied as bio-based carriers or functional modifiers that enhance the selective extraction of target analytes through host–guest interactions and surface adsorption mechanisms. For example, β-cyclodextrin-based membranes have been successfully utilized for removing endocrine-disrupting compounds (e.g., bisphenol A), phenolic pollutants, and certain pharmaceuticals from aqueous solutions [51]. Chemical modifications can be applied to CDs (cyclic compounds) to enhance their properties. For example, they can be carboxymethylated or amino-functionalized to introduce donor atoms that can coordinate with metal ions such as Pb^{2+}, Cd^{2+}, and Cu^{2+}. This coordination occurs through ion–dipole or hydrogen bonding interactions [52–55]. These modifications have been shown to improve both the selectivity and stability of the membrane phase. Cyclodextrins are highly versatile and adhere to the principles of green chemistry. They are biodegradable, renewable (as they are produced from the enzymatic degradation of starch), and have low environmental toxicity. Integrating cyclodextrins into polymer inclusion membranes (PIMs) has been shown to minimize leaching and enhance the recyclability of the membranes [56]. Moreover, cyclodextrins are highly compatible with various eco-friendly polymer matrices, such as chitosan, poly(lactic acid), and poly(vinyl alcohol). This compatibility not only improves the environmental profile of the membranes but also strengthens their structural integrity [57]. Additionally, modified tannins or polyphenols are capable of chelating heavy metals. In the study by Nowik-Zając et al., calix[4]pyrrole derivatives showed exceptional selectivity for Ag^{+} ions through anion coordination mechanisms, demonstrating fast transport kinetics and minimal leaching [58]. Similarly, calixresorcinarene-based carriers achieved over 90% removal of Pb^{2+}, exhibiting high reusability and aligning with both performance and green chemistry objectives [58,59]. The effectiveness of task-specific ionic liquids (ILs) and deep eutectic solvents (DESs) in the context of plastic-injecting molding (PIM) can be significantly improved through careful molecular design. By incorporating functional groups such as carboxylates, thiols, or phosphoryl moieties into the structures of ILs or DESs, we can achieve targeted coordination with soft or borderline metal cations. This approach allows for the fine-tuning of extraction behavior while maintaining low volatility and high thermal stability. Moreover, when coupled with biodegradable matrices such as PLA or chitosan, these carriers form fully integrated membrane systems with minimized environmental footprints. Looking ahead, research may benefit from synergistically combining bio-based carriers with natural antifouling agents. For instance, embedding DESs that contain antimicrobial organic acids or amino acids into membranes functionalized with plant-derived biopolymers could result in dual-function systems—simultaneously selective and resistant to microbial contamination. This concept, as proposed by Kandel et al., represents a forward-thinking strategy for water and wastewater treatment membranes that fulfill both operational and ecological performance metrics [48]. Table 2 provides a comparison of the most relevant carriers and their selective behavior toward different target ions. This shows the variety of options available to researchers for customizing PIM functionality for specific applications.

Table 2. Representative Task-Specific Carriers and Selectivity Profiles in PIMs.

Carrier Type	Example	Target Ion(s)	Observed Selectivity	Ref.
Imidazolium Ionic Liquid	[C_8mim][NTf_2] + Cyanex 301	Co^{2+}, Ni^{2+}	$\alpha(Co^{2+}/Li^{+}) > 30$	[27,28,60–63]
Phosphonium Ionic Liquid	[P_{66614}][Cl]	Ag^{+}, Hg^{2+}	High selectivity for soft metals	[28]
Deep Eutectic Solvent (DES)	Choline chloride + glycerol	Cd^{2+}, Zn^{2+}, Pb^{2+}	Comparable to ILs; high for Cd^{2+}, Pb^{2+}	[47,48,64]
Task-Specific Macrocycle	Calix[4]arene-phosphine oxide	La^{3+}, Eu^{3+}	$\alpha(La^{3+}/Eu^{3+}) > 100$	[49,50]
Crown Ether System	18-crown-6 + Aliquat 336	Na^{+}, K^{+}	Size-specific selectivity	[50]
Bio-based Carrier (Cyclodextrin)	Carboxymethyl-β-cyclodextrin	BPA, Phenols, Pb^{2+}, Cd^{2+}	Strong host–guest interaction	[51–57]
Task-Specific Macrocycle	Calix[4]pyrrole derivative	Ag^{+}	High affinity via anion coordination	[58]
Task-Specific Macrocycle	Calixresorcinarene derivative	Pb^{2+}	>90% removal, reusable	[59]

The future direction of carrier development for PIMs is moving toward multifunctional and stimuli-responsive extractants, including smart ILs capable of pH- or redox-controlled binding and biomimetic carriers inspired by metalloproteins. In parallel, efforts are underway to minimize carrier leaching and toxicity through covalent immobilization or encapsulation within nanoporous frameworks.

These innovations in carrier chemistry have improved the selectivity and efficiency of PIMs. They have expanded their application scope to include critical metal recovery, nutrient recycling, pharmaceutical separation, and sensing technologies, positioning PIMs as a versatile and sustainable option in modern separation science.

2.2. *Polymer Matrix Functionalization and Its Impact on Selectivity*

While the transport carrier is responsible for analyte recognition, the polymer matrix in which it is immobilized plays an equally critical role in defining the physical and chemical environment for transport. The structure and functionality of the matrix influence membrane permeability, ion–carrier interaction frequency, and even the system's long-term stability. Thus, tuning the polymer matrix is a key strategy for enhancing the selectivity of PIMs, particularly when dealing with multicomponent mixtures or low-concentration analytes.

A major direction in matrix optimization involves introducing functional groups into the polymer backbone. The sulfonation of standard matrices, such as PVC and CTA, introduces negatively charged sulfonate (-SO_3^{-}) groups, which improve electrostatic attraction toward multivalent cations and facilitate diffusion. Studies have shown that sulfonated CTA, membranes yield significantly higher fluxes and improved separation factors for metal ions such as Pb^{2+}, Cu^{2+}, and Cr^{3+} compared to their non-sulfonated analogs [65].

Another functionalization pathway is carboxylation, which is performed by grafting maleic anhydride or blending with poly(acrylic acid). This modification adds carboxylic acid (-COOH) groups with strong metal-chelating properties. Carboxylated membranes have been especially effective in the selective transport of rare-earth elements and transition metals under acidic conditions. They improve coordination and membrane stability [66,67].

In addition to chemical groups, adding nanomaterials into a polymer matrix has become an important method for modifying membrane structure and selectivity. Nanofillers such as graphene oxide (GO), titanium dioxide (TiO_2), and halloysite nanotubes increase the membrane's internal surface area and provide additional reactive sites for ion binding. For example, GO-functionalized CTA membranes showed a twofold increase in Pb^{2+} transport efficiency, attributed to better carrier dispersion, higher porosity, and reduced fouling [68]. Moreover, the nanostructure of these additives can create complex transport pathways that limit the unwanted diffusion of non-target species, improving overall selectivity. Nanomaterials are not just passive fillers; they are increasingly being used as active transport enhancers and ion carriers in electrochemical systems. Recent advancements in membrane engineering indicate that it is possible to mimic PIM-like behavior by using dual-layer electrospun membranes that are embedded with functional nanoparticles. Electrospinning allows for precise control over fiber diameter, alignment, and surface chemistry, enabling the creation of asymmetric structures where one layer provides selective ion recognition while the other promotes directional flux. For example, GO or TiO_2-loaded electrospun membranes can act as selective ion-transport channels in capacitive or electrochemical sensing systems. Their high conductivity and tunable surface chemistry facilitate charge transfer and selective electrostatic interactions with target analytes. This strategy closely parallels PIM design in terms of carrier-mediated transport, yet offers enhanced surface-to-volume ratios and greater adaptability for sensor miniaturization. A recent study by Kwak et al. (2025) showed that integrating metal–organic frameworks (MOFs) and 3D nanostructures into bipolar membranes significantly decreased the water dissociation overpotential to thermodynamic limits, highlighting the synergistic potential of nanoarchitectures in separation and sensing [69]. Such findings support the integration of functional nanomaterials into electrospun membranes as a pathway to develop PIM-inspired, hybrid sensing and separation systems with advanced performance metrics.

Another trend is the use of bio-based fillers in membrane production. Materials such as acetylated lignin, starch derivatives, and cellulose nanocrystals contribute functional binding groups (e.g., hydroxyl, phenolic, or amine) and improve the membrane's mechanical integrity and sustainability. These renewable materials reduce production costs and lessen the environmental impact. For instance, CTA membranes blended with acetylated lignin and D2EHPA efficiently extracted Ni^{2+} and Co^{2+} from wastewater while maintaining over 85% performance across multiple transport cycles [70].

An often-overlooked aspect of matrix modification is its indirect influence on the carrier's behavior. Hydrophilic modifications, such as sulfonation or nanoparticle inclusion, can increase the mobility and availability of the carrier phase. They do this by preventing clumping and improving its distribution within the membrane. Simultaneously, polar functional groups in the matrix can anchor mobile carriers, reducing leaching and prolonging the membrane's life.

These combined effects on ion–carrier interaction, membrane porosity, phase retention, and fouling resistance highlight the pivotal role of matrix design in controlling selectivity. The diverse functionalization strategies and their impact on transport behavior are conceptually illustrated in Figure 3, which summarizes how sulfonation, carboxylation, nanofiller integration, and biofiller blending shape ion migration from the source phase to the receiving phase across the membrane.

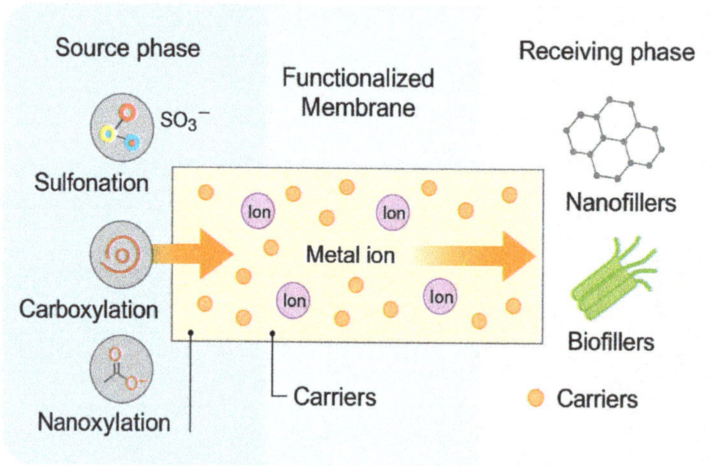

Figure 3. Matrix functionalization strategies and their effect on ion transport in PIMs.

A schematic representation of polymer matrix functionalization strategies and their impact on transport selectivity in PIMs illustrates ion migration across a functionalized membrane from the source to the receiving phase.

In summary, polymer matrix functionalization provides a complementary route to carrier design for tuning PIM selectivity. Whether by modifying surface charge, incorporating diffusion enhancers, or embedding biofunctional moieties, such alterations redefine how ions interact with the carrier and the membrane phase. These strategies not only expand the use of PIMs to more complex separation problems but also match their design with the goals of sustainability, reusability, and long-term stability, core criteria for new environmental and industrial challenges.

3. Enhancement of Mechanical and Chemical Stability

Durability under mechanical stress and chemical exposure remains a significant challenge for polymer inclusion membranes (PIMs), especially in industrial and environmental applications. Over the last five years, we have made significant progress by using better polymers and cross-linking techniques. Membrane integrity has been reinforced while performance has been preserved or enhanced.

3.1. Utilization of Advanced Polymers

The choice of base polymer matrix is critical in determining the mechanical, thermal, and chemical resistance of polymer inclusion membranes (PIMs). Traditional matrices such as cellulose triacetate (CTA) and polyvinyl chloride (PVC) have shown good compatibility with various carriers. However, their use is often limited under harsh conditions due to plasticizer leaching, swelling, and degradation in acidic or organic environments [65,71,72].

To address these challenges, recent research has increasingly focused on advanced engineering polymers such as poly(vinylidene fluoride) (PVDF), poly(vinylidene fluoride-co-hexafluoropropylene) (PVDF-HFP), and polyether ether ketone (PEEK). These materials offer enhanced mechanical stability, chemical inertness and thermal resistance, enabling membrane operation in more aggressive environments without sacrificing selectivity or permeability [29,72,73].

3.1.1. PVDF- and PVDF-HFP-Based PIMs

PVDF-HFP has emerged as one of the most promising membrane-forming polymers due to its superior flexibility, hydrophobicity, and chemical inertness. The presence of fluorine atoms confers high resistance to oxidative and acidic attack, while the amorphous HFP segments promote elasticity and improve membrane processability [72]. Recent studies have shown that PVDF-HFP-based PIMs functionalized with D2EHPA exhibit significantly longer operational lifespans and reduced carrier loss compared to CTA- or PVC-based counterparts. Kazemi and Yaftian (2024) demonstrated that PVDF-HFP membranes functionalized with D2EHPA retained over 90% of their Bi(III) extraction efficiency across five consecutive acidic transport cycles, while CTA membranes lost structural integrity under the same conditions [29]. Moreover, PVDF-HFP exhibits lower water uptake and swelling, which contribute to dimensional stability and reduced carrier loss [28]. A visual summary of these benefits is presented in Figure 4, which compares the structural behavior of PIMs based on PVC, PVDF-HFP, and PEEK under repeated acidic exposure. The superior retention of membrane structure and carrier content in PVDF-HFP and PEEK membranes is evident.

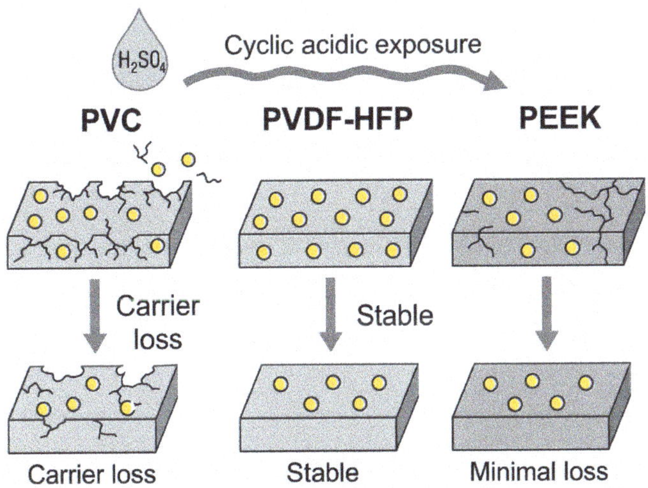

Figure 4. Schematic representation of membrane behavior under aggressive chemical conditions for different polymer matrices (PVC, PVD-HFP, PEEK), showing comparative carrier retention and structural degradation across operating cycles (●—Carrier).

3.1.2. PEEK-Based PIMs

Polyether ether ketone (PEEK) in PIMs is not as widely studied because it is complex to process and expensive. It represents a class of ultra-durable, high-temperature polymers used extensively in aerospace and biomedical applications. Its semicrystalline structure provides exceptional tensile strength (>100 MPa), thermal resistance (up to 250 °C), and chemical resistance even to strong acids and chlorinated solvents [74]. In exploratory studies, PEEK-based PIMs have shown no visible degradation after 96 h of exposure to concentrated HNO_3 and HCl, thereby maintaining their transport properties and structural integrity. These characteristics make PEEK suitable for long-term, high-pressure, or high-temperature separations in industrial wastewater treatment and analytical sensing [30].

3.1.3. Comparative Properties and Potential Applications

High-performance polymer matrices such as PVDF-HFP and PEEK provide notable benefits compared to traditional materials like PVC and CTA. Their enhanced physicochemical properties broaden the range of environmental and industrial conditions in which PIMs can function effectively.

- Wider pH Stability: PVDF-HFP and PEEK maintain stability across a broad pH range of approximately 1 to 10, making them suitable for use in both acidic and alkaline environments.
- Enhanced Thermal Resistance: PVDF-HFP can withstand temperatures up to around 120 °C, while PEEK remains stable at temperatures exceeding 250 °C. This capability allows for applications in thermally intensive processes.
- Improved Carrier Retention: Due to their lower surface energy and dense polymeric structure, these matrices reduce carrier volatilization and leaching, thereby extending membrane lifetime and selectivity.
- Antifouling Behavior: Their hydrophobic and chemically inert surfaces minimize organic and microbial adhesion, resulting in lower fouling rates and reduced cleaning frequency.

PVDF-HFP and PEEK exhibit performance features that make them highly suitable for demanding applications, such as wastewater treatment, metal recovery, and gas separation under harsh conditions. Table 3 presents a comparative overview of the key physicochemical parameters for commonly used PIM matrix polymers, including tensile strength, thermal stability, carrier retention capacity, and operational longevity.

Table 3. Comparative properties of polymer matrices used in PIMs.

Polymer Matrix	Tensile Strength (MPa)	Max Operating Temp (°C)	Acid/Base Resistance	Carrier Leaching Tendency	Reusability	Ref.
PVC	40–50	<80	Moderate (pH 3–9)	High	Low	[65,71,72]
CTA	20–40	<90	Moderate	Moderate	Medium	[65,71,72]
PVDF-HFP	50–60	120	Excellent	Low	High	[28,29,72,73]
PEEK	>100	>250	Excellent	Very Low	Very High	[30,74]

Additionally, these advanced matrices show excellent compatibility with electrospinning, solvent evaporation, and casting techniques, facilitating the development of nanostructured PIMs or multi-layered composite architectures [27].

3.2. *Cross-Linking Strategies and Stability Under Harsh Conditions*

Ensuring structural durability and carrier retention in polymer inclusion membranes (PIMs) is crucial, especially under demanding thermal, chemical, or mechanical stress. Cross-linking methods—whether UV/light-initiated, thermal, or chemical—have emerged as practical approaches to improve membrane strength without losing transport performance.

Controlled cross-linking creates interlocked polymer networks that limit plasticizer-induced swelling, reduce carrier leaching, and improve mechanical response. For example, the UV-initiated addition of poly(ethylene glycol) dimethacrylate (PEG-DMA) into PVDF-HFP membranes forms a semi-interpenetrating polymer network (semi-IPN) that significantly increases tensile modulus by 25%, lowers membrane swelling by over 50%, and maintains Pb(II) flux nearly unchanged after 10 reuse cycles [27,75]. Thermal cross-linking techniques, such as mild heating of PVDF-HFP with difunctional linkers have also shown notable improvements. Membranes subjected to 80 °C curing preserved over 90% ion-selective transport activity after exposure to 0.1 M HCl for 72 h, compared to a 40% drop in non-cross-linked counterparts [76]. Chemical cross-linkers, such as glutaraldehyde and epichlorohydrin, have been used with CTA and PVC matrices to resist organic solvents and surfactant-rich environments. These membranes displayed 20–40% improved longevity, extending workable lifetimes from ~5 cycles to more than 8 in industrial wastewater simulations [9].

Beyond chemical resistance and swelling suppression, the long-term performance of PIMs in aggressive environments depends on several interrelated factors: (i) polymer matrix stability, (ii) carrier–polymer compatibility, (iii) plasticizer migration behavior, and (iv) membrane microstructure. For example, exposure to strong oxidizing agents (such as ClO^- and H_2O_2), extreme acidic or alkaline pH levels, or high ionic strengths can speed up polymer degradation or cause carrier leaching in non-cross-linked systems. Cross-linking helps reduce these effects by limiting segmental mobility and forming a cohesive polymer network that can better withstand hydrolytic and oxidative stress. A detailed study on CTA-based membranes demonstrated that glutaraldehyde-induced cross-linking preserved 85% of Pb^{2+} transport capacity after 15 cycles in wastewater spiked with detergents and oils. In contrast, uncross-linked controls lost 50% of capacity after only six cycles [77].

Similarly, UV-cross-linked PVDF-HFP membranes retained their performance after prolonged exposure (120 h) to simulated acidic mine drainage (pH $\approx$ 2.5), maintaining stable Zn^{2+} and Cu^{2+} permeability. This was attributed to enhanced phase compatibility between ionic liquid carriers and the semi-IPN structure, which immobilized the functional phase and prevented leaching [78]. Temperature also plays a critical role. Membranes operated at $\geq$60 °C often exhibit accelerated plasticizer migration or phase separation. However, cross-linked membranes (particularly those thermally or UV-cured) showed limited deterioration in selectivity and structural integrity even after thermal cycling between 20 °C and 80 °C [79]. The use of cross-linking also improves resistance to shear stress in continuous flow or pressurized systems. For example, membranes reinforced with epoxy or multifunctional acrylates retained their tensile strength and functional flux after more than 50 h of convective flow at 2 bar, indicating their applicability in pilot-scale treatment modules [80]. Table 4 highlights the distinct impact of cross-linking routes on mechanical strength, swelling degree, ion flux stability, and operational lifespan across various polymer matrices and conditions.

Table 4. Cross-linking strategies enhancing PIM durability.

Cross-Linking Type	Polymer Base	Method	Key Benefits	Ref.
UV (PEG-DMA)	PVDF-HFP	UV + PEG-DMA	+25% modulus; −50% swelling; stable Pb(II) flux over 10 cycles	[27,75,78–80]
Thermal Cure	PVDF-HFP	Heat-curing at ~80 °C	>90% flux retention after 0.1 M HCl exposure over 72 h	[9,76]
Chemical (Glutaraldehyde)	CTA, PVC	Aqueous/vapor cross-linking	20–40% extended durability in solvent/surfactant conditions	[77,81]

Table 4. *Cont.*

Cross-Linking Type	Polymer Base	Method	Key Benefits	Ref.
Dual-stress reinforcement	PVDF-HFP, hybrid	Epoxy/acrylate reinforcement; flow and temp. cycling	Functional flux and tensile retention after >50 h of flow at 2 bar; stable under 20–80 °C thermal cycling	[78–80]

Moreover, Figure 5 schematically illustrates the difference between the structures of conventional and cross-linked membranes, showing how networked polymer structures more effectively retain carrier molecules and reduce plasticizer mobility, thereby enhancing stability in aggressive environments.

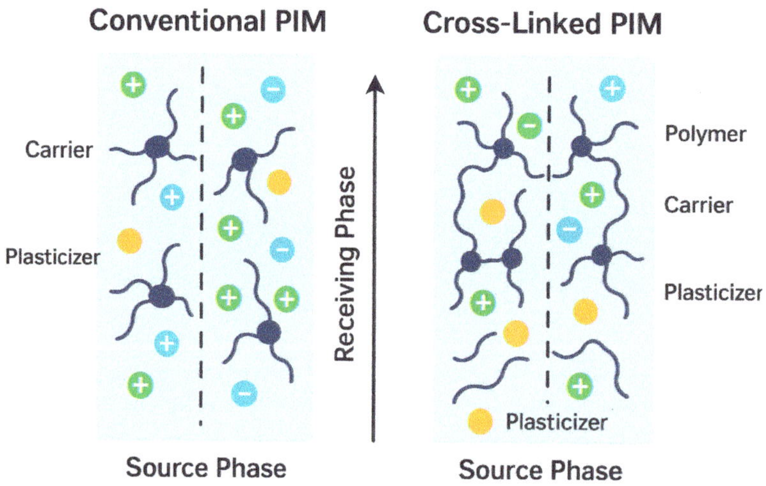

Figure 5. Schematic comparison of conventional vs. cross-linked PIM structures (●—Carrier; ●—Plasticizer; ●—Anion; ●—Cation).

Overall, strategic cross-linking strengthens membrane structure, minimizes plasticizer and carrier bleed, and creates a stable environment for facilitated transport—even in aggressive chemical or thermal scenarios. This enables PIMs to compete against conventional membranes, especially in industrial or environmental applications requiring stringent durability.

4. Environmental Applications of PIMs

Polymer inclusion membranes are increasingly used for environmental remediation, particularly for removing heavy metals and organic pollutants from water, as well as for gas separation, and carbon dioxide capture. Their tunable selectivity, combined with their structural stability, makes them highly promising for real-world applications.

4.1. Heavy Metal and Organic Pollutant Removal from Water

One of the most established and effective uses of polymer inclusion membranes (PIMs) is the selective removal of hazardous inorganic and organic pollutants from aqueous environments. These pollutants include transition metals (e.g., Pb^{2+}, Cd^{2+}, Ni^{2+}, and Cu^{2+}), toxic metalloids (e.g., Cr^{6+}, As^{5+}), and new organic contaminants such as pharmaceuticals, pesticides, and industrial dyes. PIMs provide a unique advantage due to their high ion selectivity, low carrier loss, and reusability under varying environmental conditions.

4.1.1. Mechanism of Removal

PIMs operate via facilitated transport, where the embedded carrier forms a reversible complex with the target pollutant in the source phase. This complex then diffuses through the membrane and dissociates in the receiving phase, driven by a concentration gradient or pH difference. Figure 6 illustrates this process for a representative divalent cation (e.g., Cd^{2+}).

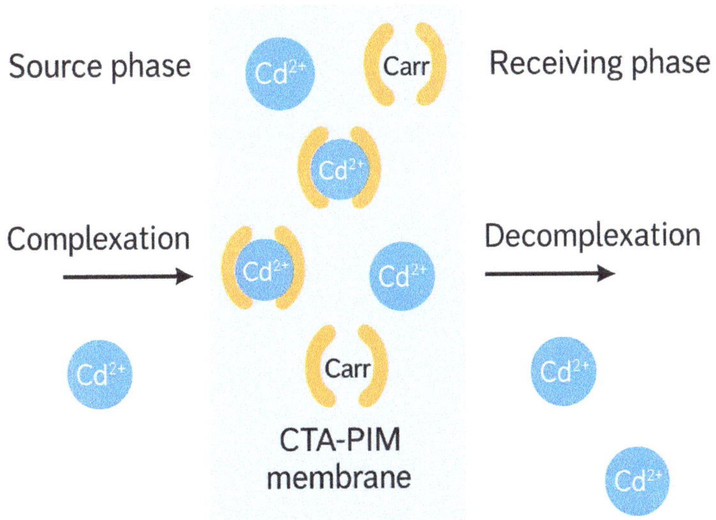

Figure 6. Schematic representation of Cd^{2+} transport across a CTA-based PIM from the source to receiving phase, showing complexation/decomplexation with the carrier.

4.1.2. Heavy Metal Removal

Recent studies demonstrate that ionic liquid (IL)-based PIMs exhibit excellent performance in selectively removing Pb^{2+} and Cd^{2+} ions from aqueous solutions. For instance, Hernández-Fernández et al. (2023) employed a CTA-based PIM containing [Bmim][PF_6] as an extractant, achieving greater than 90% Cd^{2+} removal within 2 h, with minimal leaching and stable operation across five cycles [27]. ILs also suppress carrier volatilization and improve membrane stability.

Biofiller-modified PIMs further enhance selectivity and membrane lifespan [36]. Kaczorowska et al. (2022) reported a bio-based CTA membrane blended with acetylated lignin and D2EHPA, which demonstrated consistent the removal of Ni^{2+} and Co^{2+} from real wastewater while maintaining its structural integrity under varying pH conditions [35].

4.1.3. Organic Micropollutant Removal

PIMs have also been successfully applied to emerging contaminants, particularly pharmaceuticals. Sulfamethoxazole and diclofenac are frequently detected in surface waters, posing ecological threats. Studies using Aliquat 336-loaded PVC-based PIMs revealed rejection efficiencies of 85–95% for these drugs, depending on membrane thickness and phase pH [77]. Functionalization with sulfonic acid or carboxyl groups further enhances interactions with zwitterionic drug species. Additionally, studies have reported the successful separation of dyes using PIMs modified with cellulose nanocrystals, which offer hydrogen bonding and π-π interactions with aromatic pollutants. This approach was shown to reduce methylene blue concentration by up to 90% in simulated wastewater [30].

Polymer inclusion membranes (PIMs) provide excellent selectivity and reusability, but it is important to compare their performance with other membrane technologies. Various techniques, including nanofiltration (NF), reverse osmosis (RO), ultrafiltration (UF), and hydrogel-based membranes, have commonly been used for pollutant removal. However, each of these methods has its own specific limitations. For instance, reverse osmosis and NF provide high rejection rates for heavy metals and pharmaceuticals, but they operate under high pressure, consume more energy, and are prone to membrane fouling. Ultrafiltration membranes, including polymer-enhanced ultrafiltration (PEUF), can effectively remove specific micropollutants and macromolecules; however, they often require additives (e.g., oxidized alginate) to achieve enhanced separation and encounter challenges with selectivity and membrane clogging [78]. Hydrogel membranes are known for their swelling-controlled transport and stimulus-responsiveness, but they generally lack chemical robustness under variable pH or ionic strength conditions, and long-term reusability is often limited.

In contrast, PIMs demonstrate the following:

- High ion selectivity due to the molecular specificity of the carriers;
- Low energy demands, operating under ambient pressure and temperature;
- Minimal leaching and long-term reusability, especially when incorporating ILs or bio-based matrices;
- Tailorability for specific targets, including non-ionic pharmaceuticals, nutrients, or charged dyes;
- Greater chemical resistance, particularly with CTA, PVDF-HFP, or IL-based systems.

PIMs are particularly appealing for niche applications that require selective separation under mild conditions. These include tasks such as removing specific antibiotics, pesticides, or nutrient ions like phosphate and ammonium. Additionally, their compatibility with renewable carriers and biodegradable polymers makes them an attractive option for sustainable remediation efforts.

4.1.4. Practical Benefits and Benchmarks

The principal advantages of PIM-based systems in environmental remediation include

- Exceptional selectivity toward specific target analytes, enabled by the molecular recognition capabilities of tailored carriers;
- Energy efficiency, as separation processes are typically conducted under ambient-temperature and- pressure conditions;
- Operational simplicity and renewability, with facile carrier reactivation and membrane reuse achievable through controlled pH modulation between phases;
- Enhanced sustainability, particularly when incorporating biodegradable or bio-derived fillers that reduce the environmental impact of membrane fabrication and disposal.

A performance comparison is summarized in Table 5.

Table 5. Examples of PIM applications in removing metal ions and organic pollutants from water.

Application	Membrane Composition	Target Contaminants	Removal Efficiency	Operational Stability	Ref.
Heavy metal removal	CTA + $[Bmim][PF_6]$	Pb^{2+}, Cd^{2+}	>90% in 2 h	5 reuse cycles	[27]
Biofiller-enhanced CTA	CTA + lignin + D2EHPA	Ni^{2+}, Co^{2+}	~88%	Maintained at pH 3–6	[28]

Table 5. *Cont.*

Application	Membrane Composition	Target Contaminants	Removal Efficiency	Operational Stability	Ref.
Pharmaceutical removal	PVC + Aliquat 336	Sulfamethoxazole, DCF	85–95%	3 reuse cycles	[30,77]
Dye removal	CNC-modified CTA	Methylene Blue	~90%	Stable after 4 cycles	[78]

4.2. Gas Separation and Carbon Capture

Polymer inclusion membranes (PIMs) have recently gained considerable attention in the field of gas separation, particularly for their potential application in carbon dioxide (CO_2) capture from industrial flue gases, natural gas purification, and biogas upgrading. In contrast to traditional dense polymer membranes, which rely on solution–diffusion mechanisms, PIMs can exploit facilitated transport processes, wherein the polymer matrix incorporates selective carriers (e.g., ionic liquids, amines) that interact reversibly with CO_2 molecules. This enables PIMs to achieve enhanced permeability and selectivity, even exceeding the classical Robeson upper bound for polymer membranes [79,80,82–84].

The chemical affinity between the carrier and CO_2 typically drives the facilitated transport mechanism in gas-separating PIMs. Ionic liquids (ILs), for instance, can form weakly bound complexes with CO_2, promoting selective uptake on the source phase and rapid release on the permeate side. Similarly, amine-functionalized carriers react with CO_2 via carbamate formation, facilitating directional transport. These interactions not only improve selectivity over non-reactive gases, such as N_2 or CH_4 but also contribute to increased diffusion fluxes due to the maintenance of concentration gradients across the membrane. A schematic representation of this process is shown in Figure 7, which compares IL-based and GO-enhanced membrane structures for the selective transport of CO_2.

Recent studies report outstanding performance metrics for PIMs in CO_2 separation. For example, PVA membranes incorporating aniline-based ionic liquids deposited on polysulfone supports achieved CO_2/N_2 selectivity values of approximately 30 and CO_2 permeabilities approaching 300 GPU under ambient conditions [82,83,85,86]. Similarly, mixed-matrix PIMs composed of PIM-1 and graphene oxide nanosheets demonstrated CO_2 permeabilities exceeding 6000 Barrer, coupled with CO_2/N_2 selectivities above 120 [82, 87]. These results are primarily attributed to the high free volume of PIM-1 and the additional diffusion channels introduced by the nanofillers. Another noteworthy example is the blending of PIM-1 with poly(ethylene glycol) (PEG), which not only enhances polymer–gas interaction but also improves CO_2/CH_4 separation performance, reaching permeability values of around 2000 Barrer and selectivity of approximately 39 [82,88,89]. These representative data are summarized in Table 6, offering a comparative overview of selected PIM-based gas separation systems.

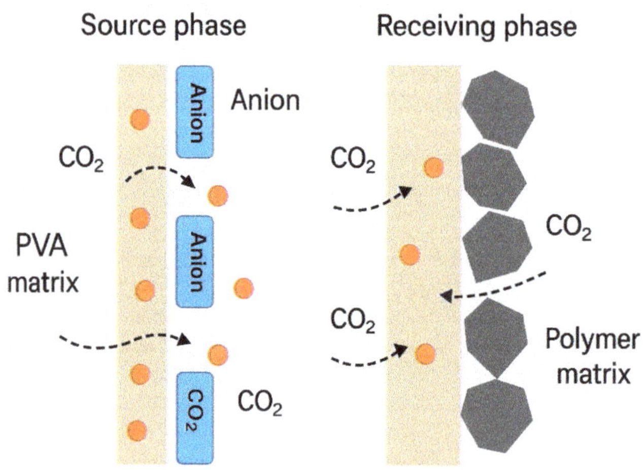

Figure 7. Schematic of CO_2 transport through IL-based and GO-enhanced PIMs (●—CO_2 molecule; ■—Graphene Oxide (GO) nanosheet or nanofiller [38].

Table 6. Performance matrix of CO_2-facilitated transport PIMs.

Membrane System	CO_2 Permeability (Barrer)	CO_2 Selectivity	Notes	Ref.
IL-PVA/aniline on PSf support	≈300 GPU	$CO_2/N_2 \approx 30$	Facilitated via IL-aniline chemistry	[82,83,86]
IL-PIM-1 + GO	~6.169	$CO_2/N_2 \approx 123.5$	Mixed-matrix, free-volume enhanced	[82,87]
PIM-1 + PEG (2.5 wt%)	1.952	$CO_2/CH_4 \approx 39$	Blend tailored for CH_4 separation	[82,88,89]

In addition to their promising transport properties, PIMs exhibit advantageous features in terms of stability and process adaptability. The incorporation of ILs such as [Bmim][PF_6] or amino-functionalized ILs contributes to thermal and chemical resistance, mitigating aging and performance loss even under humid or pressurized conditions [81,89]. Moreover, the structural reinforcement provided by inorganic fillers such as graphene oxide (GO), polyhedral oligomeric silsesquioxanes (POSS), and titanium-based nanomaterials improves stability [82,90].

The manufacturing scalability of gas-separating PIMs has also been demonstrated using established thin-film fabrication techniques, such as non-solvent-induced phase separation (NIPS), spin coating, and doctor blade casting. These methods enable the formation of selective layers as thin as 1.0–10.0 μm, supported on porous polymer substrates, which provide high gas fluxes and mechanical robustness. Notably, these membranes maintain stable performance across multiple gas cycles and varying humidity conditions, supporting their potential for real-world CO_2 removal systems [91].

In summary, gas separation and carbon capture using PIMs represent a frontier area of membrane science, combining the molecular precision of selective carriers with the process advantages of polymeric films. The integration of ionic liquids and functional nanofillers into PIM matrices provides a robust route to high-performance CO_2 separation, offering tunable selectivity, enhanced energy efficiency, and improved environmental compatibility.

5. Integration of PIMs in Sensing Technologies

Polymer inclusion membranes have transcended their traditional role in separation, emerging as versatile components in optical and electrochemical sensors. By immobilizing carriers, chromogenic reagents, or electroactive elements within a polymer matrix, PIMs combine high selectivity, signal amplification, and portability, enabling robust detection platforms for environmental monitoring and analytical applications [92–94].

5.1. Optical and Electrochemical Sensors Based on PIMs

Polymer inclusion membranes (PIMs) have evolved from mere separation barriers into highly effective sensing platforms thanks to their capacity to immobilize selective carriers, chromogenic reagents, or electroactive compounds within a planar polymer matrix. This versatility enables the development of both optical sensors (optodes) and electrochemical sensors, combining selectivity, portability, and analytical precision in one compact form.

In optical PIM-based sensors, a colorimetric reagent such as a Schiff base, azo dye, or ionic liquid complex is embedded within a cellulose triacetate (CTA) or PVC membrane. These reagents undergo visually detectable color transitions–observable by the naked eye or monitored via spectrophotometry–upon interacting with specific analytes. Optical PIM-based sensors operate through a specific chemical reaction or complexation process involving an immobilized reagent (such as a dye) and an analyte (such as a metal ion). This interaction alters the electronic structure of the reagent, causing a shift in light absorption or emission bands, which is observed as a visible color change [71,95,96]. Typically, this is the result of ligand–metal complex formation, where the reagent acts as a coordinating ligand. Depending on the type of dye and the target ion, the color change may result from tautomerism, protonation/deprotonation, or electronic transitions such as metal-to-ligand charge transfer (MLCT) [92]. The polymer matrix, typically made of CTA or PVC, is essential for stabilizing the sensing system. It provides mechanical support for the reagent, prevents leaching, and allows for the controlled diffusion of the analyte to the reactive site. Furthermore, the physical and chemical properties of the membrane, such as porosity and hydrophilicity, affect both the kinetics and selectivity of the sensor's response [92,95]. Thanks to these features, optical PIM sensors can provide a reliable signal in a short time frame (typically a few to several minutes), while maintaining reusability and reproducibility [96].

A noteworthy example, as cited by Sánchez-Ponce et al. (2021), employed 2-acetylpyridine benzoylhydrazone (2-APBH) immobilized within a PVC-PIM to selectively sense Cd^{2+}. This sensor achieved a limit of detection as low as 0.02 mg/L, demonstrated a linear response up to 1.0 mg/L, and provided results within just 20 min [92]. Additionally, the sensor maintained reliable operation over multiple cycles and proved accurate when detecting Cd^{2+} in natural water and even art paint samples.

Beyond cadmium sensors, Cu^{2+}, Zn^{2+}, Fe^{2+}, and Pb^{2+} optodes based on CTA-PIMs with chromophores like 1-(2-pyridylazo)-2-naphthol (PAN) have been reported. In one study, an optode incorporating TEHP and PAN detected metal ions within 20 min and demonstrated strong optical signal retention over a period of nine days. Techniques combining mathematical analysis methods, such as multivariate curve resolution, have further enhanced the quantitative accuracy of these sensors [92,97,98].

Electrochemical PIM sensors, on the other hand, typically involve coating an electrode—such as a screen-printed or glassy carbon electrode—with a CTA-PIM containing selective ionophores or ionic liquids. Carriers, such as calixarenes or ionophoric compounds, facilitate the specific transfer of ions, creating measurable potentiometric or amperometric responses. These systems offer remarkable detection limits in the parts-per-billion range, rapid response times (often below 30 s), and excellent stability, usually preserving functionality over 50 or more measurement cycles [99,100]. The operation of electrochemical sensors based on polymer inclusion membranes (PIMs) relies on the selective extraction and transport of target ions from the sample solution into the membrane phase. This process is facilitated by an ionophore or carrier that is embedded within the polymer matrix [101]. The selective transport of ions modifies the local electrochemical environment at the interface between the electrode and the membrane. For potentiometric sensors, this results in a change in phase boundary potential that is proportional to the logarithm of the ion activity, which is then measured as an electrical potential difference [102]. In amperometric systems, redox-active analytes are transferred into the membrane and oxidized or reduced at the electrode surface, producing a current directly related to analyte concentration [103]. The polymeric membrane provides physical separation between the sample and the electrode while allowing selective ion transport through embedded carriers. Ionic liquids or plasticizers are often incorporated into the polymeric ionic membrane (PIM) to improve membrane conductivity, enhance ion mobility, and ensure compatibility with the electrode surface. Additionally, immobilizing reactive components in a solid matrix boosts operational stability and reproducibility while reducing the leaching of active compounds, which is a common issue in traditional liquid membrane systems. This structural robustness allows the sensor to be reused for multiple measurement cycles without significant degradation in signal quality or selectivity.

What makes PIM-based sensors particularly compelling is their combination of high analytical performance, mechanical integrity, and deployment readiness—they are inexpensive, simple to fabricate, and portable. The immobilization of reactive components within a solid matrix also mitigates the leaching issues and instability often encountered in liquid membrane sensors. The working principles of both PIM-based optical and electrochemical sensors are illustrated in Figure 8.

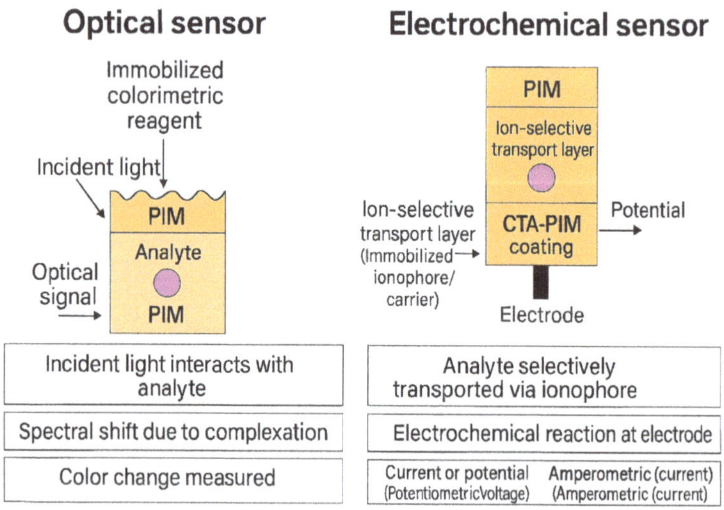

Figure 8. Schematic illustration of PIM-based sensing platforms: **left**–optical sensor with immobilized colorimetric reagent; **right**–electrochemical sensor with CTA-PIM coating on electrode.

5.2. *Performance Matrix: Sensitivity, Selectivity, and Response Time*

The analytical performance of PIM-based sensors is determined by three core parameters: sensitivity, selectivity, and response time. These factors collectively govern the practical applicability of the sensor in trace detection, environmental monitoring, and field-based diagnostics. Their values depend not only on the type of embedded reagent or carrier but also on the physical properties of the polymer matrix, the mode of signal transduction (optical or electrochemical), and the operational conditions.

Sensitivity reflects the sensor's ability to detect minute changes in analyte concentration. In PIM-based optodes, this is typically defined by the slope of the calibration curve and the limit of detection (LOD). Sensors embedding Schiff base ligands or azo dyes in plasticized PVC or CTA matrices have achieved LODs as low as 0.01 to 0.05 mg L^{-1} for divalent metal ions such as Cd^{2+}, Pb^{2+}, and Fe^{2+}. For example, the Cd^{2+} sensor developed by Sánchez-Ponce et al. achieved a linear response range from 0.02 to 1 mg L^{-1}, with an LOD of 0.02 mg L^{-1}, supporting practical deployment in environmental analysis and potable water quality control [46]. Similarly, Mancilla-Rico et al. developed a PIM-based optode capable of simultaneously detecting Cu^{2+}, Zn^{2+}, and Pb^{2+} using multivariate analysis of overlapping signals, enhancing both sensitivity and multidimensional selectivity [104].

Electrochemical sensors based on PIMs can achieve even lower detection thresholds–in the parts-per-billion range—owing to the synergistic effect of facilitated transport and enhanced charge transfer mechanisms. A study by Yacoob et al. reported the detection of Zn^{2+} and Cu^{2+} at sub-ppb levels using ionophore-modified PIMs on screen-printed electrodes [32].

Selectivity is another critical aspect, particularly when operating in complex or real-world matrices. The specificity of PIMs arises from the embedded carrier or ionophore, which is often designed to form highly stable complexes with the target ion. For instance, Suah et al. demonstrated that Fe^{2+}-selective optodes based on CTA and PAN exhibit limited cross-reactivity toward Fe^{3+} and Zn^{2+} due to the coordination geometry and the acidic pH environment favoring Fe^{2+} complexation [105].

Response time is a practical determinant of how quickly the sensor can report changes in analyte concentration. Optical systems often have response times ranging from 5 to 30 min, whereas electrochemical formats can be significantly faster. PIM-coated electrodes for Cu^{2+} and Zn^{2+} reported by Yacoob et al. stabilized in under 30 s, showing excellent reproducibility [32]. Long-term operational stability, mechanical durability, and resistance to carrier leaching are also evaluated. Sensors that retain at least 90% of their signal after 10–15 measurements or a week of storage are considered highly stable [30]. Figure 9 provides a conceptual comparison of PIM-based optical and electrochemical sensing performance, while Table 7 summarizes representative metrics for selected sensor configurations.

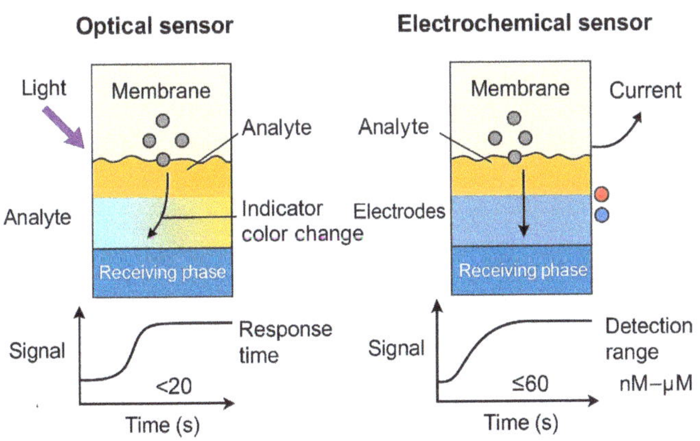

Figure 9. Comparative schematic of PIM-based optical vs. electrochemical sensor performance, showing signal generation modes, response time, and detection range (●: analyte molecules, ●●: electrons involved in electrochemical signal generation).

Table 7. Representative performance metrics of selected PIM-based sensors.

Sensor Type	Target Ion(s)	LOD	Response Time	Selectivity/Reusability	Ref.
Optical (PVC-PIM)	Cd^{2+}	0.02 mg/L	20 min	High/~10 uses	[104]
Optical (CTA-PIM)	Fe^{2+}	~0.05 mg/L	15–25 min	Moderate/9+ days	[105]
Optical (Multianalyte)	Cu^{2+}, Zn^{2+}, Pb^{2+}	<0.05 mg/L	15 min	With chemometric modeling	[30,32]
Electrochemical (PIM-SPCE)	Cu^{2+}, Zn^{2+}	ppb range	<0 s	High/>50 cycles	[32]

6. Sustainable and Green Approaches in PIM Fabrication

As environmental concerns intensify, the PIM field is shifting toward greener and more sustainable fabrication methods. Research efforts are increasingly emphasizing the use of biodegradable polymers, renewable feedstocks, and low-toxicity solvents or plasticizers. These eco-driven strategies contribute to sustainable chemistry and circular economy principles while ensuring that membrane performance is not compromised [106].

6.1. Biodegradable and Renewable Polymers

In recent years, the environmental footprint of polymer inclusion membrane (PIM) production has come under growing scrutiny, leading to substantial interest in biodegradable and renewable materials. Traditional synthetic matrices such as PVC and CTA, while effective, are not inherently biodegradable and pose long-term disposal challenges. To align with circular economy principles and the Sustainable Development Goals, researchers have explored a wide range of biodegradable polymers that offer both performance and environmental benefits [107].

Poly(lactic acid) (PLA) is one of the most widely studied biopolymers for PIM matrices. Derived from renewable resources such as corn starch or sugarcane, PLA offers industrial compostability and relatively high mechanical strength. Its main drawback—brittleness—can be mitigated through blending with poly(ethylene glycol) (PEG), plasticizers, or cellulose nanofibers. PLA-based PIMs have been employed in heavy metal ion separation, demonstrating not only selectivity but also enhanced environmental safety upon disposal after use [30].

Poly(3-hydroxybutyrate) (PHB), a member of the polyhydroxyalkanoates (PHAs), is another bio-based polymer that shows promise in membrane science. PHB is synthesized via microbial fermentation and is fully biodegradable under both aerobic and anaerobic conditions [8]. Its semicrystalline structure contributes to excellent barrier properties and chemical resistance. PIMs based on PHB exhibit notable stability in aggressive aqueous environments, making them suitable for wastewater remediation.

Poly(vinyl alcohol) (PVA) is a partially biodegradable synthetic polymer often used in combination with green additives. Its water solubility and transparency make it ideal for sensor applications. Moreover, PVA exhibits excellent film-forming capability and is highly compatible with carriers such as Aliquat 336 and various ionic liquids. PVA-based PIMs have demonstrated superior mechanical durability during repeated extraction cycles, especially when cross-linked or hybridized with biofilters [28].

Chitosan, obtained from the deacetylation of chitin, is a natural polysaccharide that is both biodegradable and biocompatible. Its amino and hydroxyl functional groups endow it with significant metal-chelating properties, allowing for the efficient removal of toxic cations like Cu^{2+}, Cr^{3+}, and Pb^{2+} [9]. Chitosan-PIMs can be fabricated using low-toxicity solvents such as dilute acetic acid, making the entire process more environmentally benign. Figure 10 illustrates the key structural features and degradation mechanisms associated with selected biodegradable PIM matrices, including PLA, PHB, PVA, and chitosan. The figure provides a comparative overview of molecular architecture (e.g., ester, hydroxyl, or amine groups), biodegradation pathways (such as microbial hydrolysis, enzymatic attack, or chemical cleavage), and the corresponding environmental fate of each polymer type. This structural and mechanistic analysis clarifies how each polymer enhances the performance of eco-friendly membranes and ensures safe end-of-life disposal.

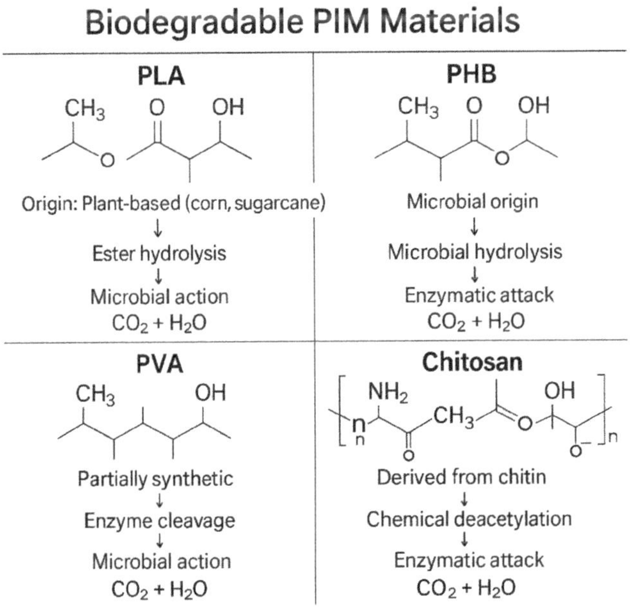

Figure 10. Comparative schematic of PIM-based optical vs. electrochemical sensor performance, showing signal generation modes, response time, and detection range.

A comparative summary of the properties of these biodegradable polymers, including their mechanical performance, thermal behavior, and carrier compatibility, is presented in Table 8.

Table 8. Properties and suitability of biodegradable polymers for use in polymer inclusion membranes (PIMs).

Plasticizer	Chemical Origin	Target Polymer(s)	Biodegradability	Key Benefit in PIMs	Ref.
Triethyl citrate	Citrate ester (natural)	PLA, PHB	High	Improves flexibility, non-toxic	[8,30]
Glycerol	Biobased polyol	PVA, Chitosan	High	Enhances water retention, compatible with hydrophilic carriers	[28]
Acetyl tributyl citrate	Citrate ester (natural)	PLA, PHB	Moderate–High	Reduces brittleness, low leaching	[30,32]

Table 8. *Cont.*

Plasticizer	Chemical Origin	Target Polymer(s)	Biodegradability	Key Benefit in PIMs	Ref.
Polyethylene glycol (PEG)	Synthetic, degradable	PLA, PVA	Moderate	Enhances elasticity and carrier dispersion	[28,30]
Sorbitol derivatives	Sugar alcohols	Chitosan, PVA	High	Maintains mechanical integrity and bio-compatibility	[9]

6.2. Green Solvents and Plasticizers

As awareness grows around the ecological footprint of membrane production, the use of green solvents and plasticizers has emerged as a critical component in the sustainable fabrication of polymer inclusion membranes (PIMs). These substances not only reduce the toxicological burden associated with traditional organic chemicals but also align with the principles of green chemistry by improving energy efficiency, reducing hazardous waste, and enhancing membrane recyclability [46,106].

In conventional PIM fabrication, chlorinated solvents such as dichloromethane (DCM), chloroform, and tetrahydrofuran (THF) have been widely used to dissolve hydrophobic polymers like cellulose triacetate (CTA). However, these solvents are classified as volatile organic compounds (VOCs) and pose significant health and environmental hazards due to their high volatility, flammability, and persistence. Therefore, a paradigm shift has occurred toward utilizing green solvents, which are non-toxic, biodegradable, and derived from renewable resources. Among the most prominent are ethanol, ethyl lactate, γ-valerolactone (GVL), and ionic liquids (ILs). These solvents facilitate polymer dissolution under milder conditions, thereby reducing the risk of carrier degradation during membrane casting. For instance, ethyl lactate, derived from lactic acid and ethanol, is a low-toxicity solvent that effectively dissolves PLA and is increasingly used in the production of eco-friendly membranes [30,108].

Additionally, ionic liquids, particularly those based on imidazolium, pyrrolidinium, or phosphonium backbones, have gained popularity due to their negligible vapor pressure and tunable solvent properties. They are particularly effective in forming PIMs when combined with biodegradable carriers and natural polymers. As illustrated in Figure 11, membranes fabricated using green solvents often exhibit more uniform morphology, enhanced interfacial compatibility between the polymer and carrier, and greater long-term stability in aqueous media compared to membranes made using conventional VOC-based solvents [30,109,110].

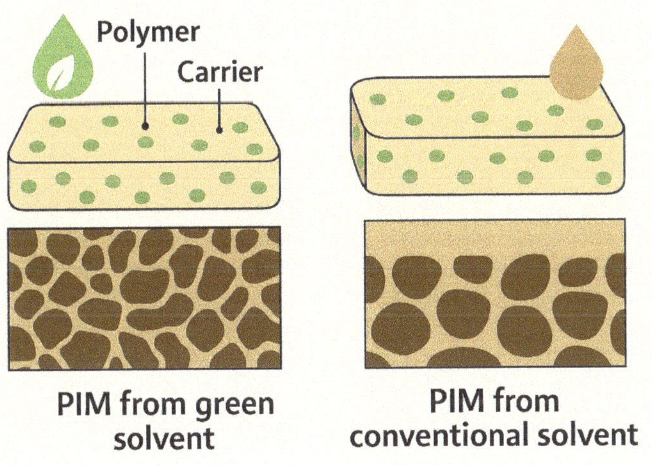

Figure 11. Influence of solvent type on the microstructural characteristics of polymer inclusion membranes.

Green plasticizers serve an equally essential role in sustainable PIM design by improving polymer flexibility, reducing brittleness, and maintaining carrier integrity during repeated separation cycles. Traditional plasticizers such as dioctyl phthalate (DOP) and tributyl phosphate (TBP) are effective softeners but pose significant health concerns, including endocrine disruption and bioaccumulation [111]. In contrast, plasticizers derived from natural sources such as triethyl citrate, acetyl tributyl citrate, glycerol, polyethylene glycol (PEG), and sorbitol derivatives demonstrate excellent biodegradability and minimal toxicity [62,64]. These compounds are particularly compatible with biopolymers such as PLA, PHB, PVA, and chitosan [8]. Their application in PIMs results in improved mechanical stability and reduced leaching of extractants into the surrounding environment. Table 9 provides an overview of the selected green plasticizers commonly used in PIM systems, highlighting their chemical origin, target compatibility with polymer matrices, and functional benefits. These materials support the development of safe, effective, and environmentally sound membranes that are suitable for applications in water treatment, sensing, and resource recovery.

Table 9. Selected green plasticizers for PIM fabrication and their properties.

Polymer	Biodegradability	Mechanical Strength	Thermal Stability	Carrier Compatibility	Notable Applications	Limitations	Ref.
PLA	High (industrial composting)	Moderate	~150 °C	Moderate	Heavy metal extraction, ion-selective PIMs	Brittle, low elongation	[8,30,108]
PHB	Excellent (natural degradation)	High	~180 °C	Good	Acid/base-stable PIMs, wastewater recovery	Rigid, high crystallinity	[8,109,110]
PVA	Partial (aqueous biodegradation)	High	~200 °C	Excellent	Optical sensors, hybrid PIMs	Water-sensitive, soluble in humid air	[18]
Chitosan	Excellent (enzymatic degradation)	Low-Moderate	~120 °C	High	Metal-binding, biocompatible membranes	Acid solubility, lower durability	[112,113]

7. Scalable Fabrication and Process Optimization

The ability to scale polymer inclusion membranes (PIMs) is crucial for their success in industrial settings. Along with selectivity and stability, modern fabrication techniques allow the production of membranes with a consistent structure and reproducible quality, which translates into lower processing costs and facilitates large-scale implementation.

7.1. Techniques: Phase Inversion, Electrospinning, and Others

The scalability of polymer inclusion membranes (PIMs) depends on how reproducible, structural integrity, and economically feasible their fabrication processes are. One popular method is the non-solvent-induced phase separation (NIPS). This technique has gained traction due to its flexibility and suitability for high-volume manufacturing. In NIPS, a polymer solution is applied to a support and then placed in a bath of non-solvent, typically water. This process starts a solvent–non-solvent exchange, which solidifies the polymer and creates a porous structure [30,35,114]. NIPS allows for precise control over porosity and membrane thickness, producing membranes that are both strong and selectively permeable [115]. NIPS works well with polymers such as cellulose triacetate (CTA), polyvinylidene fluoride (PVDF), and polyether ether ketone (PEEK), achieving porosities of up to 70% while ensuring chemical resistance.

Electrospinning is another effective method that excels at making nanofibrous PIMs. This technique uses a high voltage on a polymer solution to produce very fine fibers, typically with diameters between 100 and 500 nm. Electrospun membranes have higher surface-to-volume ratios, improved transport kinetics, and better dispersion of carriers such as ionic liquids. Electrospun PIMs are noted for their enhanced selectivity and minimal leaching, mainly when biopolymer matrices or green solvents are used [30,116]. Recent advancements such as alternating current (AC) electrospinning and needleless configurations have improved production throughput, making electrospinning increasingly viable for scale-up.

An emerging area is melt electrospinning writing (MEW), a technique that skips solvents and instead pushes molten polymers using electric fields to create microfibrous networks. This method provides precise control over fiber orientation and spacing, resulting in membranes with optimized mechanical pathways for selective ion transport. Although it is mostly used in research settings because of its lower output, MEW has shown exceptional potential in creating structured, custom-made PIMs for high-efficiency separations [29]. Figure 12 shows the structural differences between membranes made with NIPS, electrospinning, and MEW. It highlights their different shapes and the effects of each process.

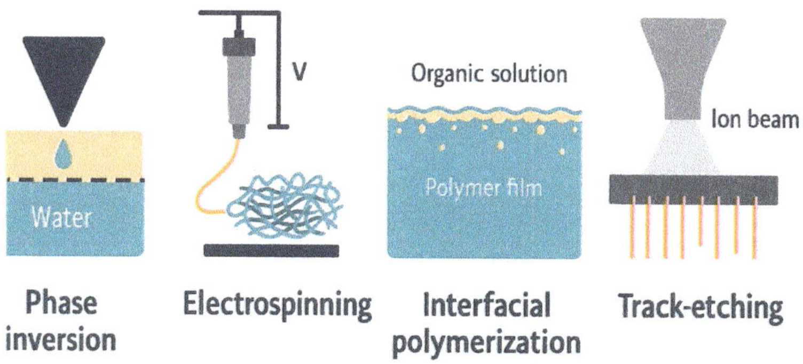

Figure 12. Comparison of PIM morphology using conventional and green solvents.

Table 10, presented below, provides a comparative overview of these techniques in terms of operational scalability, energy demands, and membrane performance metrics.

Table 10. Comparative analysis of scalable PIM fabrication techniques.

Technique	Fiber Size/Porosity	Throughput	Energy Demand	Solvent Required	Industrial Potential	Ref.
NIPS	5–50 μm/up to 70%	High	Moderate	Yes	Commercially established	[30,35,114,115]
Electrospinning	100–500 nm	Medium	High	Yes	Pilot to industrial	[30,116]
MEW	1–10 μm	Low–Medium	High	No	Emerging/R&D	[29]

7.2. Industrial Feasibility and Cost Considerations

The shift of polymer inclusion membranes (PIMs) from laboratory-scale prototypes to large-scale industrial applications depends on their cost-effectiveness, reliability during processing, and compatibility with current systems. Several techno-economic factors govern the industrial uptake of membrane technologies, including fabrication costs, operational stability under variable field conditions, reusability, and environmental compliance [117,118].

One of the primary benefits of PIM-based processes is their low energy demand, especially when compared to energy-heavy separation methods such as distillation or solvent extraction. PIMs operate efficiently under ambient pressure and temperature conditions, which removes the need for heating and vacuum systems. Energy analysis reports estimate that membrane separation processes consume nearly 90% less energy than thermal separation methods [30,119,120]. Material costs and membrane longevity are also critical. Table 11 summarizes economic indicators for typical materials used in PIM fabrication, including polymer cost per kg, number of reusability cycles, and carrier leaching resistance. Emerging biodegradable polymers like PLA and PVA provide savings due to their lower environmental disposal costs, although they may require optimization for long-term stability [29,114,120,121].

Another key factor is process integration. PIM modules can be configured in spiral-wound, hollow fiber, or cassette designs, enabling compatibility with standard water treatment and chemical recovery systems. Pilot-scale trials have shown consistent performance in removing heavy metals and recovering rare earth elements from complex mixtures. These include electroplating wastewater and acid mine drainage [122].

Table 11. Estimated economic parameters of materials used in scalable PIM fabrication [123].

Material	Approx. Cost ($/kg)	Reusability Cycles	Carrier Leaching	Environmental Impact
CTA	20–40	30–50	Moderate	Low
PVDF	60–90	50–80	Low	Moderate
PVA	10–25	15–30	High	Biodegradable
PLA	5–15	1020	Moderate	Biodegradable
PEEK	100–150	>100	Low	Low

8. Challenges and Future Perspectives

The field of polymer inclusion membranes (PIMs) has seen significant remarkable progress in recent years. There have been notable improvements in selectivity, mechanical strength, environmental compatibility, and manufacturing scalability [19,30,32,35,124]. These advancements have led to the successful application of PIMs in environmental remediation, gas separation, sensing, and analytical preconcentration. However, even with these achievements, several scientific and technological challenges still limit the complete industrial use of PIM systems [29,30,32,35,124,125]. This chapter presents at the important challenges that remain in PIM development, including material limitations, environmental risks, and engineering obstacles related to scaling up. It also looks at future strategies and research directions focused on overcoming these limitations. By identifying current bottlenecks and outlining innovative approaches—such as covalent carrier immobilization, bio-based membrane structures, hybrid nanostructures, and AI-driven design tools—this chapter sets the stage for the evolution of next-generation PIM technologies tailored for industrial and environmental resilience [19,28,30,80,126].

8.1. Remaining Limitations in Current PIMs

Despite the important progress in polymer inclusion membranes (PIMs), several limitations prevent their broad use in industry. One major challenge is carrier leaching, particularly in hydrophobic membranes where physical trapping does not hold up under dynamic flow or long-term exposure conditions [8,31,35,127]. This not only reduces membrane lifespan but also poses environmental risks. The cumulative leaching decreases membrane efficiency and may introduce secondary contamination into treated media. This issue is especially critical in membranes fabricated with biodegradable matrices such as polyvinyl alcohol (PVA) or polylactic acid (PLA), which are prone to degradation under acidic or basic pH conditions. Membranes relying solely on physically entrapped carriers–without covalent bonding or cross-linking–suffer from poor reusability. To tackle this issue, recent strategies have looked at the covalent grafting of extractants to polymer backbones or encapsulating carriers within nanoporous supports, such as mesoporous silica or cyclodextrin frameworks, to improve stability and retention [128,129].

At the same time, a significant technical challenge is the limited pH and temperature range that conventional PIMs can tolerate. Most PIMs, especially those based on cellulose triacetate (CTA), degrade or deform under aggressive industrial conditions involving high ionic strength, oxidative species, or thermal fluctuations. While engineering polymers such as polyether ether ketone (PEEK) and polyvinylidene fluoride (PVDF) provide better resistance, their use is restricted due to the need for harmful solvents, such as N-methyl-2-pyrrolidone (NMP) and dimethylformamide (DMF), along with complex and expensive processing [130,131]. Current research focuses on using green solvent systems—such as γ-valerolactone (GVL)–to process these polymers without losing their performance or sustainability [130,132].

A key issue in industrial settings is the long-term breakdown of membrane structure and function. Although many PIMs show short-term stability, but their durability over multiple transport cycles is still lacking. Repeated exposure to fluctuating pH levels, salinity, and mechanical stress can cause physical damage, such as microcracking, delamination, and carrier loss. Chemical degradation, including hydrolysis and oxidative breakdown of both the polymer and extractant, further contributes to performance decay. Research has found that CTA-based membranes can lose up to 30–40% of their transport efficiency after 10 to 15 cycles, particularly in solutions containing aggressive ions such as sulfates or chlorides [30,34].

Another ongoing issue is the limited compatibility of PIMs with complex water mixtures, like industrial wastewater or natural sources that have a lot of dissolved organic matter (DOM), humic acids, and microbial contaminants. These factors lead to fouling and pore blockage, reducing selectivity and flux. Researchers are investigating surface modifications, including PEGylation and fluorination, to improve hydrophilicity and antifouling properties. However, their long-term effectiveness in real-world conditions still needs to be clearly shown [133,134].

From a theoretical perspective, the understanding of ion transport in PIMs is still not complete. While empirical models, such as Fick's law-based equations and facilitated diffusion approximations, provide basic predictive capability, they often fail to capture the complex interplay between the polymer matrix, carrier–analyte interactions, membrane hydration, and competing species. The lack of valid transport models limits rational membrane design [135]. Recent advances in analytical techniques, such as operando FTIR spectroscopy, X-ray photoelectron spectroscopy (XPS), and molecular dynamics simulations, provide promising ways to understand these mechanisms. However, their integration into membrane engineering frameworks is still in its early stages [30,31,136].

Reproducibility and scaling up are major challenges, particularly when moving from laboratory casting to pilot or industrial production. Membrane performance can differ widely, even with the same formulations. This happens due to differences in polymer batch purity, plasticizer composition, ambient humidity, and drying conditions. Such variability affects quality control and consistency. Furthermore, the lack of standardized fabrication protocols for PIMs makes the translation to industrial use harder, especially when using new techniques like electrospinning, 3D micro-extrusion, or melt electrowriting [137,138].

Regulatory gaps and environmental concerns create major obstacles to the broad use of PIM technologies. Although these technologies are advertised as green and sustainable, many PIMs still use volatile organic solvents (VOCs), synthetic extractants such as Aliquat 336 or D2EHPA, and non-biodegradable materials. These substances can raise issues related to toxicity and environmental impact, especially in areas such as drinking water, agriculture, or pharmaceutical processing. There is no consistent international regulation setting acceptable limits of extractant leaching, biodegradability, and environmental effects of PIM components. Developing next-generation PIMs will need close cooperation with regulatory agencies to meet sustainability standards and public health needs [35,139,140].

Another bottleneck is the narrow operating pH and temperature window, which restricts the use of PIM use in aggressive industrial environments [131]. Additionally, while specific polymers such as PEEK or PVDF offer superior resistance, they often require toxic solvents or costly processing routes [141]. The scalability of advanced structures, such as electrospun nanofibers and MEW-written matrices, also faces technical and economic constraints. PEEK- or PVDF-based membranes offer promising alternatives due to their chemical resilience; however, their fabrication often relies on hazardous solvents, such as NMP and DMF. Efforts to adapt green solvent systems (e.g., γ-valerolactone) to dissolve engineering polymers without degrading their performance are ongoing. Moreover, uniform membrane casting at an industrial scale, especially via electrospinning or micro-extrusion, demands precise control over solution rheology, temperature, and phase separation dynamics [131,142,143].

8.2. Future Reasearch Directions for PIMs

As PIM technologies evolve, the next generation of research aims to overcome current limitations by incorporating innovative, adaptive, and multifunctional features into membrane systems. One major direction involves the development of stimuli-responsive membranes that can modulate their transport properties in response to external cues such as pH, temperature, or electric fields. For example, using pH-sensitive carriers, such as benzimidazole derivatives or thermoresponsive polymers like PNIPAm can provide dynamic control over ion transport. This control could greatly benefit biomedical and environmental applications [144,145].

Another promising area is combining nanotechnology with PIM architecture. Adding metal–organic frameworks (MOFs), covalent organic frameworks (COFs), or functionalized nanoparticles (e.g., TiO_2, ZnO, Fe_3O_4) into polymer matrices can improve selectivity and permeability. This approach also brings in photocatalytic or antimicrobial functions. These hybrid materials could pave the way for PIMs capable of simultaneously separating and degrading of organic micropollutants [146–148].

In structural engineering, new fabrication methods like additive manufacturing, including direct ink writing, melt electrowriting, as well as nanoscale patterning, are being studied to create membranes with specific porosity, different channels orientations, or layered architectures. These approaches aim to optimize both flux and selectivity while reducing resistance to mass transfer [149–152].

Bioinspired designs also offer considerable potential. Mimicking biological membranes using phospholipid analogs, aquaporin-like channels, or self-assembling peptide carriers can lead to ultra-selective and energy-efficient transport. Moreover, the use of artificial ion channels or gated nanopores may introduce new paradigms for molecular recognition and selective separation [153,154]. Figure 13 illustrates the multidisciplinary innovations driving the next-generation development of PIMs, including the integration of nanotechnology, innovative materials, and bioinspired designs.

Digital tools are becoming increasingly integral to PIM research. Machine learning algorithms predict membrane performance using compositional and processing parameters. When combined with high-throughput screening and multi-objective optimization, these methods should speed up material discovery and process design [155,156].

From a sustainability perspective, next-generation PIMs will focus more on circular economy principles. This includes using waste-derived polymers, recyclable membranes, and closed-loop manufacturing systems. Combining these materials with life cycle analysis (LCA) and techno-economic analysis (TEA) will be important for measuring their environmental impact and practicality [30,157,158].

To support these improvements, multidisciplinary collaboration between chemists, material scientists, process engineers, and computational experts will be vital. Integrating new fields, such as green chemistry, artificial intelligence, and synthetic biology, could change the future of membrane science. This integration would lead to scalable, high-performance PIMs with low environmental impact [159–163]. Table 12 offers a comparison of emerging research directions, their goals, and related technical challenges.

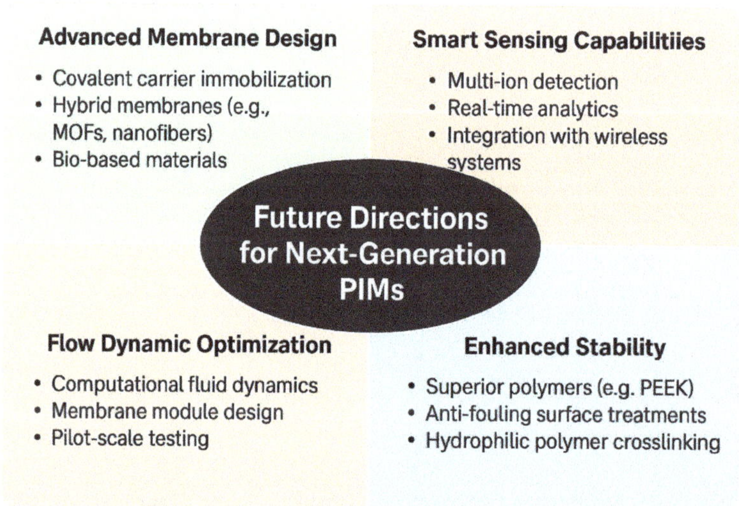

Figure 13. Comparative overview of future research directions.

Table 12. Comparative Overview of Future Research Directions in PIMs.

Research Direction	Objective	Key Challenges
Green Polymer Matrices	Develop fully biodegradable and non-toxic matrices	Limited mechanical stability and low compatibility with conventional carriers
Advanced Carrier Anchoring	Enhance carrier stability through covalent attachment or encapsulation	Synthesis complexity and limited scalability
Stimuli-Responsive PIMs	Enable selective transport triggered by pH, light, or temperature	Slow response time and material fatigue under cycling
Multiscale Modeling and Simulation	Predict transport performance under complex conditions	High computational cost and lack of experimental validation
Hybrid Membrane Structures	Combine PIMs with MOFs, COFs, or electrospun layers	Interfacial incompatibility and manufacturing difficulties

9. Conclusions

The development of polymer inclusion membranes (PIMs) has become an important area in separation science. PIMs offer a unique mix of selectivity, structural flexibility, and environmental compatibility. In the last twenty years, there has been significant progress in material design, membrane engineering, and the range of applications. The findings presented in this manuscript, spanning fundamental chemistry, functionalization strategies, fabrication techniques, environmental applications, and prospects, highlight the multidisciplinary potential of PIM-based systems.

PIMs are highly effective in selective ion transport, particularly for heavy metals and rare-earth elements. This is thanks to their specific carrier-polymer design and improved transport methods. Developments in polymer matrix engineering, like sulfonation, carboxylation, nanofiller incorporation, and blending with biofillers, have significantly improved membrane stability, permeability, and reusability. The use of ionic liquids and smart extractants has also improved the specificity and efficiency of metal separation processes. From a sustainability standpoint, the move towards biodegradable polymers and green solvents demonstrates an increasing awareness of environmental impacts and regulatory requirements. This shift has been supported by using green plasticizers and solvent-free fabrication methods, allowing for safer membrane production without lowering performance. Scalability and industrial feasibility remain critical milestones. The successful transfer of laboratory-scale fabrication techniques, such as phase inversion, spin coating, and electrospinning, to pilot and full-scale production environments demonstrates the readiness of PIM technology. Nonetheless, standardization and quality control still pose technical challenges, particularly in achieving reproducibility and consistent membrane quality. PIMs are also gaining popularity in analytical chemistry and sensor development. Optical and electrochemical sensors based on PIM technology can quickly, selectively, and accurately detect trace analytes. This highlights their value in environmental monitoring and medical diagnostics.

Despite this progress, some limitations, such as carrier leaching, reduced lifetime in aggressive environments, and gaps in regulations, highlight the need for ongoing innovation. Chapter 8 discussed strategic research directions, including stimuli-responsive membranes, bioinspired architectures, and AI-guided membrane creation, which could help overcome current obstacles.

In conclusion, PIMs are evolving into scalable, sustainable, and multifunctional membrane solutions. The combination of green chemistry, nanotechnology, and computational tools is expected to shape the next generation of high-performance PIM systems that can tackle global challenges in clean water, resource recovery, and environmental health.

Author Contributions: Conceptualization, A.N.-Z.; validation, A.N.-Z.; writing—original draft preparation, A.N.-Z.; writing—review and editing, A.N.-Z.; visualization, A.N.-Z.; supervision, V.S. All authors have read and agreed to the published version of the manuscript.

Funding: This research received no external funding.

Institutional Review Board Statement: Not applicable.

Informed Consent Statement: Not applicable.

Data Availability Statement: Not applicable.

Conflicts of Interest: The authors declare no conflicts of interest.

References

1. Kagaya, S.; Maeno, T.; Ito, K.; Gemmei-Ide, M.; Catrrrall, R.W.; Kolev, S.D. Improvement of Chromium(VI) extraction from acidic solutions using a poly(vinyl chloride)-based polimer inclusion membranę with Aliquat336 as the carrier. *Anal. Sci.* **2017**, *33*, 643–646. [CrossRef] [PubMed]
2. Kagaya, S.; Ryokan, Y.; Tohda, T.; Cattrall, R.W.; Kolev, S.D. Stability Studies of Poly(vinyl chloride)-Based Polymer Inclusion Membranes Containing Aliquat 336 as a Carrier. *Sep. Purif. Technol.* **2012**, *101*, 69–75. [CrossRef]
3. Uğur, A.; Sener, I.; Hol, A.; Alpoguz, H.K.; Elci, L. Facilitated transport of Zn(II) and Cd(II) ions through polymer inclusion membranes immobilized with calix[4]resorcinarene derivative. *J. Macromol. Sci. A* **2014**, *51*, 611–618. [CrossRef]
4. Shannon, M.A.; Bohn, P.W.; Elimelech, M.; Georgiadis, J.G.; Marinas, B.J.; Mayes, A.M. Science and technology for water purification in the coming decades. *Nature* **2008**, *452*, 301–310. [CrossRef] [PubMed]
5. Suksaroj, C.; Héran, M.; Allègre, C.; Persin, F. Treatment of textile plant effluent by nanofiltration and/or reverse osomosis for water reuse. *Desalination* **2005**, *178*, 333–341. [CrossRef]
6. Vera, R.; Anticó, E.; Fontàs, C. The Use of a Polymer Inclusion Membrane for Arsenate Determination in Groundwater. *Water* **2018**, *10*, 1093. [CrossRef]
7. Fontàs, C.; Vera, R.; Anticó, E. New Insights on the effects of water on polimer inclusion membranes containing Aliquat 336 Derivatives as Carriers. *Membranes* **2022**, *12*, 192. [CrossRef]
8. Gherasim, C.V.; Cristea, M.; Grigoras, V.C.; Bourceanu, G. New polymer inclusion membrane. Preparation and characteryzation. *Dig. J. Nanomater. Biostructures* **2011**, *6*, 1499–1508.
9. Almeida, M.I.G.S.; Cattrall, R.W.; Kolev, S.D. Recent trends in extraction and transport of metal ions using polymer inclusion membranes (PIMs). *J. Membr. Sci.* **2012**, *415–416*, 9–23. [CrossRef]
10. Al-Haddabi, M.; Vuthaluru, H.; Ahmed, M.; Znad, H. Use of Ceramic Membrane Technology for Sustainable Management of Oil Production Water: A Review. In *Recent Progress in Desalination, Environmental and Marine Outfall Systems*; Springer Nature: Berlin/Heidelberg, Germany, 2015; pp. 11–23.
11. Hakami, M.W.; Alkhudhri, A.; Al-Batty, S.; Zacharof, M.-P.; Maddy, J.; Hilal, N. Ceramic Microfiltration Membranes in Wastewater Treatment: Filtration Behavior, Fouling and Prevention. *Membranes* **2020**, *10*, 248. [CrossRef]
12. Zielińska, M.; Galik, M. Use of Ceramic Membranes in a Membrane Filtration Supported by Coagulation for the Treatment of Dairy wastewater. *Water Air Soil Pollut.* **2017**, *228*, 173. [CrossRef]
13. Asif, M.B.; Zhang, Z. Ceramic membranę technology for water and wastewater treatment: A critical review of performance, fullscale applications, membranę fouling and prospects. *Chem. Eng. J.* **2021**, *418*, 129481. [CrossRef]
14. Almeida, M.I.; Cattrall, R.W.; Kolev, S. Polymer Inclusion Membranes: Concept and applications. *Procedia Eng.* **2012**, *44*, 681–682. [CrossRef]
15. Płotka-Wasylka, J.; Namiesnik, J. *Green Analytical Chemistry Past, Present and Perspectives*; Springer Nature: Berlin/Heidelberg, Germany, 2019; ISBN 978-981-13-9104-0.
16. Sgarlata, C.; Arena, G.; Longo, E.; Zhang, D.; Yang, Y.; Bartsch, R.A. Heavy metal separation with polimer inclusion membranes. *J. Membr. Sci.* **2008**, *323*, 444–451. [CrossRef]
17. Ouchn, R.; Chaouqi, Y.; Oukkass, S.; Santoro, S.; Avci, A.H.; Curcio, E.; Hlaibi, M. Advancing Sustainable Direct Contact Membrane Distillation: Performance and Stability of Novel Polymer Inclusion membranes. *Sep. Purif. Technol.* **2025**, *354*, 129056. [CrossRef]
18. Musa, A.A.; Bello, A.; Adams, S.M.; Onwualu, A.P.; Anye, V.C.; Bello, K.O.; Obianyo, I.I. Nano-Enhanced Polymer Composite Materials: A Review of Current Advancements and Challenges. *Polymers* **2025**, *17*, 893. [CrossRef]
19. Nitti, F.; Selan, O.T.E.; Hoque, B.; Tambaru, D.; Djunaidi, M.C. Improving the Performance Inclusion Membranes in Separation Process Using Alternative Base Polymers: A review. *Indones. J. Chem.* **2022**, *22*, 284–302. [CrossRef]

20. Zante, G.; Boltoeva, M.; Masmoudi, A.; Barillon, R.; Trébouet, D. Supported ionic liquid and polimer inclusion membranes for metal separation. *Sep. Purif. Rev.* **2022**, *51*, 100–116. [CrossRef]
21. Xiao, T.; Zhu, Z.; Li, L.; Shi, J.; Li, Z.; Zuo, X. Mebrane fouling and cleaning strategies in microfiltration/ultrafiltration and dynamic membrane. *Sep. Purif. Technol.* **2023**, *318*, 123977. [CrossRef]
22. Zulkefli, N.F.; Alias, N.H.; Jamaluddin, N.S.; Abdullah, N.; Manaf, S.F.A.; Othman, N.H.; Marpani, F.; Mat-Shayuti, M.S.; Kusworo, T.D. Recent Mitigation Strategies on Membrane Fouling for Oily Wastewater Treatment. *Membranes* **2021**, *12*, 26. [CrossRef]
23. Shukla, B.K.; Sharma, P.K.; Yadav, H.; Singh, S.; Tyagi, K.; Yadav, Y.; Rajpoot, N.K.; Rawat, S.; Verma, S. Advanced membrane technologies for water treatment: Utilization of nanomaterials and nanoparticles in membranes fabrication. *J. Nanoparticle Res.* **2024**, *26*, 222. [CrossRef]
24. Sellami, F.; Senhadji, O.K.; Marais, S.; Fatyeyeva, K. PVC/EVA-based polymer inclusion membranes with improved stability and Cr(VI) extraction capacity: Water plasticization effect. *J. Hazard. Mater.* **2022**, *436*, 129069. [CrossRef] [PubMed]
25. Ovejero-Pérez, A.; Nakasu, P.Y.S.; Hopson, C.; Costa, J.M.; Hallett, J.P. Challenges and opportunities on the utilisation of ionic liquid for biomass pretreatment and valorisation. *Mater. Sustain.* **2024**, *2*, 7. [CrossRef]
26. Wong, K.C.; Gog, P.S.; Ismail, A.F.; Kang, H.S.; Guo, Q.; Jiang, X.; Ma, J. The State-of-the-Art Functionalized Nanomaterials for Carbon Dioxide Separation Membrane. *Membranes* **2022**, *12*, 186. [CrossRef]
27. Hernández-Fernández, A.; Iniesta-López, E.; Ginestá-Anzola, A.; Garrido, Y.; Pérez de los Ríos, A.; Quesada-Medina, J.; Hernádez-Fernádez, F. Polymeric Inclusion Membranes Based on Ionic Liquids for Selective Separation of Metal Ions. *Membranes* **2023**, *13*, 795. [CrossRef]
28. Adigun, B.; Thapaliya, B.P.; Luo, H.; Dai, S. Ionic Liquid-Based Extraction of Metal Ions via Polymer Inclusion Membranes: A Critical Review. *RSC Sustain.* **2024**, *2*, 2768–2780. [CrossRef]
29. Kazemi, D.; Yaftian, M.R. PVDF-HFP-Based Polymer Inclusion Membrane Functionalized with D2EHPA for the Selective Extraction of Bismuth(III) from sulfate media. *Sci. Rep.* **2024**, *14*, 11622. [CrossRef]
30. Senila, M. Polymer Inclusion Membranes (PIMs) for Metal Separation-Toward Environmentally Friendly Production and Applications. *Polymers* **2025**, *17*, 725. [CrossRef] [PubMed]
31. Keskin, B.; Zeytuncu-Gökoğlu, B.; Koyuncu, I. Polymer inclusion membrane applications for transport of metal ions: A critical review. *Chemosphere* **2021**, *279*, 130604. [CrossRef] [PubMed]
32. Yaacob, S.F.F.; Olasupo, A.; Suah, F.B.M. Polymer Inclusion Membranes Based Optode: Recent Advances and Perspectives. *TrAC—Trends Anal. Chem.* **2024**, *171*, 117498. [CrossRef]
33. Bloch, R.; Kedem, O.; Vofsi, D. Ion Specific Mambrane. *Nature* **1963**, *199*, 802–803. [CrossRef]
34. Kolev, S.D.; Ines, M.I.G.; Cattrall, R.W. Polymer Inclusion Membranes. In *Handbook of Membrane Separations*; CRC Press: Boca Raton, FL, USA, 2023.
35. Kaczorowska, M. The use of polymer inclusion membranes for the removal of metal ions from aqueous solutions—The Latest Achievements and potential industrial applications: A review. *Membranes* **2022**, *12*, 1135. [CrossRef]
36. Latos-Brozios, M.; Rułka, K.; Masek, A. Review of Bio-fillers Dedicated of Polymer Compositions. *Chem. Biodivers.* **2025**, *22*, e202500406. [CrossRef]
37. Maiphetlho, K.; Shumbula, N.; Motsoane, N.; Chimuka, L.; Richards, H. Evaluation of Silver-Nanocomposite PIMs for Trace Metal Transports. Selectivity and stability studies. *J. Water Process Eng.* **2020**, *37*, 101527. [CrossRef]
38. Negul, E.A.; Fontas, C.; McKelvie, I.D.; Cattrall, R.W.; Kolev, S.D. The use of a polymer inclusion membrane for separation and preconcentration of orthophosphate in flow analysis. *Anal. Chim. Acta* **2013**, *803*, 82–90. [CrossRef] [PubMed]
39. Nghiem, D.; Mornane, P.; Potter, I.D.; Perera, J.M.; Cattrall, R.W.; Kolev, S.D. Extraction and transport of metal ions and small organic compounds using polymer inclusion membranes (PIMs). *J. Membr. Sci.* **2006**, *281*, 7–41. [CrossRef]
40. O'Rourke, M.; Cattrall, R.W.; Kolev, S.D.; Potter, I.D. Electrochemical Impedance Spectroscopy-A Simple Method for the Characterization of Polymer Inclusion Membranes Containing Aliquat 336. *Membranes* **2011**, *1*, 132–148. [CrossRef]
41. Gherasim, C.V.; Bourceanu, G.; Timpu, D. Experimental and modeling studies of lead(II) sorption onto a polyvinyl-chloride inclusion membrane. *Chem. Eng. J.* **2011**, *172*, 817–827. [CrossRef]
42. Kolev, S.D.; St John, A.M.; Cattrall, R.W. Mathematical modelling of the extraction of uranium(VI) into a polymer inclusion membrane composed of PVC and di-(2-ethylhexyl) phosphoric acid. *J. Membr. Sci.* **2013**, *425–426*, 169–175. [CrossRef]
43. Garcia-Rodríguez, A.; Matamoros, V.; Kolev, S.D.; Fontàs, C. Development of a polimer inclusion membrane (PIM) for the preconcentration of antibiotics in environmental water samples. *J. Membr. Sci.* **2015**, *492*, 32–39. [CrossRef]
44. Matsumoto, M.; Takemori, S.; Tahara, Y. Lactic acid permeation through deep eutectic solvents-based polymer inclusion membranes. *Membranes* **2020**, *10*, 244. [CrossRef]
45. Kończyk, J.; Ciesielski, W. Calixresorcin[4]arene-Mediated Transport of Pb)II) Ions through Polymer Inclusion Membrane. *Membranes* **2021**, *11*, 285. [CrossRef] [PubMed]

46. Sánchez-Ponce, L.; Casanueva-Marenco, M.J.; Diaz-de-Alba, M.; Galindo-Riano, M.D.; Granada-Castro, M.D. A novel polymer inclusion membrane-based green optical sensor for selective determination of Iron: Design, characterization, and analytical applications. *Polymers* **2023**, *15*, 4082. [CrossRef]
47. Taghizadeh, M.; Taghizadeh, A.; Vatanpour, V.; Ganjali, M.R.; Saeb, M.R. Deep Eutectic Solvents in Membrane science and technology: Fundamental, preparation, application and future perspective. *Sep. Purif. Technol.* **2020**, *258*, 118015. [CrossRef]
48. Kandel, D.R.; Kwak, D.; Lee, S.; Lim, Y.J.; Subedi, S.; Lee, J. Harnessing natural antifouling agents for enhancing water and wastewater treatment membranes. *Sep. Purif. Technol.* **2025**, *359*, 130254. [CrossRef]
49. Fang, H.; Han, Y.; Wang, J.; Ding, Y.; He, M.; Lan, J.; Ding, X.; Shi, W.; Liu, X. Calix[4]arene—Decorated polymers of intrinsic microporosity for lithium isotopes separation. *Chem. Eng. J.* **2024**, *500*, 156916. [CrossRef]
50. Yaftian, M.R.; Burgard, M.; Dieleman, C.B.; Matt, D. Rare-earth metal-ion separation using a supported liquid membrane mediated by a narrow rim phosphorylated calix[4]arene. *J. Membr. Sci.* **1998**, *144*, 57–64. [CrossRef]
51. Syeda, S.E.Z.; Nowacka, D.N.; Khan, M.S.; Skwierawska, A. Recent Advancements in Cyclodextrin-Based Adsorbants for the Removal of Hazardous pollutants from Waters. *Polymers* **2022**, *14*, 2341. [CrossRef]
52. Kozlowski, C.A.; Walkowaik, W.; Girek, T. Modified cyclodextrin polymers as selectiveion carriers for Pb(II) separation across plasticized membranes. *J. Membr. Sci.* **2008**, *310*, 312–320. [CrossRef]
53. Liu, D.; Huang, Z.; Li, M.; Sun, P.; Yu, T.; Zhou, L. Novel porous magnetic nanophores funcionalized by ß-cyclodextrin polimer and its application in organic pollutants from aqueous solution. *Environ. Pollut.* **2019**, *250*, 639–649. [CrossRef]
54. Miyah, Y.; El Messaoudi, N.; Benjelloun, M.; Georgin, J.; Franco, D.S.P.; El-Habacha, M.; Ali, O.A.; Acikbas, Y. A comprehensive review of ß-cyclodextrinpolymer nanocomposities exploration for heavy metal removal from wastewater. *Carbohydr. Polym.* **2025**, *350*, 122981. [CrossRef]
55. Liu, Y.; Huang, S.; Zhao, X.; Zhang, Y. Fabrication of three-dimensional β-cyclodextrin/chitosan functionalized Graphene oxide hydrogel for methylene blue removal from aqueous solution. *Colloids Surf. A* **2019**, *539*, 1–10. [CrossRef]
56. Moulahcene, L.; Skiba, M.; Bounoure, F.; Benamor, M.; Milon, N.; Hallouard, F.; Lahiani-Skiba, M. New Polymer Inclusion membrane Containing β-cyclodextrin Polymer: Application for Pharmaceutical Pollutant Removal from Wastewater. *Int. J. Environ. Res. Public Health.* **2019**, *16*, 414. [CrossRef] [PubMed]
57. Edward, K.; Yuvaraj, K.M.; Kapoor, A. Chitosan-blende membranes for heavy metal removal from aqueous systems: A review of synthesis, separation, mechanism, and performance. *Int. J. Biol. Macromol.* **2024**, *279*, 134996. [CrossRef] [PubMed]
58. Nowik-Zajac, A.; Zawierucha, I.; Kozlowski, C. Selective removal of silver(I) using polymer inclusion membranes containing calixpyrroles. *RSC Adv.* **2019**, *9*, 31122–31132. [CrossRef]
59. Zawierucha, I.; Nowik-Zajac, A.; Kozlowski, C. Removal of Pb(II) ions using polymer inclusion membranes containing calix[4]resorcinarene derivative as ion carrier. *Polymers* **2019**, *11*, 2111. [CrossRef]
60. Rout, A.; Binnemans, K. Influence of the ionic liquid cation on the solvent extraction of trivalent rare-earth ions by mixtures of Cyanex 923 and ionic liquids. *Dalton Trans.* **2015**, *44*, 1379–1387. [CrossRef] [PubMed]
61. Wang, L.Y.; Lee, M.-S. Separation of Co(II) and Ni(II) from chloride leach solution of nickel laterite ore by solvent extraction with Cyanex 301. *Int. J. Miner. Process.* **2017**, *166*, 45–52. [CrossRef]
62. Skoronski, E.; Fernandes, M.; Malaret, F.J.; Hallett, J.P. Use of phosphonium ionic liquids for highly efficient extraction of phenolic compounds from water. *Sep. Purif. Technol.* **2020**, *248*, 117069. [CrossRef]
63. Fraser, K.J.; MacFarlane, D.R. Phosphonium-based Ionic Liquids: An Overview. *Aust. J. Chem.* **2009**, *62*, 309–321. [CrossRef]
64. Zhu, Y.; Qi, B.; Liang, X.; Luo, J.; Wan, Y. Comparison of Corn Stover Pretreatments with Lewis Acid Catalyzed Choline Chloride, Glycerol and Choline Chloride-Glycerol Deep Eutectic Solvent. *Polymers* **2021**, *13*, 1170. [CrossRef]
65. Kozłowski, C.; Zawierucha, I. Polymer Inclusion Membranes Based on Sulfonic Acid Derivatives as Ion Carriers for Selective Separation of Pb(II) Ions. *Membranes* **2025**, *15*, 146. [CrossRef]
66. Bashiri, A.; Nikzad, A.; Maleki, R.; Asadnia, M.; Razmjou, A. Rare earth elements recovery using selective membranes via extraction and rejection. *Membranes* **2022**, *12*, 80. [CrossRef]
67. Croft, C.F.; Nagul, E.A.; Almeida, M.I.G.S.; Kolev, S.D. Polymer-based extracting materials in the green recycling of rare earth elements: A review. *ACS Omega* **2024**, *9*, 40315–40328. [CrossRef]
68. Husna, S.M.; Yusoff, A.H.; Mohan, M.; Azmi, N.A.; Ter, T.P.; Shoparwe, N.F.; Sulaiman, A.Z. Graphene Oxide (GO) Membrane in Removing Heavy Metals from Wastewater: Effect of Graphene Oxide on the Properties of Polymer Inclusion Membranes for Gold Extraction from Acidic Solution. *Membranes* **2022**, *12*, 996. [CrossRef] [PubMed]
69. Kwak, D.; Tran, H.M.; Kandel, D.R.; Lee, J. Synergistic effects of MOF catalysts and a 3D junction enable thin bipolar membranes to reach the thermodynamic minimum potential for water dissociation. *Chem. Eng. J.* **2025**, *507*, 160573. [CrossRef]
70. Ncib, S.; Barhoumi, A.; Bouguerra, W.; Larchet, C.; Dammak, L.; Hamrouni, B.; Elaloui, E. Preparation and Characterization of Cellulose Triacetate Polymer Inclusion Membrane Blended with Acetylated Kraft Lignin: Effect of AKL and Application to the Extraction of Copper(II) from Acidic Media. *Desalination Water Treat.* **2018**, *104*, 263–272. [CrossRef]

71. Almeida, M.I.G.S.; Cattrall, R.W.; Kolev, S.D. Polymer inclusion membranes (PIMs) in chemical analysis—A review. *Anal. Chim. Acta* **2017**, *987*, 1–14. [CrossRef]
72. O'Bryan, Y.; Truong, Y.B.; Cattrall, R.W.; Kyratzis, I.L.; Kolev, S.D. Electrospun polystyrene/Aliquat 336 for preconcentration and determination of thiocyanate in flow analysis. *Electrospinning* **2017**, *1*, 100–110. [CrossRef]
73. Geng, Z.; Yang, X.; Boo, C.; Zhu, S.; Lu, Y.; Fan, W.; Huo, M.; Elimelech, M.; Yang, X. Self cleaning anti fouling hybrid ultrafiltration membranes via side chain grafting of poly(aryl ether sulfone) and titanium dioxide. *J. Membr. Sci.* **2017**, *529*, 1–10. [CrossRef]
74. Wang, J.; Sun, Y.; Bi, W.; Jiang, Z.; Zhang, M.; Pang, J. High-strength corrosion resistant membranes for the separation of oil/water mixtures and immiscible oil mixtures based on PEEK. *J. Membr. Sci.* **2020**, *616*, 118418. [CrossRef]
75. O'Bryan, Y.; Truong, Y.B.; Cattrall, R.W.; Kyratzis, I.L.; Kolev, S.D. A New Generation of Highly Stable and Permeable Polymer Inclusion Membranes (PIMs) with Their Carrier Immobilized in a Cross-Linked Semi-Interpenetrating Polymer Network. Application to the transport of thiocyanate. *J. Membr. Sci.* **2017**, *529*, 55–62. [CrossRef]
76. Aristizábal, S.L.; Chisca, S.; Pulido, B.A.; Nunes, S.P. Preparation of PEEK Membranes with Excellent Stability Using Common Organic Solvents. *Ind. Eng. Chem. Res.* **2020**, *59*, 5218–5226. [CrossRef]
77. Ncib, S.; Brahmi, K.; Othmen, K.; Lasaad, D.; Elaloui, E.; Bouguerra, W. Enhanced Ni(II) recovery via D2EHPA-cellulose triacetate-based polymer inclusion membranes: Optimized extraction and stability analysis. *Euro-Mediterr. J. Environ. Integr.* **2025**, *10*, 2245–2263. [CrossRef]
78. Chang, S.; Kandal, D.R.; Lee, U.; Tran, H.M.; Lee, J. Improving separation efficiency of various micropollutants from water by polymer-enhanced ultrafiltration using oxidized alginate featuring less viscous and low filtration resistance. *J. Environ. Chem. Eng.* **2024**, *12*, 114753. [CrossRef]
79. Wang, H.; He, S.; Qin, X.; Li, C.; Li, T. Interfacialo Engineering in metal organic framework-based mixed matrix membranes using covalently grafted polyimide Brushes. *J. Am. Chem. Sci.* **2018**, *140*, 17203–17210. [CrossRef]
80. Astorino, C.; De Nardo, E.; Lattieri, S.; Ferraro, G.; Pirri, C.F.; Bocchini, S. Advancements in gas separation for energy applications: Exploring the potential of polymer membranes with intrinsic microporosity (PIM). *Membranes* **2023**, *13*, 903. [CrossRef]
81. Dong, X.; Lu, D.; Harris, T.A.; Escobar, I.C. Polymers and solvents used in membrane fabrication: A review focusing on sustainable membrane development. *Membranes* **2021**, *11*, 309. [CrossRef]
82. Fan, Y.; Yu, W.; Wu, A.; Shu, W.; Zhang, Y. Recent progress on CO_2 separation membranes. *RSC Adv.* **2024**, *14*, 20714. [CrossRef]
83. Li, Y.-H.; Wang, D.-M.; Hsien, T.-Y.; Chan, K.-Y.; Lai, J.-Y. Polymer Inclusion Membranes with Strip Dispersion. *Water* **2017**, *9*, 399. [CrossRef]
84. Radzyminska-Lenarcik, E.; Pyszka, I. New polymer inclusion membrane in the separation of nonferrous metal ion from aqueous solutions. *Membranes* **2020**, *10*, 385. [CrossRef] [PubMed]
85. Heo, J.; Choi, M.; Chang, J.; Ji, D.; Kang, S.W.; Hong, J. Highly permeable graphene oxide/polyelectrolytes hybrid thin films for enhanced CO_2/N_2 separation performance. *Sci. Rep.* **2017**, *7*, 456. [CrossRef]
86. Lee, H.J.; Kang, S.W. CO_2 Separation with polymer/aniline composite membranes. *Polymers* **2020**, *12*, 1363. [CrossRef] [PubMed]
87. Fatima, N.; Waheed, H.; Raza, A.; Mukhtar, A.; Ahmad, F. A Critical Review on membrane Technology and Its application in CO_2 Capturing. *Chem. Methodol.* **2024**, *8*, 662–698.
88. Tong, Z.; Sekizkardes, A.K. Recent developments in high-performance membranes for CO_2 separation. *Membranes* **2021**, *11*, 156. [CrossRef] [PubMed]
89. Yahia, M.; Refaat, D.; Coronas, J.; Tellez, C. Enhancing CO_2/CH_4 separation performance in PIM-1 based MXene nanosheets mixed matrix membranes. *Sep. Purif. Technol.* **2025**, *356*, 129825. [CrossRef]
90. Ferraro, G.; Astorino, C.; Bartoli, M.; Martis, A.; Lattieri, S.; Pirri, C.F.; Bocchini, S. Ionic Liquid-Polymer of Intrinsic Microporosity (PIMs) Blend Membranes for CO_2 Separation. *Membranes* **2022**, *12*, 1262. [CrossRef]
91. Zeng, H.; Qu, X.; Xu, D.; Luo, Y. Porous adsorption materials for carbon dioxide capture in industrial flue gas. *Front. Chem.* **2022**, *10*, 939701. [CrossRef]
92. Sánchez-Ponce, L.; Galindo-Riano, M.D.; Casanueva-Marenco, M.J.; Granada-Castro, M.D.; Diaz-de-Alba, M. Sensing Cd(II) using a disposable optical sensor based on a Schiff base immobilization on a polymer-inclusion membrane. Applications in water and art paint samples. *Polymers* **2021**, *13*, 4414. [CrossRef]
93. Kuswandi, B.; Nitti, F.; Almeida, M.I.G.S.; Kolev, S.D. Water monitoring using polymer inclusion membranes: A review. *Environ. Chem. Lett.* **2019**, *18*, 129–150. [CrossRef]
94. Alshehri, R.F.; Amin, A.S.; Aish, M. PVC-DOS membrane immobilized with 2-amino-4-(3-chloro-phenylozo) pyridine-3-ol for environmentally friendly detection of Cd(II) ions. *Chem. Data Collect.* **2023**, *48*, 101098. [CrossRef]
95. Suah, F.B.M.; Ahmad, M. Preparation and characterization of polimer inclusion membrane based optode for determination of Al^{3+} ion. *Anal. Chim. Acta* **2017**, *951*, 133–139. [CrossRef]
96. Alharbi, K.H. A Review on Organic Colorimetric and Fluoroscent Chemosensors for the Detection of Zn(II) Ions. *Crit. Rev. Anal. Chem.* **2023**, *53*, 1472–1488. [CrossRef]

97. Alizadeh, N.; Moemeni, A.; Shamsipur, M. Poly(vinyl chloride)-membrane ion-selective bulk optode based on 1, 10-dibenzyl-1, 10-diaza-18-crown-6 and 1-(2-pyridylazo)-2-naphthol for Cu^{2+} and Pb^{2+} ions. *Anal. Chim. Acta* **2002**, *464*, 187–196. [CrossRef]
98. Meng, X.; Jiang, X.; Long, Y.; Chen, J.; Wang, L.; Zhang, Y. Optical sensing membrane for determination of trace cadmium(II), zinc(II) and cooper(II) based on immobilization of 1-(2-pyridylazo)-2-naphthol on polymer inclusion membrane. *Microchem. J.* **2021**, *162*, 105767. [CrossRef]
99. Lenar, N.; Piech, R.; Paczosa-Bator, B. A new planar potentiometric sensor for in situ measurements. *Sensors* **2024**, *24*, 2492. [CrossRef] [PubMed]
100. Fathy, M.A.; Bühlmann, P. Next-generation potentiometric sensors: A review of flexible and wearable technologies. *Biosensors* **2025**, *15*, 51. [CrossRef]
101. Apostu, M.; Vieriu, M.; Bibire, N.; Panainte, A.; Tantaru, G. Design and Study of Electrochemical Sensors Based on Polymer Inclusion Membranes Containing Polyoxometalate. *Mater. Plast.* **2019**, *56*, 429–433. [CrossRef]
102. Lyu, Y.; Gan, S.; Bao, Y.; Zhong, L.; Xu, J.; Wang, W.; Liu, Z.; Ma, W.; Yang, G.; Niu, L. Solid-contact Ion-Selective Electrodes: Response Mechanisms, Transducer Materials and Wearable Sensors. *Membranes* **2020**, *10*, 128. [CrossRef]
103. Sulthana, S.F.; Iqbal, U.M.; Suseela, S.B.; Chinthaginjala, R.; Chittathuru, D. Electrochemical Sensors for Heavy Metal Ion Detection in Aqueous Medium: A Systematic Review. *ACS Omega* **2024**, *9*, 25493–25512. [CrossRef]
104. Mancilla-Rico, A.J. Simultaneous determination of Cu(II), Zn(II), and Pb(II) from aqueous solutions using a polymer inclusion membrane (PIM) based-sensor with 1-(2-pyridylazo)-2-naphthol (PAN) as chromophore and chemometric methods. *Front. Anal. Sci.* **2022**, *2*, 971352. [CrossRef]
105. Suah, F.B.M. Preparation and Characterisation of a Novel Co(II) Optode Based on Polymer Inclusion Membrane. *Anal. Chem. Res.* **2017**, *12*, 40–46. [CrossRef]
106. Carner, C.A.; Croft, C.; Kolev, S.; Almeida, M.I. Green Solvents for the fabrication of polymer inclusion membranes (PIMs). *Sep. Purif. Technol.* **2020**, *239*, 116486. [CrossRef]
107. Maiphetlho, K.; Chimuka, L.; Tutu, H.; Richards, H. Technical design and optimization of polymer inclusion membranes (PIMs) for sample pretreatment and passive sampling—A review. *Sci. Total Environ.* **2021**, *799*, 149483. [CrossRef] [PubMed]
108. Xie, W.; Li, T.; Tiraferri, A.; Drioli, E.; Figoli, A.; Crittenden, J.C.; Liu, B. Toward the Next Generation of Sustainable Membranes from Green Chemistry Principles. *ACS Sustain. Chem. Eng.* **2021**, *9*, 50–75. [CrossRef]
109. Zhao, S.; Samadi, A.; Wang, Z.; Pringle, J.M.; Zhang, Y.; Kolev, S.D. Ionic liquid-based polymer inclusion membranes for metal ions extraction and recovery: Fundamentals, considerations, and prospects. *Chem. Eng. J.* **2024**, *481*, 148792. [CrossRef]
110. Baczyńska, M.; Waszak, M.; Nowicki, M.; Prządka, D.; Borysiak, S.; Regel-Rosocka, M. Characteryzation of polymer inclusion membranes (PIMs) containing phosphonium ionic liquids as Zn(II) carriers. *Ind. Eng. Chem. Res.* **2018**, *57*, 5070–5082. [CrossRef]
111. Cindradewi, A.W.; Bandi, R.; Park, C.-W.; Park, J.-S.; Lee, E.-A.; Kim, J.-K.; Kwon, G.-J.; Han, S.-Y.; Lee, S.-H. Preparation and characterization of cellulose acetate film reinforced with cellulose nanofibril. *Polymers* **2021**, *13*, 2990. [CrossRef]
112. Mena-Prado, I.; Fernández-Garcia, M.; Blázquez-Blázquez, E.; Muñoz-Bonilla, A.; Del Campo, A. Plasticizing PLA with Biobased fatty esters: Comprehensive study on film properties. *J. Polym. Environ.* **2025**, *33*, 1337–1352. [CrossRef]
113. Cocco, S.; Resende Assis Silva, R.; Veloso De Oliveira, T.; Stringheta, P.; Moacir Ribeiro Pinto, M. Glycerol and triethyl citrate plasticizer effects om molecular, thermal, mechanical, and barrier properties of cellulose acetate films. *Food Biosci.* **2021**, *42*, 101202.
114. Rachid, O.; Youssf, C.; Louafy, R.; Avci, A.H.; Santoro, S.; Cherkaoui, O.; Hlaibi, M. Polymer inclusion membranes with long term-stability in desalination via membrane distillation. *Chem. Eng. Process.* **2023**, *191*, 109442.
115. Cheng, X.; Bae, J. Recent advancements in fabrication, separation, and purification of hierarchically porous polymer membranes and their applications in next-generation electrochemical energy storage devices. *Polymers* **2024**, *16*, 3269. [CrossRef]
116. Ghosal, K.; Agatemor, C.; Tucker, N.; Kny, E.; Thomas, S. Chapter 1: Electrical Spinning to Electrospinning: A Brief History. In *Electrospinning: From Basic Research to Commercialization*; Royal Society of Chemistry: London, UK, 2025; pp. 1–23. ISBN 978-1-78801-491-5.
117. Gkika, D.A.; Filiz, V.; Rongou, S.; Kyzas, G.K.; Mitropoulos, A.C. Cost profile of membranes that use polymers of intrinsic microporosity (PIMs). *Membranes* **2022**, *12*, 433. [CrossRef]
118. Caliskan, E.; Shishatskiy, S.; Abetz, V.; Filiz, V. Pioneering the preparation of porous PIM-1 membranes for enhanced water vapor flow. *RSC Adv.* **2024**, *14*, 9631–9645. [CrossRef]
119. Goh, L.M.; Thong, Z.; Li, W.P.; Ooi, T.S.; Esa, F.; Ng, K.S.; Dhalla, A.; Gudipati, C. development and Industrial-scale Fabrication of Next Generation Low-Energy membranes for Desalination. *Membranes* **2022**, *12*, 540. [CrossRef]
120. Dischinger, S.M.; Miller, D.J.; Vermaas, D.A.; Kingsbury, R.S. Unifying the Conversation: Mebrane Separation Performance in Energy, Water, and Industrial Applications. *ACS EST Engg.* **2024**, *4*, 277–289. [CrossRef]
121. Jiang, L.; Yan, C.; Zhai, Y.-M.; Luo, F.; Meng, H.; Yin, B.; Zhang, k.; Yang, M.-B. Highly-toughened PLLA/PVA biodegradable blends: Graft copolymer tailored crystallization and phase morphology. *Polymer* **2024**, *312*, 127606. [CrossRef]
122. Zawierucha, I.; Nowik-Zajac, A.; Malina, G. Selective Removal of As(V) Ions from Acid Mine Drainage Using Polymer Inclusion Membranes. *Minerals* **2020**, *10*, 909. [CrossRef]

123. Ihueze, C.C.; Ekhibise, N.; Onwurah, U.O.; Okafor, C.E. Process Parameters Optimization in Plastic manufacturing Industry using machine Learning. *J. Innov. Res. Eng. Sci.* **2024**, *5*, 572–589.
124. Kian, L.K.; Saba, N.; Jawaid, M.; Sultan, M.T.H. A Review on Processing Techniques of Bast Fibers Nanocellulose and Its Polylactic Acid (PLA) Nanocomposites. *Int. J. Biol. Macromol.* **2019**, *121*, 1314–1328. [CrossRef] [PubMed]
125. Kaczorowska, M.; Bożejewicz, D.; Witt, K. The Application of Polymer inclusion membranes for the Removal of Emerging Contaminants and Synthetic Dyes from Aqueous Solutions—A Mini Review. *Membranes* **2023**, *13*, 132. [CrossRef]
126. Nitti, F.; Boliona, A.A.; Nitbani, F.O.; Naat, J.N.; Lapailaka, T.; Kadang, L.; Pingak, R.K.; Abadi, A.; Kiswandono, A.A. Fabrication of Polymer Inclusion Membrane Using Recycled Polyvinyl Chloride as Sustainable Alternative Support Polymer for the Extraction of Zn(II) from Water. *J. Appl. Polym. Sci.* **2025**, *142*, e57365. [CrossRef]
127. Kunene, P.; Akinbami, O.; Motsoane, N.; Tutu, H.; Chimuka, L.; Richards, H. Feasibility of Polysulfone as Base Polymer in a Polymer Inclusion Membrane: Synthesis and Characrerisation. *J. Membr. Sci. Res.* **2020**, *6*, 203–210.
128. Coscarello, M.; Nardi, M.; Alipieva, K.; Bonacci, S.; Popova, M.; Procopio, A.; Scarpelli, R.; Simeonov, S. Alternative Assisted Extraction Methods of Phenolic Compounds Using NaDESs. *Antioxidants* **2023**, *13*, 62. [CrossRef] [PubMed]
129. Richard, E.; Fatyeyeva, K.; Marais, S. Ionic liquid-based polymer inclusion membranes for heavy metal ions removal from water: Achievements and challenges. *Chem. Eng. J.* **2025**, *505*, 158916. [CrossRef]
130. Kerkel, F.; Markiewicz, M.; Stolte, S.; Müller, E.; Kunz, W. The green platform molecule gamma-valerolactone—Ecotoxicity, biodegradability, solvent properties, and potential applications. *Green Chem.* **2021**, *23*, 2962–2976. [CrossRef]
131. Rasool, M.A.; Vankelecom, I.F.J. γ-Valerolactone as Bio-Based Solvent for Nanofiltration Membrane Preparation. *Membranes* **2021**, *11*, 418. [CrossRef]
132. Jose, E.S.; Rosa de la Viuda, M.; Carmona, J.; Soto, C.M.; Palacio, L.; Pradanos, P.; Hernandez, A.; Tena, A. Green Dipolar aprotic Solvents for the Dynamic Polycondensation of High-Performance Polyimide Membranes. *Green Chem.* **2024**, *26*, 11984–12007. [CrossRef]
133. Li, Y.; Chen, L.; Wang, K.; Zhang, M. A PEGylated PVDF antifouling membrane prepared by grafting of mPEGA. *Membranes* **2022**, *11*, 16.
134. Zhang, X.; Zhang, Y.; Gao, J.; Li, Y. Investigation of anti-fouling and UV-cleaning properties of PVDF/TiO_2 mixed-matrix membranes. *Membranes* **2021**, *11*, 16.
135. Nikonenko, V.; Nebavsky, A.; Mareev, S.; Kovalenko, A.; Urtenov, M.; Pourcelly, G. Modeling of Ion Transport in Electromembrane Systems: Impacts of membrane Bulk and Surface Heterogeneity. *Appl. Sci.* **2019**, *9*, 25. [CrossRef]
136. Di Salvo, J.L.; de Luca, G.; Cipollina, A.; Micale, G. Effect of ion exchange capacity and water uptake on hydroxide transport in PSU-TMA membranes: A DAFT and molecular dynamics study. *J. Membr. Sci.* **2020**, *599*, 117837. [CrossRef]
137. Witzleben Von, M.; Stoppe, T.; Strauwitz ne Ahlfeld Von, T.; Bernhardt, A.; Polk, M.-L.; Bornitz, M.; Neudert, M.; Gelinsky, M. Biomimetic Tympanic Membrane Replacement Made by Melt Electrowriting. *Adv. Healthc. Mater.* **2021**, *10*, 2002089. [CrossRef]
138. Bonakdar, M.A.; Rodrigue, D. Electrospinning: Processes, Structures and Materials. *Macromol* **2024**, *4*, 58–103. [CrossRef]
139. Vazquez, M.I.; Romero, V.; Fontas, C.; Antico, E.; Benavente, J. Polymer inclusion membranes (PIMs) with the ionic liquid (IL) Aliquat 336 as extractant: Effect of base polymer and IL concentration on their physical-chemical and elastic characteristics. *J. Membr. Sci.* **2014**, *455*, 312–319. [CrossRef]
140. Sellami, F.; Marais, S.; Ounissa, S.-K.; Kobzar, Y.; Fatyeyeva, K. Poly(vinyl-chloride)-based advanced polymer inclusion membranes with Aliquat 336 and inorganic filler for efficient Cr(VI) removal. *Chem. Eng. J.* **2024**, *493*, 152056. [CrossRef]
141. Ismail, N. *Sustainable Membrane Fabrication Using Greener Solvents*; Cityprint I Norr AB: Umea, Sweden, 2023; ISBN 978-91-8070-134-1.
142. O'Neill, K.; Dalton, P.D. A decade of melt electrowriting. *Small Methods* **2023**, *7*, 2201589. [CrossRef] [PubMed]
143. Wang, Y.; Dai, M.; Luo, G.; Fan, J.; Clark, J.H.; Zhang, S. Preparation and application of green sustainable solvent Cyrene. *Chemistry* **2023**, *5*, 2322–2346. [CrossRef]
144. Uredat, S.; Gujare, A.; Runge, J.; Truzzolillo, D.; Oberdisse, J.; Hellweg, T. A review of stimuli-responsive polymer-based gating membranes. *Phys. Chem. Chem. Phys.* **2024**, *26*, 2732–2744. [CrossRef] [PubMed]
145. Jurak, I.; Ramos, S.; Caldeira, B.; Domingues, C.; Veiga, F.; Figueiras, A. The Many Faces of Cyclodextrins within Self Assembling Polymer Nanovehicles: From Inclusion Complexes to Valuable Structural and Functional Elements. *Int. J. Mol. Sci.* **2024**, *25*, 9516. [CrossRef]
146. Chen, Q.; Tang, Y.; Ding, Y.-T.; Jiang, H.-Y.; Zhang, Z.-B.; Li, W.-X.; Liu, M.-L.; Sun, S.-P. Synergistic Construction of Sub-Nanometer Channel Membranes through MOF-Polymer Composites: Strategies and Nanofiltration Applications. *Polymers* **2024**, *16*, 1653. [CrossRef]
147. Wu, Z.; Chen, Z.; Chen, J.; Ning, X.; Chen, P.; Jiang, H.; Qiu, H. Enhanced adsorption and synergistic photocatalytic degradation of tetracycline by MOF-801/Go composite via solvothermal synthesis. *Environ. Sci. Nano* **2022**, *9*, 4609–4618. [CrossRef]
148. Chabalala, M.B.; Gumbi, N.N.; Mamba, B.B.; Al-Abri, M.Z.; Nxumalo, E.N. photocatalytic Nanofiber membranes for Degradation of Micropollutants and their antimicrobial Activity: Recent Advances and Future Prospects. *Membranes* **2021**, *11*, 678. [CrossRef]

149. Kade, J.C.; Dalton, P.D. Polymers for Melt Electrowriting. *Adv. Healthc. Mater.* **2020**, *10*, 2001232. [CrossRef]
150. Zhang, G.; Xu, Y.; Sun, Y.; Ye, X.; Du, L.; Xu, H. Fabrication of janus Composite Membranes with a Gradient Pore Structure based on Melt Electrowriting and Solution Electrospinning for Directional water transport. *Langmuir* **2023**, *39*, 7891–7900. [CrossRef]
151. Polychronopoulos, N.D.; Brouzgou, A. Direct Ink Writing for electrochemical Device fabrication: A review of 3D-Printed Electrodes and Ink Rheology. *Catalysts* **2024**, *14*, 110. [CrossRef]
152. Chen, Q.; Chen, Z.; Wang, Y.; Tian, E.; Mo, J. Hierarchical diffusion pathways into VOC adsorption films by direct ink writing and ammonium carbonate treatment. *Chem. Eng. J.* **2023**, *471*, 144560. [CrossRef]
153. Liang, Z.; Yun, Y.; Wang, M.; Liu, G.; Lu, P.; Yang, W.; Li, C. Performance evaluation of interfacial polymerization-fabricated aquaporin-based biomimetic membranes in forward osmosis. *RSC. Adv.* **2019**, *9*, 19163. [CrossRef]
154. Ahuja, R.; Shivhare, V.; Konar, A.D. Recent Advances in Smart Self-Assembled Bioinspired Hydrogels: A Bridging Weapon for Emerging Health Care applications from Bench to Bedside. *Macromol. Rapid Commun.* **2024**, *45*, 2400255. [CrossRef]
155. Osman, A.I.; Nasr, M.; Farghali, M.; Bakr, S.S.; Eltaweil, A.S.; Rashwan, A.K.; Abd El-Monaem, E.M. Machine learning for membrane design in energy production, gas separation, and water treatment: A review. *Environ. Chem. Lett.* **2024**, *22*, 505–560. [CrossRef]
156. Ren, E.; Guilbaud, P.; Coudert, F.-X. High-throughput computational screening of nanoporous materials in targeted applications. *Digit. Discov.* **2022**, *1*, 355–374. [CrossRef]
157. Khanzada, N.K.; Al-Juboori, R.A.; Khatri, M.; Ahmed, F.E.; Ibrahim, Y.; Hilal, N. Sustainability in membrane Technology: Membrane Recycling and fabrication Using Recycled waste. *Membranes* **2024**, *14*, 52. [CrossRef]
158. Goh, W.H.D.; Lau, H.S.; Yong, W.F. An integrated life cycle assessment and techno-economic analysis: Evaluation on the production of polymers of intrinsic microporosity (PIM-1) and UiO-66-NH_2 as membrane materials. *Sci. Total Environ.* **2023**, *892*, 164582. [CrossRef] [PubMed]
159. Yin, H.; Xu, M.; Luo, Z.; Bi, X.; Li, J.; Zhang, S.; Wang, X. Machine learning for membrane design and discovery. *Green Energy Environ.* **2024**, *9*, 54–70. [CrossRef]
160. Sharma, B.; Moghimianavval, H.; Hwang, S.W.; Liu, A.P. Synthetic Cell as a Platform for Understanding Membrane-Membrane Interactions. *Membranes* **2021**, *11*, 912. [CrossRef] [PubMed]
161. Ignacz, G.; Bader, L.; Beke, A.; Ghunaim, Y.; Shastry, T.; Vovusha, H.; Carbone, M.R.; Ghanem, B.; Szekely, G. Machine learning for the advancement of membrane science and technology: A critical review. *J. Membr. Sci.* **2024**, *713*, 123256. [CrossRef]
162. Maleki, R.; Shams, S.M.; Chellehbari, Y.M.; Rezvantalab, S.; Jahromi, A.M.; Asadnia, M.; Abbassi, R.; Aminabhavi, T.; Razmjou, A. Materials discovery of ion-selective membranes usinf artificial intelligence. *Commun. Chem.* **2022**, *5*, 132. [CrossRef]
163. Coronado, S.; Herrera, J.; Pino, M.G.; Martin, S.; Ballesteros-Rueda, L.; Cea, P. Advancements in Engineering Planar Model Cell Membranes: Current Techniques, Applications, and Future Perspectives. *Nanomaterials* **2024**, *14*, 1489. [CrossRef]

membranes

MDPI

Article

Impact of Long-Term Alkaline Cleaning on Ultrafiltration Tubular PVDF Membrane Performances

Marek Gryta * and Piotr Woźniak

Faculty of Chemical Technology and Engineering, West Pomeranian University of Technology in Szczecin, Piastów Ave. 17, 70-310 Szczecin, Poland; piotr.wozniak@zut.edu.pl
* Correspondence: marek.gryta@zut.edu.pl

Abstract: The application of an ultrafiltration (UF) process with periodic membrane cleaning with the use of alkaline detergent solutions was proposed for the recovery of wash water from car wash effluent. In order to test the resistance of the membranes to the degradation caused by the cleaning solutions, a pilot plant study was carried out for almost two years. The installation included an industrial module with FP100 tubular membranes made of polyvinylidene fluoride (PVDF). The module was fed with synthetic effluent obtained by mixing foaming agents and hydrowax. To limit the fouling phenomenon, the membranes were cleaned cyclically with P3 Ultrasil 11 solution (pH = 11.7) or Insect solution (pH = 11.5). During plant shutdowns, the membrane module was maintained with a sodium metabisulphite solution. Changes in the permeate flux, turbidity, COD, and surfactant rejection were analysed during the study. Scanning electron microscopy (SEM), atomic force microscopy (AFM), differential scanning calorimetry (DSC), and Fourier-transform infrared spectroscopy (FTIR) analysis were used to determine the changes in the membrane structure. As a result of the repeated chemical cleaning, the pore size increased, resulting in a more than 50% increase in permeate flux. However, the quality of the recovered wash water did not deteriorate, as an additional separation layer was formed on the membrane surface due to the fouling phenomenon.

Keywords: car wash wastewater; ultrafiltration; fouling; alkaline cleaning; PVDF membrane; membrane endurance

1. Introduction

Washing cars requires large amounts of water; thus, many car washes use water recycling systems [1,2]. Today, more than 50% of car washes have recycling systems that use sedimentation and sand filtration supported by water aeration [3–5]. In addition, membrane recycling systems are increasingly being developed [6–9].

The applied type of wastewater treatment system strongly depends on the size of the car wash. Large automated car washes use multi-stage systems that produce high-quality wash water, but this is expensive [3]. The average car wash customer spends $15 per visit, and the global car wash market is estimated to be worth $41 billion by 2025 [2]. Such revenues allow for investment, and 85% of professional car washes use water reclamation systems [10].

The financial situation is different for small car washes. Such car washes are generally supplied with municipal water and discharge their effluent into the sewerage system. In Poland, car wash owners pay $5–7 per 1 m^3 of water for such use. Such low prices do not provide an incentive to invest in expensive equipment to reduce fresh water consumption.

Most of these car washes have a reverse osmosis (RO) system (pump, filter, and membrane module) that produces pure water, resulting in a spot-free final car rinse. Owners of such car washes expect a similarly simple ultrafiltration (UF) system to be used for water recycling. The development of this solution is the focus of the research presented here. Such a simple system for recovering wash water has been proposed using ultrafiltration tubular polyvinylidene fluoride (PVDF) membranes [11]. PVDF membranes are commonly used in the manufacture of industrial microfiltration and ultrafiltration modules [12,13]. However, car wash wastewater treatment by UF causes the rapid fouling of the membranes [3,7,14,15].

It is well-known that the fouling phenomenon is a fundamental obstacle to the industrial application of membrane processes [16,17]. The fouling layer formed on the membrane surface can be removed by physical methods such as backwashing or flushing [18]. Unfortunately, such methods are often insufficient for wastewater separation and, consequently, the chemical cleaning of the membrane is required [12,19]. Cleaning agents containing sodium hydroxide (NaOH) and sodium hypochlorite ($NaClO_3$) are used for membrane washing [12,13]. Such solutions are aggressive, and, therefore, their frequent and prolonged application causes damage of the polymeric membranes [12,20–22]. However, alkaline detergents, such as P3 Ultrasil 11, are commonly used in membrane plants due to their high cleaning efficiency [13,23–25]. To limit membrane damage, companies generally specify their use [13,26]. However, for new technologies, it is necessary to test the resistance of the membranes to such detergents. The results of studies involving filtration for a maximum of a few days are mostly presented, and only a few papers present longer studies, such as 120-day tests of PVDF membranes [27]. Short-term tests do not provide satisfactory information; hence, in this study, tests lasting several months were carried out.

The popular spiral wound UF modules cannot be used for car wash wastewater separation due to the high turbidity [3,14]. Their application requires the use of a pre-filtration stage [3,7], which is unfavourable for small car washes as it increases the size (cost) of the treatment system. For these reasons, tubular PVDF membranes have been used to treat car wash wastewater [11,28]. It has been shown that the UF permeate obtained from the filtration of the wastewater can be successfully used for car washing [29]. In this study, PVDF FP100 tubular membranes, installed in the B1 module manufactured by PCI, were successfully used to produce washing water. A permeate flux of 50–60 L/m^2h was achieved during the UF of real wastewater treatment at a pressure differential of 0.1 MPa [11]. For such a process' yield, the UF plant consisting of two to five B1 modules would be suitable for small car washes.

One limitation of the presented solution is the considerable fouling caused by the compounds of car wash wastewater. In this case, the fouling of the UF membranes was reduced by membrane washing with cleaning agents containing NaOH, such as P3 Ultrasil 11 and an agent used in car washes to remove insects [11,14,15]. Membranes can also be cleaned with $NaClO_3$ solutions; however, these cause much more damage to the PVDF membrane [12]. Previous studies have obtained results for FP100 membranes in tests involving a small module lasting only a few weeks [11]. While these results are promising, they are insufficient to convince washer manufacturers to modify their designs and incorporate UF plants into their washing systems. What is required is proof that the proposed method of cleaning the UF membranes will prevent them from failing quickly. To obtain such data, a two-year study of an industrial B1 module was conducted.

The performed UF studies showed that PVDF membranes are not fully resistant to NaOH solutions. As a result, it has been observed that cyclic filtration and cleaning has altered the stability of PVDF membranes [12,30]. On the other hand, it has been observed that fouling and the presence of surfactants reduce the negative effect of alkaline solutions [31], which may reduce the degradation of PVDF membranes during car wash wastewater separation. However, the degradation process can be very slow and its deleterious effects can significantly damage the membranes after several months of membrane module operation. Following several weeks of testing FP100 membranes, it was found that alkaline cleaning caused minor damage to the membranes' active layer [11]. This raises the question of whether these membranes could withstand several years of exposure in car washes. Due to the lack of such data, a pilot scale study was undertaken in this work.

For industrial implementation, commercially available membrane modules are required. Hence, the B1 module from PCI with FP100 membranes were chosen for the study. This module has been successfully used for several years in studies on oily wastewater separation [32]. However, the feed solutions tested in this work, such as bilge water, were not alkaline. For this reason, the industrial PVDF membrane module tested in this study was fed with a mixture of car wash chemicals (pH = 8.5–9), which is similar in composition to the wastewater generated at car washes. Moreover, the membranes were cleaned cyclically using alkaline agents with pH 11–12. During shutdowns, the membrane module was maintained with a sodium metabisulfite solution. The effect of the repeated chemical cleaning of fouled PVDF membranes on the permeate flux and surfactant rejection was analysed. Scanning electron microscopy (SEM), atomic force microscopy (AFM), differential scanning calorimetry (DSC) and Fourier-transform infrared spectroscopy (FTIR) analysis were used to analyse the changes in the membrane structure. The main objective of the study was to determine whether the selected tubular module would remain undamaged over the long term and whether the quality of the permeate obtained would not deteriorate significantly.

2. Materials and Methods

2.1. UF Installation

The UF process was investigated using the UF pilot plant shown schematically in Figure 1. The unit was equipped with a PCI B1 tubular module with 18 FP100 (100 kDa) polyvinylidene fluoride membranes (PCI Membranes, Kostrzyn Wielkopolski, Poland) mounted inside (Table 1). The diameter and length of the tubular membrane were 1.25 cm and 120 cm, respectively. The feed tank (200 L) was connected to a VNR-8 centrifugal pump (Grundfos, Bjerringbro, Denmark). The feed flowed inside the tubular membranes at a velocity of 1.9 m/s, and the UF tests were performed at a transmembrane pressure (TMP) of 0.1 MPa. For all types of solutions tested, the feed temperature was 298 ± 2 K. Temperature control was provided by a tubular heat exchanger (cooled by tap water), through which the feed flowed. The UF tests were carried out for 7–8 h a day, with the feed remaining in the system overnight. Overnight breaks also occur in most car washes, as 85% of car wash customers prefer to visit car washes during daylight hours [10].

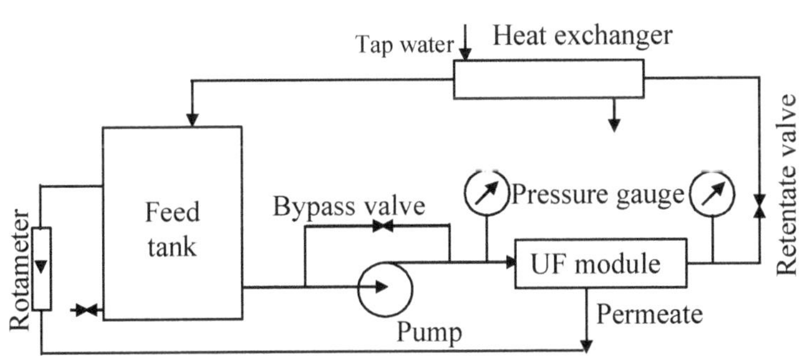

Figure 1. Experimental set-up of the pilot-scale UF unit.

Table 1. Tubular module and FP100 membrane characteristics.

Parameter	Unit	B1 Module	FP100
Length	mm	1220	1200
Diameter	mm	100	1.25
Tube	-	18	1
Material	-	AISI 316 Stainless Steel	PVDF
Operating pressure	bar	64	10
Operating temperature	°C	80	80
Volume shroud-side	L	6.7	-
Volume tube-side	L	2.8	0.147
Nominal retention	kDa	-	100
pH range	-	1.5–12	1.5–12

The permeate flux [L/m^{2}h] was obtained based on the following equation:

$$J = \frac{V}{S \cdot t} \quad (1)$$

where V [L]—cumulative permeate volume; S [m^2]—effective membrane surface area; and t [h]—filtration duration.

The rejection efficiency R [%] was determined as follows:

$$R = 1 - \frac{C_P}{C_F} \cdot 100\% \quad (2)$$

where C_P [mg/L] and C_F [mg/L] are the measured concentrations of the permeate and feed, respectively.

The UF tests were conducted at constant feed concentrations, with the permeate returned to the tank. Initially, the feed tank was filled with 100 L of the test solutions.

2.2. Feed Solutions

The permeate from the nanofiltration (NF) process, fed with tap water, was used to flush/rinsing the UF unit and to prepare test solutions. The NF permeate obtained had a conductivity in the range of 70–90 μS/cm and its composition has been reported in previous work [32,33].

In car washes, cars are first washed with various foam solutions and, then, after rinsing with water, their surfaces are treated with hydrophobic coatings (hydrowax). Synthetic car wash wastewater has been produced from a mixture of these agents with a composition similar to that of a car wash. Detergents (Euro Foams and Hydrowax) produced by EuroEcol (Łódź, Poland) were used. The foam solutions contained surfactants, diethylene glycol butyl ether, benzene sulphonic acid, and polymers. The exact composition of the commercial concentrates is given in [14,15]. The mixture used for the UF tests contained 0.5 vol.% foaming agent and 0.2 vol.% Hydrowax solution (EuroEcol, Łódz, Poland). In addition to Euro Turbo Foam without dyes ("White"), foam agent concentrates with dyes were also used: Euro Turbo Foam Colour Green ("Green") and Euro Turbo Foam Colour Blue ("Blue"). The latter contained indigo carmine dye, which increased the fouling of PVDF membranes [34]. The parameters of the synthetic wastewater are shown in Table 2.

Before weekend shutdowns, the feed was removed, and the plant was flushed with water (200 L) and filled with NF permeate. For longer shutdowns, additional maintenance was performed by flushing the system with 0.25 wt.% sodium metabisulphite ($Na_2S_2O_5$) solution (ChemLand, Stargard, Poland). This preservative solution was recirculated for

15 min, after which the system was drained. The resumption of UF operation after the break was started by rinsing the UF installations twice with NF permeate (2 × 100 L).

Table 2. Parameters of prepared synthetic wastewater. Anionic and nonionic—surfactants.

Wastewater	COD [mg/L]	Anionic [mg/L]	Nonionic [mg/L]	pH [-]	Cond. [μS/cm]	Turbidity [NTU]
White	2275 ± 378	463 ± 26	25 ± 8	8.5 ± 0.2	155 ± 1	11.3 ± 1.4
Green	2150 ± 70	420 ± 20	21 ± 6	8.8 ± 0.2	162 ± 2	12.5 ± 5.2
Blue	2391 ± 115	393 ± 29	40 ± 14	8.6 ± 0.1	167 ± 2	83.2 ± 14.7

Separation of the wastewater caused fouling and a consequent reduction in permeate flux. The membranes were, therefore, chemically cleaned at regular intervals. PCI company recommends P3 Ultrasil 11 from Hankel Ecolab (Suturamed, Szczecin, Poland) for washing the FP100 membranes. This agent contains surfactants and tetra sodium salt of EDTA in addition to NaOH [23]. A 0.1 wt.% solution with pH = 11.7 was prepared for washing the UF plant. A similar pH value is found in the 0.5 vol.% solution of Insect (EuroEcol, Łódź, Poland). This detergent contains NaOH and is used in commercial car washes to remove insects from car bodies. It has also been shown to be effective in removing deposits formed on membranes during the filtration of car wash wastewater [11,14]. The prepared cleaning solutions (100 L) were recirculated for 30 min (TMP = 0.02 MPa). After chemical cleaning, the UF system was flushed twice with NF permeate (100 L) and the maximum permeate flux was determined (TMP = 0.1 MPa).

2.3. Analytical

The Hach cuvette tests were used to determine the concentration of surfactants (LCK 333—nonionic, and LCK 432—anionic) and chemical oxygen demand—COD (LCK 1014). The values were recorded automatically with the use of a Hach Lange DR2800 spectrophotometer (Hach Lange, Düsseldorf, Germany).

The membrane surface was examined by atomic force microscopy. A multi-Mode 8 AFM apparatus equipped with a Nanoscope V converter from Bruker (Billerica, MA, USA) characterised the membrane roughness in the scanasyst mode. The Ra and Rq parameters were evaluated on a basis of at least five AFM images (10 μm × 10 μm).

The turbidity of tested solutions was measured with a portable turbidity meter model 2100 AN IS with a detection limit of 0.01 NTU (Hach Company, Loveland, CO, USA).

Differential scanning calorimetry analysis (DSC) was performed in a NETZSCH STA 449 F3 Jupiter (Erich NETZSCH GmbH & Co. Holding KG, Selb, Germany). The samples were heated from room temperature to 523 K at a heating rate of 2 K/min.

3. Results and Discussion

3.1. Car Wash Wastewater Ultrafiltration

The PVDF module installed in the UF plant was tested for almost two years, separating different types of car wash wastewater (Green, White, and Blue—Table 2). Changes in the maximum permeate flux after chemical cleaning are shown in Figure 2. The installation operated for five days a week, totalling almost 300 working days in the study. In addition to the weekend breaks, there were also several 7- to 55-day shutdowns periods, before which the installation was maintained with a sodium metabisulfite solution. It is worth noting that a similar situation can occur with small car washes, as their owners also do not work on holidays. Figure 2 shows the numbers indicating longer shutdown periods.

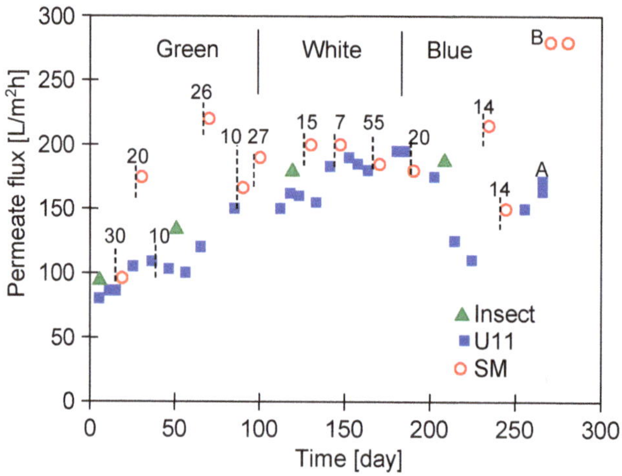

Figure 2. Changes in the maximum permeate flux during UF pilot plant exploitation. NF permeate used as a feed. Flux determined after membranes cleaning. Cleaning solutions: Insect—0.5 vol.% of Insect agent concentrate (pH = 11.5), U11—0.1 wt.% P3 Ultrasil 11 (pH = 11.7), and SM—0.25 wt.% sodium metabisulfite. Numbers—days of shutdowns periods. Point A—membranes washed with hot 0.1 wt.% P3 Ultrasil 11 solution. Point B—installation rinsed with 0.25 wt.% sodium metabisulfite solution.

As shown in Figure 2, repeated washing of the membrane with NaOH solutions increased the permeate flux, possibly due to an increase in the membranes' pore size [12]. The largest changes occurred during the initial 40 days' (excluding shutdowns) exploitation period, which was also observed in other pilot studies [13]. After this period, the maximum permeate flux increased from 95 to 175 L/m^{2}h. Subsequent to the third maintenance of the plant with an $Na_2S_2O_5$ solution, the permeate flux increased to 220 L/m^{2}h. Lower maximum permeate flux values were obtained after washing the membrane with a P3 Ultrasil 11 solution. The conducted membrane washing with this agent probably did not remove all contaminants, especially those in the largest pores [11]. These pores were cleaned to a greater extent by the sodium metabisulphite solution and, as a result, the highest values of maximum permeate flux were obtained after plant shutdowns (points SM). It has been found that, after one year of module operation (Figure 2, after 140 days), this effect disappeared. This could be due to a better cleaning of the fouling layer (generated during White wastewater separation) also by the P3 Ultrasil 11 solution. Differences again occurred in the separation of Blue wastewater, as it contained components that were more difficult to remove by alkaline cleaning solutions. These wastewaters were found to contain indigo carmine dye, which was found to adsorb strongly onto the surface of PVDF membranes [34].

A higher flux recovery for membranes contaminated with Blue wastewater was achieved by washing the module twice (for 30 min each time) with a hot (50 °C) solution of P3 Ultrasill 11 (Figure 2, point A), after which the UF plant was rinsed with a sodium metabisulphite solution. Consequently, the maximum permeate flux increased to 275 L/m^{2}h (point B). This indicates that this procedure allowed us to clean the membranes and led to an increase in their pore size.

After one week of ultrafiltration testing, the plant was rinsed again with a sodium metabisulphite solution; however, this did not result in any further increase in the permeate flux. This suggests that the $Na_2S_2O_5$ solution did not enlarge the pores of the tested membranes, but, rather, cleaned them thoroughly. While the membrane manufacturer permits the use of hot P3 Ultrasil 11 solution for cleaning purposes, it should be noted that this will accelerate the adverse effects of NaOH on the PVDF membrane matrix.

Obviously, increasing the pore size increases permeate flux; however, this also worsens rejection [35]. However, in ongoing UF studies, it is important to note that the results presented in the current study demonstrated that no significant change in rejection (Figure 3) was observed despite the increase in permeate flux (Figure 2). In addition, the separation of the Blue wastewater improved the rejection (Figure 3, after 200 days). The Blue wastewater had a significantly higher turbidity (Table 2). This increases the formation of a filter cake, which acts as an additional separation layer and, consequently, improves the separation [36]. Chemical cleaning removes most of the filter cake from the membrane surface, and, as a result, a deterioration in rejection was observed (Figure 3, points A).

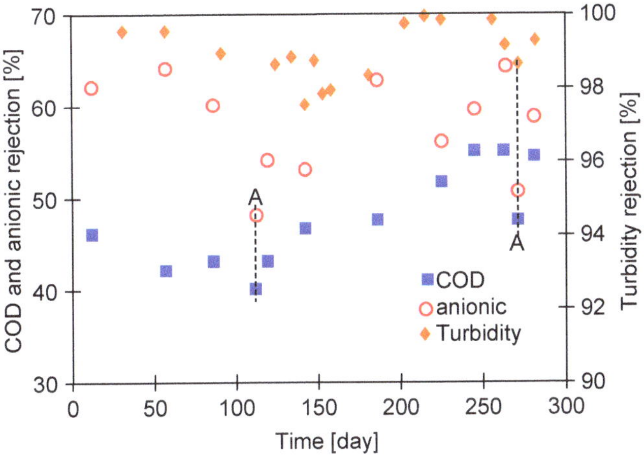

Figure 3. Changes in rejection of COD, turbidity, and anionic surfactants during UF studies (samples collected after one-day UF of wastewater). Point A—parameter determined direct after membrane cleaning (cleaning solutions: 0.25 wt.% sodium metabisulfite).

The positive influence formation of the fouling layer on the rejection is also confirmed by the results shown in Figure 4. The separation results obtained immediately after membrane chemical cleaning (COD and Anion) and the results after 2–3 days of wastewater filtration (COD UF and Anion UF) can be seen. The results obtained after chemical cleaning of the membrane deteriorate with the time of filtration of the UF module, especially for the removal of anionic surfactants, can also be seen. This finding confirms that the repeated chemical cleaning of the membrane with alkaline agents caused an increase in pore size. However, during the filtration of the effluents, due to fouling, the separation improves and remains at a similar level throughout the UF test period.

The changes in the membrane structure induced by the chemical cleaning also affected the filtration of the car wash wastewater. Figure 5 shows an example of the changes in permeate flux obtained during two consecutive days of UF testing at different times of the module exploitation. During the initial period for the new membranes, the flux decreased rapidly and stabilised at 50 L/m^{2}h. Similar 50% permeate flux decreases after 60 min of UF have been reported in other work [7,11]. The repeated cleaning of the membranes with NaOH solutions resulted in an increase in membrane permeability (Figure 2) and, consequently, an increase in permeate flux during wastewater separation.

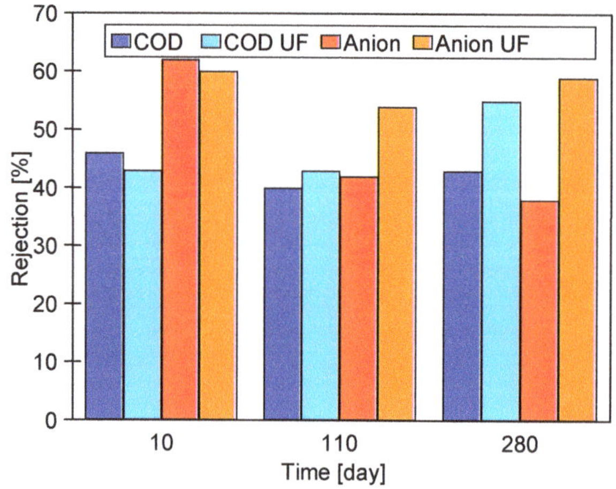

Figure 4. Changes in the rejection during performed UF studies. Membranes after chemical cleaning (COD and Anion) and after 2–3 days of UF process covered by fouling layer (COD UF and Anion UF).

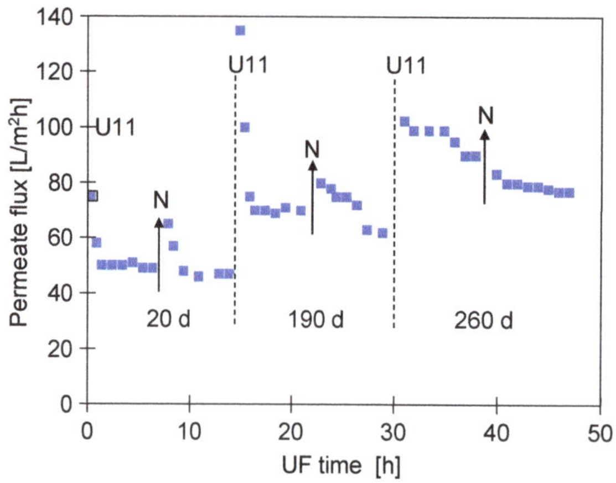

Figure 5. Effect of UF module lifetime (d—days) on permeate flux changes during car wash wastewater separation. Points: U11—membranes washed with P3 Ultrasil 11; N—overnight stops, installation filled with NF permeate.

In addition, the permeate flux also increased after overnight stops (Figure 5, point N). This was due to the decompression of the membrane and sediment (TMP = 0) and, also, to the phenomenon of osmotic washing (module filled with NF permeate). The pore size increased with the time of use and this effect had practically disappeared after one year of operation (Figure 5, after 30 h). Importantly, the larger pores favour internal fouling, which is more difficult to wash off during overnight soaking. Therefore, an increase in permeate flux was only achieved after intensive chemical cleaning (Figure 2, after 250 h). It was found that rinsing the system with sodium metabisulphite solution not only maintained the membranes but also resulted in significant cleaning. As a result, a significant increase in permeate flux was observed after each system maintenance (Figure 2, SM).

3.2. Membrane Morphology

On completion of the UF tests, the membrane was removed from the module and rinsed with deionised water. The membrane surface was brown in colour, indicating that iron oxides were also present in the resulting fouling layer [32]. After washing with a 30 wt.% HCl solution, the membranes regained their white colour. The AFM images of the surface of such washed membranes and the new membrane are shown in Figure 6.

Figure 6. AFM images of the membranes surface: (**a**) new membrane; (**b**) membrane used for separation wastewater; and (**c**) membrane after HCl rinsing.

The AFM analysis confirmed that the membrane surface changed during the wastewater separation and cyclic membrane washing with the alkaline detergents. Despite intensive washing, large amounts of contaminants remained on the membrane surface (Figure 6b). HCl rinsing removed most of these sediments, but the sediment residues are still visible on the membrane surface (Figure 6c). Due to the formation of deposits, the roughness parameters for the tested membranes increased significantly during the UF of car wash wastewater. The values of Rq increased from 35.1 to 223.7 nm, and Ra from 28.2 to 179.2 nm (Table 3, after UF tests). The roughness parameters after rinsing with HCl solution decreased, e.g., for Rq, from 223.7 to 114.5 nm; however, these were still significantly higher than the values set for the new membranes.

Table 3. Membranes' surface roughness.

Parameter	New	After UF Tests	Rinsed
R_q [nm]	35.1 ± 1.4	223.7 ± 12.8	114.5 ± 17.4
R_a [nm]	28.2 ± 4.1	179.2 ± 9.4	81.9 ± 18.4

The AFM images show that the sediment structure was not dense and that the numerous voids visible facilitated the transport of the filtrate to the membrane surface. As a result, the formation of a filter cake of considerable thickness did not lead to a decrease in permeate flux (Figure 5).

The areas of the membrane surface exposed after washing with HCl solution show no damage (Figure 6c), and, in these areas, the image is similar to that obtained for a new membrane (Figure 6a). However, the SEM observation revealed that, after the UF tests, the number of larger pores on the surface of the active layer increased compared to the new membrane (Figure 7). Consequently, the membrane's permeability increased, resulting in an increase in the permeate flux over the course of the module operation (Figures 2 and 5). The obtained SEM results indicate that the repeated cleaning of the membrane with an alkaline agent can cause minor changes to the active layer's structure. This also confirms that PVDF membranes are not fully resistant to the prolonged exposure to NaOH solutions.

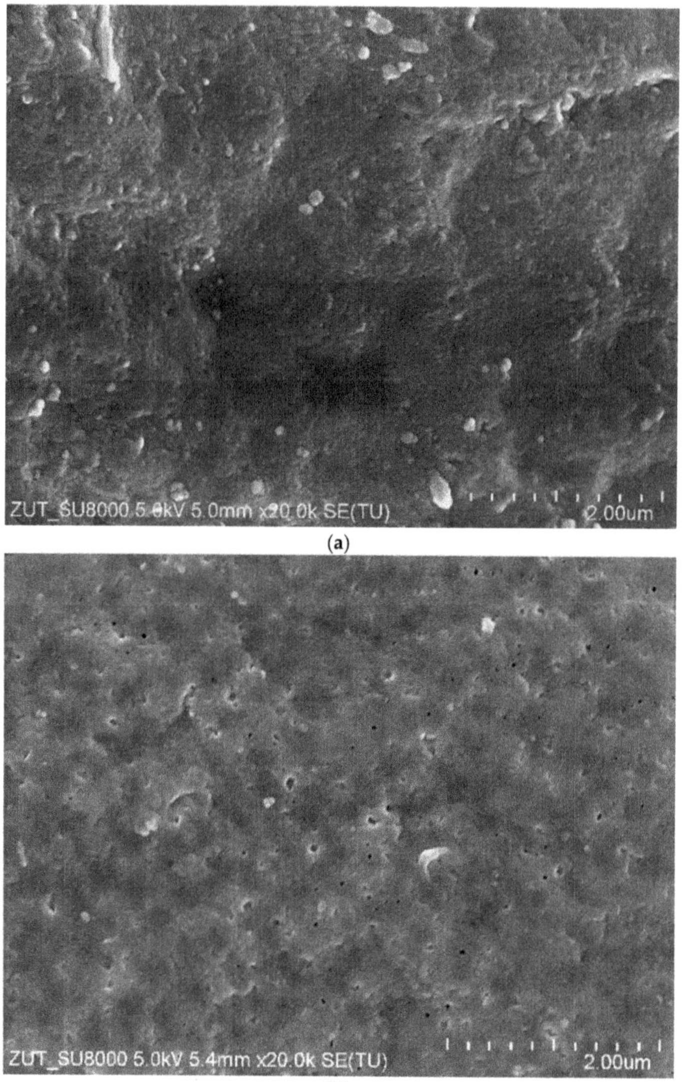

Figure 7. SEM images of FP100 membrane surface: (**a**) new membrane, and (**b**) membrane after 2 years of UF tests.

These results raise the question of whether the proposed alkaline washing method would damage the PVDF membranes to such an extent that they could not be used to obtain wash water from car wash wastewater. As demonstrated in Figures 3 and 4, the thorough cleaning of the pores worsens the retention of wastewater constituents. However, once wastewater filtration begins, these pores quickly become blocked by sludge (Figure 6b), which improves the rejection degree (Figure 4). Consequently, despite the formation of larger pores, the tested membranes still effectively filter out turbidity (Figure 3) and bacteria [11], and the UF permeate obtained from the wastewater was successfully used for car washing [29].

In addition to enlarging the pore size, the long-term use could have caused changes to the matrix structure of the PVDF membranes. For example, this could result in membrane cracking. PVDF forms five crystalline polymorphs, the most common of which are the α and β forms, which affect membrane performance differently [37]. Transitioning from the β to α phase can negatively impact the membrane's flexibility [37,38]. The FTIR spectra of the studied membranes revealed several absorption bands that can be attributed to the α and β conformations of PVDF (Figure 8).

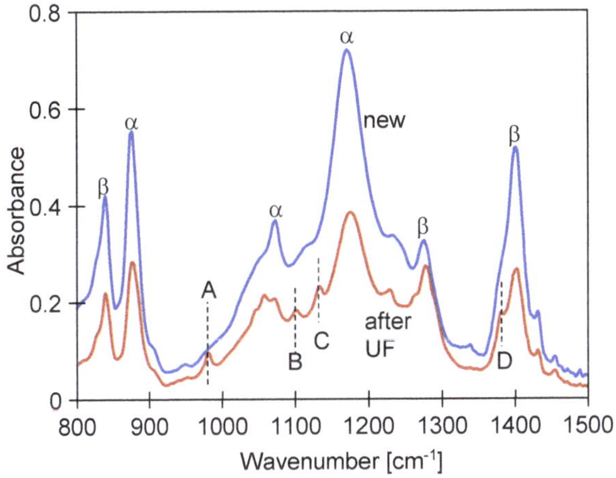

Figure 8. FTIR analysis of new FP100 and after UF test membrane samples.

Characteristic bands appeared at 875, 976, 1070, and 1181 cm^{-1}, indicating the α conformation, while the 841, 1279, and 1403 cm^{-1} results belong to the β phase [12,37,38]. An FTIR analysis showed that the intensity of these peaks decreased significantly as a result of the UF tests (Figure 8). New peaks also appeared in the membranes after two years of testing and are labelled A, B, C, and D in Figure 8. Peak A, at 978 cm^{-1} (assigned to CH_2 twisting), is attributed to the α-conformation [12]. Similarly, peak D (1383 cm^{-1}) indicates the occurrence of phase α [37]. However, the intensity of these peaks is low, suggesting that a significant conversion of phase β to the α conformation does not occur in the studied membranes.

The changes in the structure of the PVDF membranes can be revealed by the melting point and enthalpy of the phase transformation measurements in the DSC analysis [39]. The DSC signal of the new membrane shows an endothermic effect (peak temperatures of 157.1 °C and 168.2 °C) starting from an extrapolated initial temperature of 137.9 °C (Figure 9). The endothermic DSC peaks are due to the melting of the sample and the calculated enthalpy was 53.05 J/g. The results obtained for the sample after the UF tests showed that the second peak decreased, while the peak temperatures were 158.3 °C and 168.4 °C for the first and second peaks, respectively. In this case, the enthalpy decreased from 53.05

to 50.75 J/g. All of the determined melting temperatures are below 179 °C, indicating that phase α and β are the only phases of PVDF present in the studied samples [37].

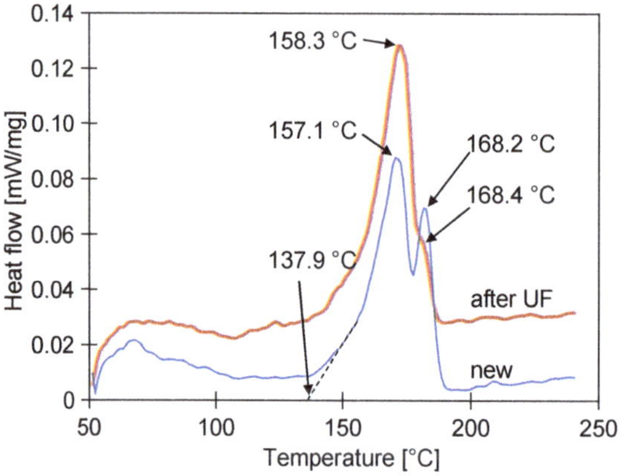

Figure 9. DSC analysis of new FP100 membrane sample, and membrane sample collected from module after two years of UF tests.

Moreover, the results obtained indicate that the almost two years of the module operation induced some changes in the polymeric structure of the PVDF membranes studied. This finding confirms the results obtained from the FTIR analysis. However, the small differences in the determined melting temperatures indicate that there is little degradation of the membrane matrix.

The performed SEM and AFM examination of the samples after the UF test did not reveal any cracks in the membrane and the rejection value during wastewater separation was close to the initial value (Figure 4); thus, it can be assumed that the membranes tested retained good separation properties despite some changes in the polymeric structure.

4. Conclusions

UF tests conducted over a period of almost two years showed that tubular PVDF membranes could be successfully used to separate car wash wastewater of varying composition and turbidity close to 100 NTU. However, filtration of the wastewater caused severe fouling, resulting in a decrease in permeate flux of more than 50% after 2–3 h of UF operation.

The intensity of the fouling phenomenon was reduced by cyclic membrane washing with detergent agents containing NaOH (pH = 11.5–11.7). Satisfactory cleaning efficiency was also demonstrated by a 0.25 wt.% sodium metabisulphite solution, which was used to maintain the UF module during plant shutdown periods.

For a car wash installation with UF, the membrane cleaning programme would include washing the FP100 membrane with a 0.1 wt.% P3 Ultrasil 11 solution. This should be performed at least every five days for 30 min. Additionally, it is recommended to flush the installation with a solution of 0.25 wt.% sodium metabisulphite every fortnight.

After one year of module exploitation with cyclical alkaline washing, the PVDF membranes showed a deterioration in the degree of separation and an increase in permeate flux, indicating that the alkali caused partial pore growth. However, after two years of membrane exploitation, the performed AFM and SEM studies showed no significant damage, although FTIR and DSC studies revealed some changes in the membrane matrix structure. This finding confirms that NaOH solutions have a negative effect on PVDF membranes. However, the resulting changes in the membrane structure do not significantly worsen membrane rejection during wastewater separation. This is due to the fact that the resulting filter cake reduces permeate flux but, at the same time, improves separation.

The results of the current study indicated that periodic cleaning of FP100 membranes (used in car washes) with alkaline agents enables the B1 tubular module to be successfully used for recycling wash water in car washes.

Author Contributions: Conceptualisation, M.G. and P.W.; methodology, M.G.; investigation, P.W.; writing—original draft preparation, M.G. and P.W.; visualisation, M.G.; supervision, M.G.; review and editing, M.G. and P.W. All authors have read and agreed to the published version of the manuscript.

Funding: The publication was financed from the Polish budget within a framework of the programme of the Minister of Education and Science in Poland entitled "Science for Society", project No. NdS/538617/2021/2022.

Institutional Review Board Statement: Not applicable.

Data Availability Statement: The original contributions presented in the study are included in the article; further inquiries can be directed to the corresponding author.

Conflicts of Interest: The authors declare no conflicts of interest.

References

1. Kuan, W.-H.; Hu, C.-Y.; Ke, L.-W.; Wu, J.-M. A Review of On-Site Carwash Wastewater Treatment. *Sustainability* **2022**, *14*, 5764. [CrossRef]
2. 39 Car Wash Industry Statistics for 2025. Available online: https://www.keevee.com/car-wash-industry-statistics (accessed on 27 May 2025).
3. Sarmadi, M.; Foroughi, M.; Saleh, H.N.; Sanaei, D.; Zarei, A.A.; Ghahrchi, M.; Bazrafshan, E. Efficient technologies for carwash wastewater treatment: A systematic review. *Environ. Sci. Pollut. Res.* **2020**, *27*, 34823–34839. [CrossRef] [PubMed]
4. Mark VII AquaPur. Available online: https://www.markvii.net/wasserrueckgewinnung/mark-vii-aquapur/ (accessed on 20 May 2025).
5. Car Wash Global Market Report 2025. Available online: https://www.thebusinessresearchcompany.com/report/car-wash-global-market-report (accessed on 17 May 2025).
6. System Overview—Hoffman innovateIT. Available online: https://innovateitcarwash.com/docs/ro/manual/system-overview/ (accessed on 20 May 2025).
7. Uçar, D. Membrane processes for the reuse of car washing wastewater. *J. Water Reuse Desalination* **2018**, *8*, 169–175. [CrossRef]
8. Boluarte, I.A.R.; Andersen, M.; Pramanik, B.K.; Chang, C.Y.; Bagshaw, S.; Farago, L.; Shu, L. Reuse of car wash wastewater by chemical coagulation and membrane bioreactor treatment processes. *Int. Biodeterior. Biodegrad.* **2016**, *113*, 44–48. [CrossRef]
9. Moazzem, S.; Wills, J.; Fan, L.; Roddick, F.; Jegatheesan, V. Performance of ceramic ultrafiltration and reverse osmosis membranes in treating car wash wastewater for reuse. *Environ. Sci. Pollut. Res.* **2018**, *25*, 8654–8668. [CrossRef]
10. Car Wash Industry Statistics. Available online: https://gitnux.org/car-wash-industry-statistics/ (accessed on 20 May 2025).
11. Woźniak, P.; Gryta, M. Application of Polymeric Tubular Ultrafiltration Membranes for Separation of Car Wash Wastewater. *Membranes* **2024**, *14*, 210. [CrossRef]
12. Rabuni, M.F.; Sulaiman, N.M.N.; Aroua, M.K.; Chee, C.Y.; Hashim, N.A. Impact of in situ physical and chemical cleaning on PVDF membrane properties and performances. *Chem. Eng. Sci.* **2015**, *122*, 426–435. [CrossRef]
13. Menon, S.; Bansode, K.; Nandi, S.; Kalyanraman, V. Impact of cleaning agents on properties of tubular polyvinylidene fluoride (PVDF) membrane. *Mater. Today Proc.* **2021**, *47*, 1466–1471. [CrossRef]
14. Gryta, M.; Woźniak, P. Polyethersulfone membrane fouling mitigation during ultrafiltration of wastewaters from car washes. *Desalination* **2024**, *574*, 117254. [CrossRef]
15. Tomczak, W.; Gryta, M. The Application of Polyethersulfone Ultrafiltration Membranes for Separation of Car Wash Wastewaters: Experiments and Modelling. *Membranes* **2023**, *13*, 321. [CrossRef]
16. Zulkefli, N.F.; Alias, N.H.; Jamaluddin, N.S.; Abdullah, N.; Abdul Manaf, S.F.; Othman, N.H.; Marpani, F.; Mat-Shayuti, M.S.; Kusworo, T.D. Recent Mitigation Strategies on Membrane Fouling for Oily Wastewater Treatment. *Membranes* **2021**, *12*, 26. [CrossRef] [PubMed]
17. Shi, X.; Tal, G.; Hankins, N.P.; Gitis, V. Fouling and Cleaning of Ultrafiltration Membranes: A Review. *J. Water Process Eng.* **2014**, *1*, 121–138. [CrossRef]

18. Lau, W.J.; Ismail, A.F.; Firdaus, S. Car wash industry in Malaysia: Treatment of car wash effluent using ultrafiltration and nanofiltration membranes. *Sep. Purif. Technol.* **2013**, *104*, 26–31. [CrossRef]
19. Palanisamy, T.; Tabatabai, S.A.A.; Zhang, T.; Leiknes, T.O. Role of surfactants in cleaning of PVDF ultrafiltration membranes fouled by emulsified cutting oil. *J. Water Process Eng.* **2021**, *40*, 101923. [CrossRef]
20. Chheang, M.; Hongprasith, N.; Ratanatawanate, C.; Lohwacharin, J. Effects of Chemical Cleaning on the Ageing of Polyvinylidene Fluoride Microfiltration and Ultrafiltration Membranes Fouled with Organic and Inorganic Matter. *Membranes* **2022**, *12*, 280. [CrossRef]
21. Hashim, N.A.; Liu, Y.; Li, K. Stability of PVDF hollow fibre membranes in sodium hydroxide aqueous solution. *Chem. Eng. Sci.* **2011**, *66*, 1565–1575. [CrossRef]
22. Li, K.; Su, Q.; Li, S.; Wen, G.; Huang, T. Aging of PVDF and PES Ultrafiltration Membranes by Sodium Hypochlorite: Effect of Solution pH. *J. Environ. Sci.* **2021**, *104*, 444–455. [CrossRef]
23. Weis, A.; Bird, M.R.; Nyström, M. The chemical cleaning of polymeric UF membranes fouled with spent sulphite liquor over multiple operational cycles. *J. Membr. Sci.* **2003**, *216*, 67–79. [CrossRef]
24. Weis, A.; Bird, M.R. The influence of multiple fouling and cleaning cycles upon the membrane processing of lignosulphonates. *Food Bioprod. Process.* **2001**, *79*, 184–187. [CrossRef]
25. Falsanisi, D.; Liberti, L.; Notarnicola, M. Ultrafiltration (UF) Pilot Plant for Municipal Wastewater Reuse in Agriculture: Impact of the Operation Mode on Process Performance. *Water* **2009**, *1*, 872–885. [CrossRef]
26. DuPont™ IntegraTec™ P Series PES-UF Modules T-Rack™ and Modules for Open Platform. Process and Design Manual. Version 6. November 2022. Available online: https://www.google.com/url?sa=t&source=web&rct=j&opi=89978449&url=https://www.dupont.com/content/dam/water/amer/us/en/water/public/documents/en/UF-IntegraTec-P-Series-PES-IO-T-Rack-Assembly-Manual-45-D02230-en.pdf&ved=2ahUKEwjzo6iYx8WNAxUIA9sEHUy0KjMQFnoECB4QAQ&usg=AOvVaw1465VsDPd3Rb8PgX6K0FZa (accessed on 20 May 2025).
27. Cai, X.; Wang, Z.; Qian, Y.; Wang, A.; Yang, Y.; Xia, S. Comprehensively understanding fouling properties of cake and bulk sludge in an anammox membrane bioreactor: Focusing on the composition, interfacial thermodynamics and microbial community. *J. Environ. Chem. Eng.* **2022**, *10*, 108612. [CrossRef]
28. Li, J.-X.; Sanderson, R.D.; Chai, G.Y. A focused ultrasonic sensor for in situ detection of protein fouling on tubular ultrafiltration membranes. *Sens. Actuators B Chem.* **2006**, *114*, 182–191. [CrossRef]
29. Woźniak, P.; Gryta, M. Influence of Reclaimed Water on the Visual Quality of Automotive Coating. *Materials* **2024**, *17*, 5382. [CrossRef]
30. Ali, S.; Rana, U.A.; Zahid, M.; Hussain, A. Effect of NaOH concentration on morphology and performance of PVDF membranes for desalination. *Sep. Purif. Technol.* **2019**, *209*, 46–56. [CrossRef]
31. Wen, G.; Chen, K.; Zhang, Y.; Zhou, Y.; Pan, J.; Wang, Q.; Sun, S.; Wang, Z. Preparation of alkali-resistant PVDF membranes via immobilization of sodium lauryl sulfate (SDS) on surface. *Appl. Water Sci.* **2021**, *11*, 55. [CrossRef]
32. Tomczak, W.; Gryta, M. Long-Term Performance of Ultrafiltration Membranes: Corrosion Fouling Aspect. *Materials* **2023**, *16*, 1673. [CrossRef] [PubMed]
33. Gryta, M.; Tomczak, W. Changes in Tubular PVDF Membrane Performance During Initial Period of Pilot Plant Operation. *Membranes* **2025**, *15*, 119. [CrossRef] [PubMed]
34. Tomczak, W.; Woźniak, P.; Gryta, M. Ultrafiltration of Car Wash Wastewater: Pilot-Scale Studies. *Water* **2025**, *17*, 213. [CrossRef]
35. Gan, X.; Lin, T.; Jiang, F.; Zhang, X. Impacts on characteristics and effluent safety of PVDF ultrafiltration membranes aged by different chemical cleaning types. *J. Membr. Sci.* **2021**, *640*, 119770. [CrossRef]
36. Khoiruddin, I.G.W.; Hakim, A.N.; Aryanti, P.T.P.; Rova, N. Long-Term Performance of a Pilot Scale Combined Chemical Precipitation-Ultrafiltration Technique for Waste Brine Regeneration at Chevron Steam Flooding Plant. *J. Eng. Technol. Sci.* **2020**, *52*, 501–513. [CrossRef]
37. Cui, Z.; Hassankiadeh, N.T.; Zhuang, Y.; Drioli, E.; Lee, Y.M. Crystalline polymorphism in poly(vinylidenefluoride) membranes. *Prog. Polym. Sci.* **2015**, *51*, 94–126. [CrossRef]
38. Ali, B.T.I.; Romadiansyah, T.Q.; Lestari, W.C.; Kusumawati, Y.; Ermavitalini, D.; Widiastuti, N. Modification of PVDF ultrafiltration membrane for high concentration of nannochloropsis as a raw material for bioethanol: Computations and experiments. *S. Afr. J. Chem. Eng.* **2023**, *46*, 42–55. [CrossRef]
39. Kong, Y.; Hay, J.N. The measurement of the crystallinity of polymers by DSC. *Polymer* **2002**, *43*, 3873–3878. [CrossRef]

membranes

Article

Determining Accurate Pore Structures of Polypropylene Membrane for ECMO Using FE-SEM Under Optimized Conditions

Makoto Fukuda [1,2,*], Yoshiaki Nishite [1], Eri Murata [2], Koki Namekawa [3], Tomohiro Mori [4], Tsutomu Tanaka [5] and Kiyotaka Sakai [6]

1 Department of Biomedical Engineering, Kindai University, 930 Nishimitani, Kinokawa-City 649-6493, Wakayama, Japan; menisite@waka.kindai.ac.jp
2 Graduate School of Biology-Oriented Science and Technology, Kindai University, 930 Nishimitani, Kinokawa-City 649-6493, Wakayama, Japan
3 Department of Clinical Engineering, Juntendo University, 6-8-1 Hinode, Urayasu-City 279-0013, Chiba, Japan; k.namekawa.at@juntendo.ac.jp
4 Industrial Technology Center of Wakayama Prefecture, 60 Ogura, Wakayama-City 649-6261, Wakayama, Japan; tomohiro_mori@wakayama-kg.jp
5 Osaka Research Institute of Industrial Science and Technology, 2-7-1 Ayumino, Izumi-City 594-1157, Osaka, Japan; t_tanaka@orist.jp
6 School of Advanced Science and Engineering, Waseda University, 3-4-1 Okubo, Shinjuku-ku 169-8555, Tokyo, Japan; kisakai@waseda.jp
* Correspondence: fukuda@waka.kindai.ac.jp; Tel.: +81-736-77-0345 (ext. 4308); Fax: +81-736-77-4754

Abstract: Long-term ECMOs are expected to be put into practical use in order to prepare for the next emerging severe infectious diseases after the novel coronavirus pandemic in 2019–2023. While polypropylene (PP) and polymethylpentene (PMP) are currently the mainstream materials for the hollow fiber membranes of ECMO, the PP membrane coated with a silicone layer on the outer surface has also been commercialized. In this study, we sought a method to accurately observe the detailed pore morphologies of the PP membrane by suppressing irreversible changes in the morphology in SEM observation, which is a general-purpose observation with higher resolution. As a result, the convex surface morphologies of the PP membrane, which was a non-conductive porous structure, were confirmed in detail by utilizing the lower secondary electron image (LEI) mode (FE-SEM, JSM-7610F, JEOL Ltd., Tokyo, Japan) at low acceleration voltage, low magnification, and long working distance, to minimize morphological alterations caused by osmium (Os) sputtering. On the other hand, although the sputter-coating on non-conductive samples is mandatory for imaging morphologies with SEM, the non-sputtering method is also worthwhile for porous and fragile structures such as this sample to minimize morphological alterations. Furthermore, we propose a method to confirm the morphology of the deep part of the sample by utilizing the secondary electron image (SEI) mode at an appropriate acceleration voltage and high magnification with higher resolution.

Keywords: extracorporeal membrane oxygenator; pore structure; polypropylene; field emission scanning electron microscopy; sputter coating

1. Introduction

The extracorporeal membrane oxygenator (ECMO) is an artificial organ that has attracted worldwide attention after the novel coronavirus pandemic in 2019–2023 [1]. The

 https://doi.org/10.3390/membranes15060174

requirements for the membrane have both high gas permeability and low water vapor permeability, while the development of ECMO for long-term usage is anticipated [2–5].

There was concern that SARS-CoV-2 in plasma penetrated through the pore in the hollow fiber membrane and diffused as an aerosol from the gas outlet port of ECMO during the treatment of COVID-19 severely ill patients with ECMO [6]. We previously used atomic force microscopy (AFM) to clarify the pore structure of ECMO membranes by using our approach and theoretically validate the risk of SARS-CoV-2 permeation [7]. However, the surface roughness of the membranes was larger than the amplitude of the probes, and we did not obtain appropriate data for polypropylene (PP) membranes for ECMO [7]. Therefore, field emission scanning electron microscopy (FE-SEM) has been used to clarify the pore structures of the membranes, and theoretically determine the permeability of SARS-CoV-2 through the pores [7]. Furthermore, the pore structures of the walls of the cross-sectional membranes made from polymethypentene (PMP) were clarified utilizing the ion-milling (IM) method [8,9], which did not apply physical stress to the porous structure when cutting the wall part of the membranes [10]. The phenomena of blockage of voids in the porous structures were confirmed while using the FE-SEM, and by elevated temperature during sputter-coating of platinum (Pt) or gold (Au). Notably, the observed images of membrane morphology made from polypropylene (PP) with a lower melting point showed differences depending on the acceleration voltage and other conditions [7].

Along with the evolution of microscopic technologies such as FE-SEM [10,11], research and development for membrane science can progress by taking advantage of those techniques. The advantages and disadvantages of microscopy and spectroscopy techniques for polymeric membranes were summarized in detail [12]. In particular, the preparation of cross-sectional membrane samples has been comprehensively studied [13]. These methods have led to the evolution of practical studies of membrane porous structure–property relationship, especially in artificial organs [4].

After the novel coronavirus pandemic, studies on the structure–function relationship or antithrombotic properties of ECMO membranes have been published worldwide [14–18]. In addition to the typical materials such as PP and poly(4-methyl-1-pentene) [16–18], the composite membrane of PP and poly(4-methyl-1-pentene) [17] or the PP membrane with gradient pore structure [18] have been developed. A novel membrane based on the hollow fiber membrane made from PP coated with a silicone layer on the outer surface [5,19] has also been developed.

This study focuses on the methods to observe the pore structure of the PP membrane more precisely and in detail by utilizing an FE-SEM with different observation modes and sputter-coating methods. We explore the methods of conductive materials sputter-coating and the conditions for observation by FE-SEM, and verify the differences in the images of pore morphologies. We chose the sputtering of osmium (Os) compared to the sputtering of Au because the sputtering of Os was considered less damaging to the sample. The quantitative differences are examined in the pore diameters and distributions, and surface porosities.

2. Experimental Section

2.1. Materials

MERA NHP (SENKO MEDICAL INSTRUMENT Mfg. Co., Ltd., Tokyo, Japan) is approved as an extracorporeal membrane lung and assisted circulation/assisted respiration membrane lung for assisting respiration. The outer surface of the porous hollow fiber membrane made from PP is coated with thin silicone layer (thickness: 0.2 μm) [19], which is the extremely rare silicone-composite membrane for ECMO in the world. Inner diameter of the lumen is 246 ± 3 μm (n = 30), membrane thickness is 27 ± 1 μm (n = 30), sterilization method is EOG (ethylene oxide gas). Heparin compound is coated on the outer surface of the hollow fiber membrane for blood flow channel as an antithrombogenic material.

2.2. Observation of the Inner and Outer Surfaces of the Hollow Fiber Membrane Utilizing Field Emission Scanning Electron Microscopy

An FE-SEM (JSM-7610F, JEOL Ltd., Tokyo, Japan) was utilized to observe the inner and outer surfaces of the hollow fiber membrane at an accelerating voltage of 1.0, 2.0 kV, and 5.0 kV, a working distance of 15.0 or 6.0 mm, and an illumination current of 37.5 pA in observation fields of a magnification of 30,000, 50,000, and 100,000.

SEI (secondary electron image) and LEI (lower secondary electron image) modes were utilized. They differ in the position of the detector and the type of secondary electrons that are detected, and the characteristics of the images are different. In the LEI mode, since the detector is located at the same height as the sample, the LEI mode detects electrons emitted from the sample at a low angle. The depth of the field to be imaged is increased by the longer working distance (WD) making it suitable for imaging a three-dimensional structure [10]. On the other hand, in the SEI mode, since the detector is positioned above the sample, it detects low-energy secondary electrons with a shorter WD. The information on deeper parts of the sample can be obtained at high resolution by irradiating the sample with acceleration voltages from 1.0 to 30 kV. Higher acceleration voltages make it easier to obtain high-resolution images, but irreversible destruction or charge-up of the sample must be avoided.

2.3. Sputter-Coating of Os or Au Particles on the Inner and Outer Surfaces of the Hollow Fiber Membrane

Nonsputter-coating samples and samples with osmium (Os) or Au sputter-coating on the membranes were observed accurately and in high-resolution imaging. Various combinations of sputter-coating and observation modes were used to prevent irreversible changes in original morphologies of the samples.

Tenant 20 (MEIWAFOSIS Co., Ltd., Tokyo, Japan) was used for Os sputter-coating and Os thickness was set at 5 nm (5 mA). JFC-1500 (JEOL Ltd., Tokyo, Japan) was used for Au sputter-coating and Au thickness was set at 7 nm.

2.4. Determination of the Pore Diameter and Surface Porosity of the Hollow Fiber Membrane

Pore areas were measured in observation fields of magnification of 30,000 or 50,000 in size by the analysis of digital imagery utilizing Image J software (Ver.1.53, National Institute of Health, Bethesda, MD, USA). Equivalent pore diameters and surface porosities of the images were calculated using the values from pore areas. Data on the samples were treated using a paired t-test ($p < 0.05$).

3. Results and Discussion

3.1. FE-SEM Images of the Inner Surfaces of the PP Membrane with the Different Sputter-Coating Methods and Observation Modes

Figure 1 shows the overview of FE-SEM images of the morphologies of the inner and outer surfaces of the PP capillary membranes under different observation modes and sputter-coating methods. The morphologies of the Os or Au sputter-coated samples were also compared to those without any sputter-coating. For example, (c) is the image of an inner surface of sputter-coating with Os in LEI mode, accelerating voltage of 1.0 kV, working distance of 15.0 mm, and magnification of 30,000. Three samples were at least prepared and observed for each condition.

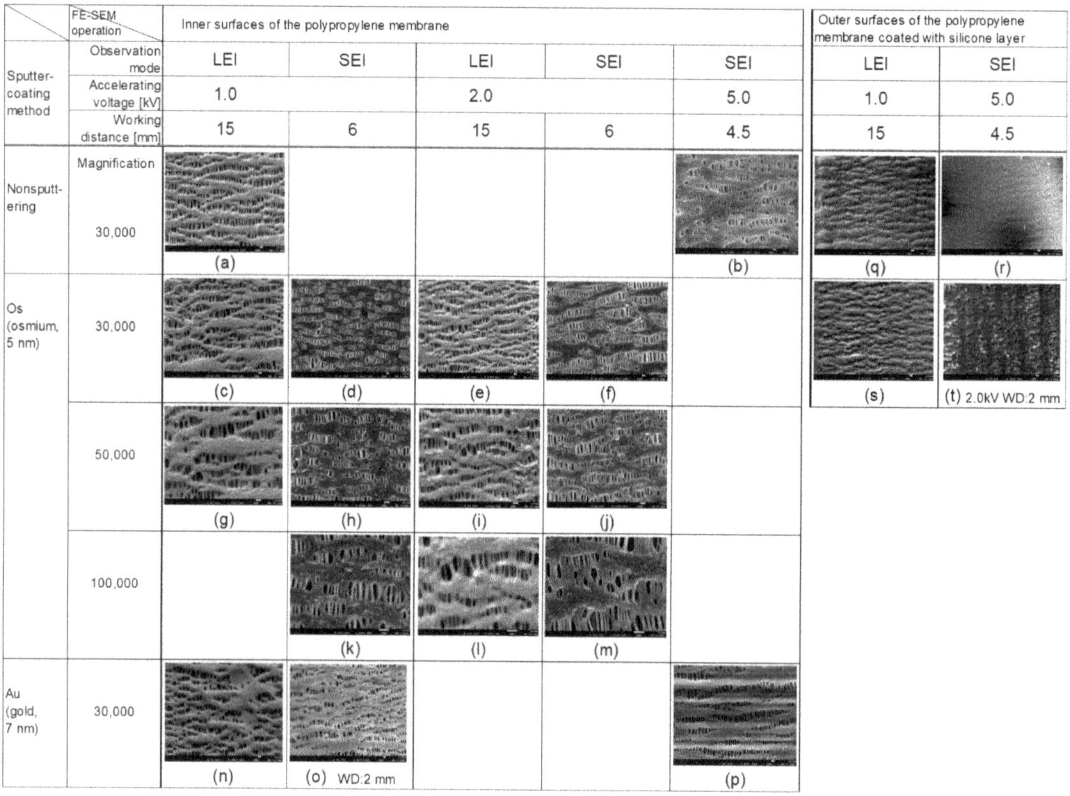

Figure 1. Outline of the morphologies of the inner and outer surfaces of the capillary membrane made of polypropylene coated with silicone layer on the outer surface using an FE-SEM with the different observation modes and sputter-coating methods. For example, (c) is the image of an inner surface of sputter-coating with Os in LEI mode, accelerating voltage of 1.0 kV, working distance of 15.0 mm, and magnification of 30,000.

Figure 2(1) shows three images each of (a), (c) and (n). (a) is a nonsputter-coated sample, (n) is a sputter-coating with Au particles, and the other observation conditions are the same as in (c). Elongated pore structures along the longitudinal hollow fiber membrane and convex morphologies of the inner surfaces are confirmed in all images, which are characteristic of PP membranes manufactured by the stretching mechanism. These images are similar to those of the cross-sectional longitudinal structure of the PP membrane [17]. While observing the nonsputter-coated sample (a), reducing the number of secondary electrons that were detected, made it difficult to obtain the images stably. The figures of the Os sputter-coated sample (c) were stably imaged, but some of the pores were found to be occluded, and large areas composed of fibrils were specifically confirmed due to morphological alterations caused by metal sputtering. On the other hand, the Au sputter-coated images (n) showed an overall loss of convex morphology and elongated pores, and areas of the PP fibrils were large. Because of the higher conductivity of Au than Os, the morphological changes during sputtering were larger.

The differences in the images (a) (c) (n) were determined to be sputtered differences rather than differences in the micro-fiber lengths of different sections of the melt-spun stretched membranes. In terms of quality uniformity, at this frequency of sampling numbers, manufacturing of melt-spun stretched membranes (PP) does not permit the morphology in (c) and (n).

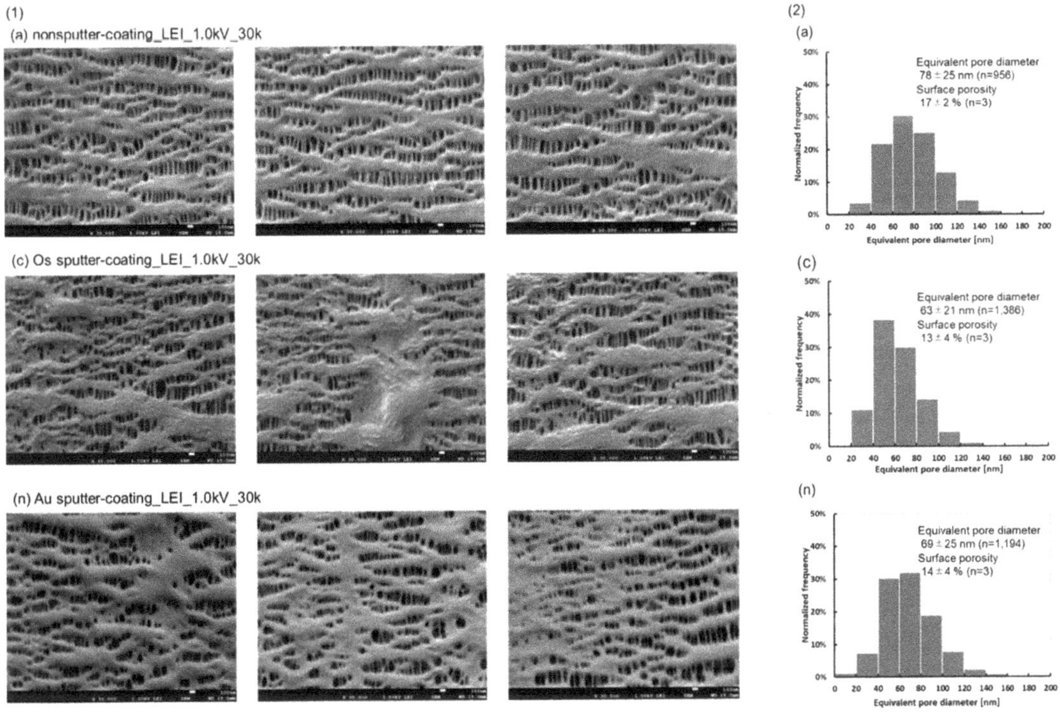

Figure 2. (**1**) FE-SEM images of the samples (a), (c), (n), inner surfaces of the capillary membrane, magnification; 30,000. (**2**) Distribution of the equivalent pore diameter of the samples, inner lumen of the capillary membrane determined via FE-SEM, (a), (c), and (n).

These images are different from those of the PP membrane surfaces [18]. The PP pieces were quenched in liquid nitrogen and coated with Pt using an ion sputter coater (JEC-3000FC, JEOL Ltd., Tokyo, Japan), SEM imaging at an accelerating voltage of 3 kV was used [18].

Figure 2(2) shows the distributions of the equivalent pore diameters in (a), (c) and (n). The distribution of (a) shifted slightly larger than that of (c) and (n). Significant differences in the equivalent pore diameters were found between (c) and (a) ($p = 0.008$) and between (a) and (n) ($p = 0.021$). Significant differences in the surface porosities were found between (c) and (a) ($p = 0.013$) and between (a) and (n) ($p = 0.045$).

Although charge-up occurred and the images were difficult to stabilize while observing the nonsputter-coated sample (Figure 2(1) (a)), the equivalent pore diameter and surface porosity of the observed images were significantly higher than those of (c) and (n). The slightly smaller equivalent pore diameter and surface porosity of (c) compared to those of (a) are due to the blockages of the Os sputter-coating, which reduced the size of individual pores and increased the area of the fibril. In the comparison of samples (a) and (c), the difference in the graphs shows that the average pore diameter is reduced by 15 nm on average due to the Os (5 nm) coating, which suggests that larger diameter pores are more likely to be occluded. Therefore, this can be interpreted as partial occlusion due to the Os coating.

These results demonstrate that the images of (a) in the LEI mode and at an accelerating voltage of 1.0 kV better reflect the morphology and size of the original individual pores. The relationship between different sputter-coating methods on the porous membrane made from PP and the observed images is clarified, as well as the differences in the distribution of equivalent pore diameter and surface porosity. Although the sputter-coating on the non-conductive samples is mandatory for imaging the morphologies by SEM, the non-sputtering method is worthwhile for the porous and fragile structure to minimize morphological alterations caused by metal sputtering.

Figure 3 shows the example of actual analyzed images of Figure 2(a), (c). The openings on the innermost surface of the membrane were visually checked and numbered one by one to determine if they were pores or not. Pores were differentiated from convex morphologies based on the difference in depth. The area and the circle equivalent diameter were calculated by trimming off the areas that could be recognized as pores one by one. Contrasts are slightly adjusted. The inner and outer surfaces were observed randomly, so there are no cross-sectional edges.

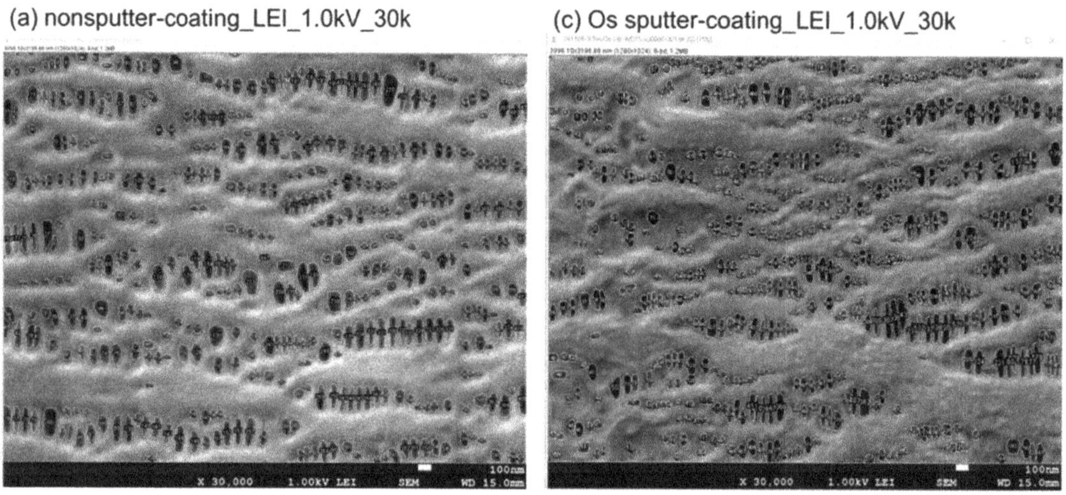

Figure 3. (a) Image analysis of the samples (a). (c) Image analysis of the samples (c).

3.2. Comparison of LEI and SEI Modes

Figure 4 shows the images in SEI mode (b)(p), compared to the nonsputter-coated sample imaged in LEI mode (a) and the Au sputter-coated sample imaged in LEI mode (n).

The morphologies of (a) and (b) were quite different. The (b) was affected by the accelerating voltage of 5.0 kV, and almost all of the elongated pores were deformed during the observation by FE-SEM, and the convex surfaces of the PP fibril chains also disappeared.

The mean values of the major axis and minor axis of (b) were 166 $\pm$ 51 nm (n = 50), and 53 $\pm$ 15 nm (n = 50), respectively [8], so the remaining pores were larger and had a larger equivalent pore diameter due to the denaturation of the pore structure, compared to the equivalent diameter of (a) (78 $\pm$ 25 nm) (n = 956). The Au sputter-coating image (p) also showed a characteristic morphology, differing from (n) or (a). Notably, thin PP chains were severed, pores shrunk or disappeared, and the convexity of the PP fibril chains also disappeared during FE-SEM observation. A large difference of (p) from the image in (b) was confirmed because sample (p) was only sputtered with Au with high conductivity. The results in Figure 4 indicate undesirable physical effects on the membrane, but they are remarkable outcomes.

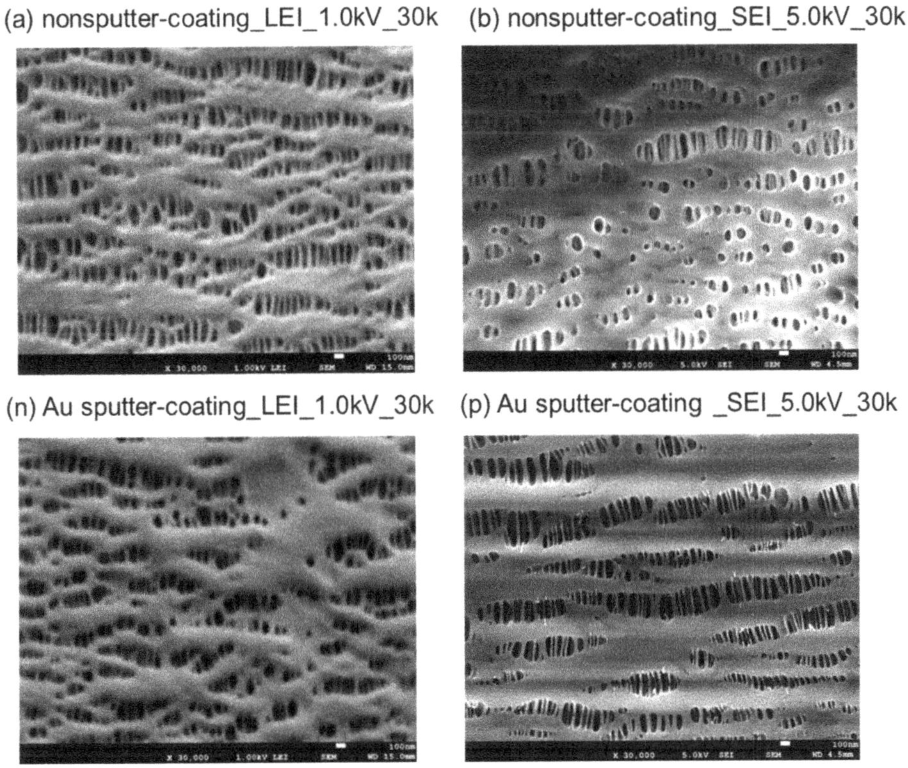

Figure 4. FE-SEM images of the samples ((a), (b), (n), and (p)) and inner surfaces of the capillary membrane, observed in different sputter-coating methods, modes, and acceleration voltages, magnification; 30,000.

3.3. Appropriate Accelerating Voltages in LEI and SEI Modes (Os Sputter-Coating)

Figure 1 compares images observed in LEI mode (c)(e)(g)(i) with those in SEI mode (d)(f)(h)(j) at accelerating voltages of 1.0 kV or 2.0 kV, for Os sputter-coating samples. From the results of Figures 1 and 4, an upper limit of 2.0 kV was set as the accelerating voltage at which irreversible changes in the sample did not occur while observing the sample in SEI mode.

The images of (c) and (d) are completely different, and (d) is quite different from the typical image of the PP membrane ((a), nonsputter-coated sample). The deep morphology of the sample was not confirmed at magnification; 30,000 (d).

The images obtained by changing the accelerating voltage from 1.0 kV to 2.0 kV in each of (c) and (d) are shown in (e) and (f). The images of (e) and (f) have higher resolution than those of (c) and (d). In addition, the deep morphologies of the samples are well-defined in (f) (SEI mode) than in (e) (LEI mode). (c), (d), (e), and (f) are the ones in which the magnification factor of 30,000X, and those for (g), (h), (i) and (j) are changed to magnification; 50,000. The images of (g) and (i) (magnification; 50,000) had the same level of detail compared to (c) and (e) (LEI mode, magnification; 30,000), but (g) and (i) were slightly more indistinct.

On the other hand, (h) and (j) (SEI mode, magnification; 50,000) showed the same high resolution and detailed structure compared to (d) and (f) (SEI mode, 30,000X). In addition, the accelerating voltage of (j) was higher than that of (h), so the accurate pore morphology in the deep part of (j) was confirmed.

Figure 5(1) shows a comparison of the enlarged figures of (g), (i), and (j), and Figure 5(2) shows a comparison of the distributions of their equivalent pore diameters. No significant differences between (g) and (i), and between (i) and (j) were found in either case.

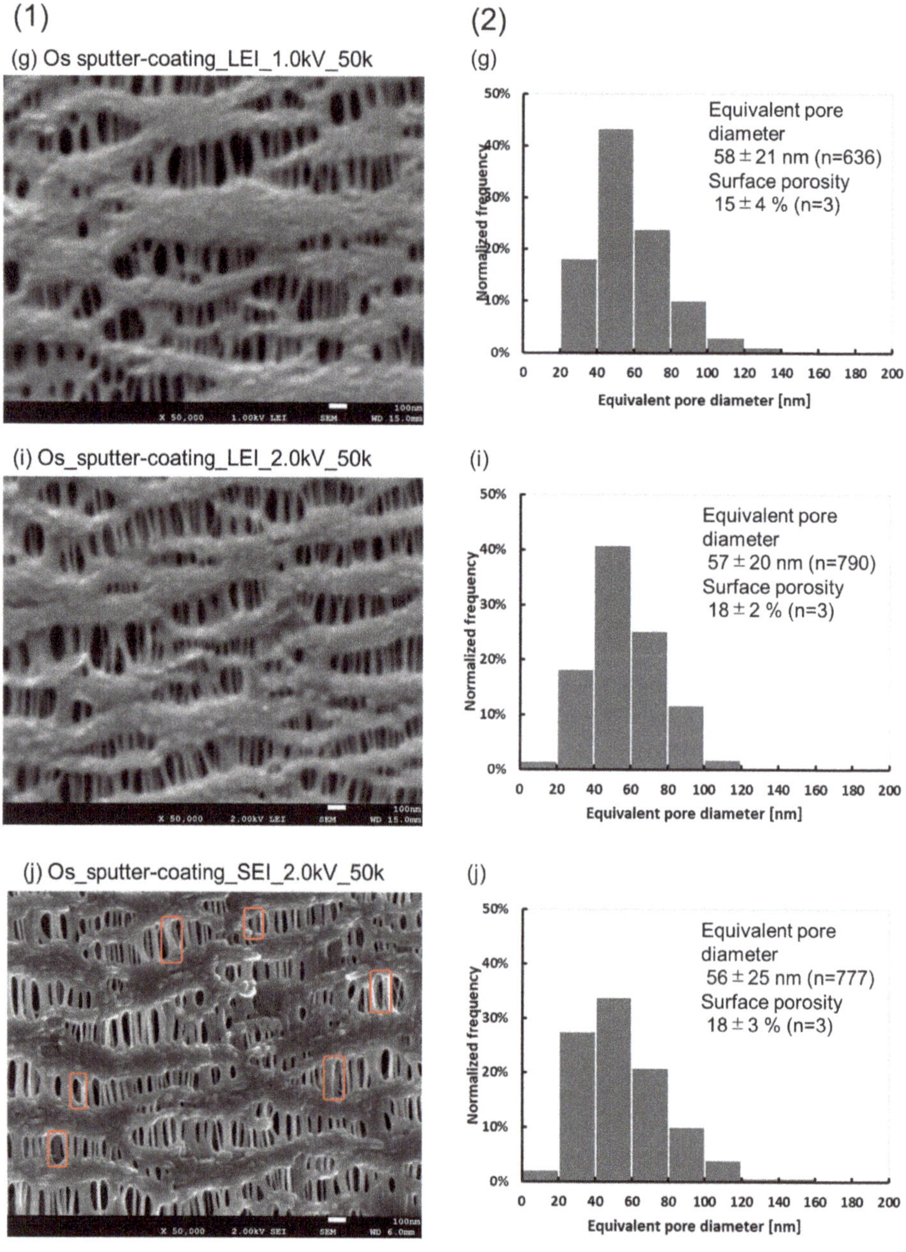

Figure 5. (**1**) FE-SEM images of the sample ((g), (i), and (j)), inner surfaces of the capillary membrane, observed in LEI or SEI mode, the acceleration voltages of 1.0 kV or 2.0 kV, and magnification; 50,000. (**2**) Distribution of the equivalent pore diameter of the samples, inner lumen of the capillary membrane determined via FE-SEM, (g), (i), and (j). red box: another fibril is recognized deeper than the top surface of the image.

The differences were obvious when comparing (i) with (j) (accelerating voltage 2.0 kV, magnification; 50,000) observed in LEI and SEI mode at a more magnified view. The image of (i) was a little blurred, while (j) had a higher resolution and more detailed structure, and the deep morphology was also confirmed. Although it is difficult to measure the depth, since another fibril is recognized deeper than the top surface of the image (red box), its depth is considered to be deeper than the thickness of the fibril of the top surface (e.g., 50 nm, 100 nm, etc.). The mean value of the surface equivalent pore diameter of (j) was 56 ± 25 nm (n = 777) and the surface porosity was 18 ± 3% (n = 3), which were not significantly different from those of (i), indicating that the surface morphology was equivalent. These results demonstrate that the deep morphology and precise structure are clearly confirmed with high resolution without irreversible changes if the accelerating voltage is appropriately set in SEI mode. Since the images of (j) were quite different from the previous findings, the reproducibility of these images was carefully validated.

Figure 6 shows the example of actual analyzed images of Figure 5(1) (j). The openings on the innermost surface of the membrane were visually checked and numbered one by one to determine if they were pores or not. The area and the circle equivalent diameter were calculated by trimming off the areas that could be recognized as pores one by one.

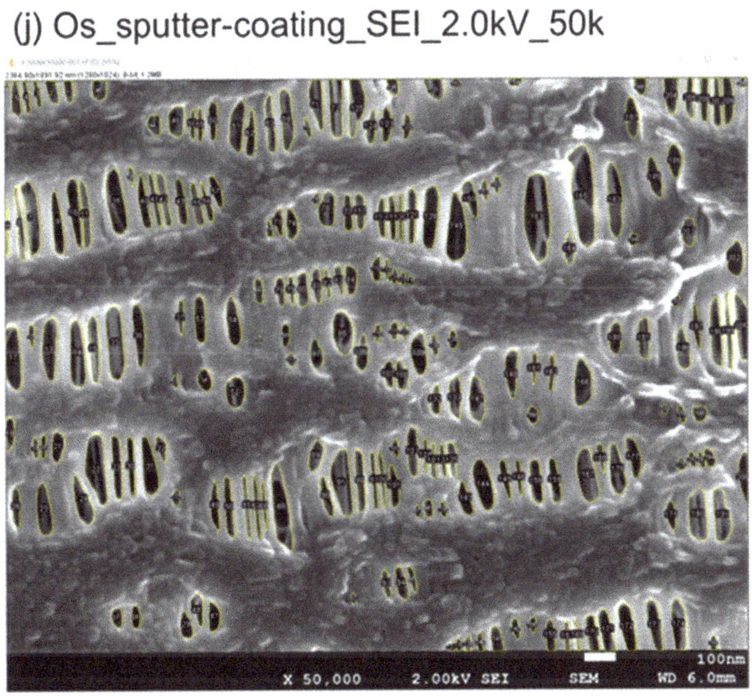

Figure 6. Image analysis of the sample (j).

3.4. *Outer Surface of the Membrane Coated with Silicone Layer*

Figure 1(q), (r), (s) and (t) shows the images of the outer surface of the PP membrane coated with the silicone layer. (q) is an image of a nonsputter-coated sample in LEI mode, and (s) is an image of a sputter-coated sample with Os in LEI mode.

In both cases, the pores on the outer surface of the membrane were found to be covered by the silicone layer [7]. The convexity was observed in the (s) image more than in the (q) image.

(r) and (t) were observed by SEI mode, (r) was taken with the nonsputter-coating and an acceleration voltage of 5.0 kV [7], and (t) was taken with the Os sputter-coated sample at an accelerating voltage of 2.0 kV. (r) was smoother than the others. The features of (t) were also different from those of (s), this was thought to be because it represented a more deeper view of the sample.

3.5. *Significance of This Study*

It is important to precisely design the anisotropy of the pore structure in the cross-section of the capillary membrane to have highly controlled anisotropic functions [5,9]. Microscopic observation technology such as FE-SEM systems have evolved independently for each individual system [10,11], and it is necessary to utilize these techniques according to research needs. The results of this study are expected to further progress the prototyping and evaluation of such related membranes., and it is expected that next-generation gas exchange membranes and ECMOs with novel antithrombogenic materials will be put into practical use.

Although multiple analytical tools should be utilized to achieve a comprehensive structural analysis, it is noteworthy that the authors are able to reflect on the findings of the various sample preparations and modes in the previous study, understand the causes, and apply countermeasures, making this study highly practical. Even though we do not have other analytical instruments such as permporometry or environmental SEM (ESEM), the relationship between pore structure and gas permeability of several gases in hollow fiber membranes have been investigated which are the outcome of this study or by FIB-SEM [20,21] in future. The goal is to clarify the anisotropic gas permeation behavior in the anisotropic pore structure of ECMO membranes based on the behavior of Knudsen flow [22] and Poiseuille flow, and how it depends on the viscosity or molecular weight of the gas.

When using SEM to observe non-conductive samples, it is difficult to minimize morphological alterations caused by metal sputtering. Since PP has a lower melting point and PP hollow fiber membranes for ECMO are porous, the observed images of membrane morphology made from PP showed differences depending on the acceleration voltage and other conditions by using an FE-SEM. Therefore, we have investigated methods to minimize morphological alterations caused by Os sputtering or without sputtering. This is contrary to common sense and has not been seen in previous studies, and utilizing this method will advance research on a daily basis. Non-sputtered LEI imaging may be applicable to other porous polymer membranes with similar fragility.

4. Conclusions

In this study, we attempted to observe the original morphology of the porous PP membranes for ECMO. The unique LEI mode of the FE-SEM reveals the surface convex morphology of the porous PP membrane for ECMO. Although it is possible to obtain stable images by the Os sputter-coating on non-conductive materials, the non-sputtering method is also worthwhile for the porous and fragile structure of the PP membrane with a lower melting point. On the other hand, the morphology of the deep structure of the porous PP membrane is confirmed by utilizing the SEI mode at an appropriate accelerating voltage. It is noteworthy that the authors are able to reflect on the findings of the previous study, understand the causes, and apply countermeasures, making this study highly practical. The achievements and approaches of this study are being applied to the development of next-generation gas exchange membranes.

Author Contributions: M.F., Y.N., K.N.: Conceptualization, Methodology, Validation, Visualization, Writing—original draft, Writing-review and editing. E.M., T.M.: Methodology, Software, Validation, Visualization. T.T.: Conceptualization, Methodology. K.S.: Conceptualization, Methodology, Writing—review, and editing. All authors have read and agreed to the published version of the manuscript.

Funding: The corresponding author would like to express our appreciation for the financial support of the JSPS KAKENHI (Grant-in-Aid for Scientific Research-C (No. 23K08487)).

Institutional Review Board Statement: Not applicable.

Data Availability Statement: The data that support the findings of this study are available.

Acknowledgments: The authors acknowledge Naomi Backes Kamimura (Kindai University) for helpful suggestions during the preparation of the manuscript.

Conflicts of Interest: The authors declare no conflicts of interest.

References

1. Barbaro, R.P.; MacLaren, G.; Boonstra, P.S.; Combes, A.; Agerstrand, C.; Annich, G.; Diaz, R.; Fan, E.; Hryniewicz, K.; Lorusso, R.; et al. Extracorporeal membrane oxygenation for COVID-19: Evolving outcomes from the international Extracorporeal Life Support Organization Registry. *Lancet* **2021**, *398*, 1230–1238. [CrossRef] [PubMed]
2. Arens, J.; Grottke, O.; Haverich, A.; Maier, L.S.; Schmitz-Rode, T.; Steinseifer, U.; Wendel, H.P.; Rossaint, R. Toward a Long-Term Artificial Lung. *ASAIO J.* **2020**, *50*, 491–497. [CrossRef] [PubMed]
3. Akiyama, D.; Katagiri, N.; Mizuno, T.; Tsukiya, T.; Takewa, Y.; Tatsumi, E. Preclinical biocompatibility study of ultra-compact durable ECMO system in chronic animal experiments for 2 weeks. *J. Artif. Organs* **2020**, *23*, 335–341. [CrossRef] [PubMed]
4. Duy Nguyen, B.T.; Nguyen Thi, H.Y.; Nguyen Thi, B.P.; Kang, D.-K.; Kim, J.F. The Roles of Membrane Technology in Artificial Organs: Current Challenges and Perspectives. *Membranes* **2021**, *11*, 239. [CrossRef] [PubMed]
5. Fukuda, M. Evolutions of extracorporeal membrane oxygenator (ECMO): Perspectives for advanced hollow fiber membrane. *J. Artif. Organs* **2024**, *27*, 1–6. [CrossRef] [PubMed]
6. Ogawa, T.; Uemura, T.; Matsuda, W.; Sato, M.; Ishizuka, K.; Fukaya, T.; Kinoshita, N.; Nakamoto, T.; Ohmagari, N.; Katano, H.; et al. SARS-CoV-2 leakage from the gas outlet port during extracorporeal membrane oxygenation for COVID-19. *ASAIO J.* **2021**, *67*, 511–516. [CrossRef] [PubMed]
7. Fukuda, M.; Furuya, T.; Sadano, K.; Tokumine, A.; Mori, T.; Saomoto, H.; Sakai, K. Electron microscopic confirmation of anisotropic pore characteristics for ECMO membranes theoretically validating the risk of SARS-CoV-2 permeation. *Membranes* **2021**, *11*, 529. [CrossRef] [PubMed]
8. Kaneko, A.; Takasu, H. Introduction of ion milling system and its application. *J. Surf. Finish. Soc. Jpn.* **2015**, *66*, 581–585. (In Japanese) [CrossRef]
9. Fukuda, M.; Sadano, K.; Maeda, T.; Murata, E.; Miyashita, N.; Tanaka, T.; Mori, T.; Saito, A.; Sakai, K. Characterization of anisotropic pore structure and dense selective layer of capillary membranes for long-term ECMO by cross-sectional ion-milling method. *J. Artif. Organs* **2024**, *28*, 50–60. [CrossRef] [PubMed]
10. Jeol Ltd. Available online: https://www.jeol.co.jp/products/scientific/sem/ (accessed on 10 February 2025).
11. Jeol Ltd. Available online: https://www.jeol.co.jp/words/semterms/20121024.022959.html#gsc.tab=0 (accessed on 10 February 2025).
12. Alqaheem, Y.; Alomair, A.A. Microscopy and spectroscopy tech niques for characterization of polymeric membranes. *Membranes* **2020**, *10*, 33. [CrossRef] [PubMed]
13. Ren, H.; Zhang, X.; Li, Y.; Zhang, D.; Huang, F.; Zhang, Z. Preparation of cross-sectional membrane samples for scanning electron microscopy characterizations using a new frozen section technique. *Membranes* **2023**, *13*, 634. [CrossRef] [PubMed]
14. He, T.; He, J.; Wang, Z.; Cui, Z. Modification strategies to improve the membrane hemocompatibility in extracorporeal membrane oxygenator (ECMO). *Adv. Compos. Hybrid Mater.* **2021**, *4*, 847–864. [CrossRef] [PubMed]
15. Feng, Y.; Wang, Q.; Sun, S.; Zhao, W.; Zhao, C. Advanced hemocompatible polyethersulfone composite artificial lung membrane with efficient CO_2/O_2 exchange channel constructed by modified carbon nanotubes network. *J. Mater. Sci. Technol.* **2023**, *160*, 181–193. [CrossRef]
16. Zhang, T.-Q.; Jia, Z.-Q.; Peng, W.; Li, S.; Wen, J. Preparation of 4-methyl-1-pentene membranes via non-solvent induced phase separation (NIPS). *Eur. Polym. J.* **2022**, *178*, 111480. [CrossRef]
17. Wang, Y.; Liu, Y.; Han, Q.; Lin, H.; Liu, F. A novel poly (4-methyl-1-pentene) / polypropylene (PMP/PP) thin film composite (TFC) artificial lung membrane for enhanced gas transport and excellent hemo-compatibility. *J. Membr. Sci.* **2022**, *649*, 120359. [CrossRef] [PubMed]

18. Tang, Y.; Zhang, F.; Wu, F.; Wang, L.; Feng, A.; Yu, L.; Wu, H.; Lin, Y.; Wang, X. A novel tunable polypropylene hollow fiber membrane with gradient structure for extracorporeal membrane oxygenation applications. *J. Membr. Sci.* **2024**, *693*, 122325. [CrossRef]
19. Nakanishi, H.; Nishitani, Y.; Kuwana, K.; Tahara, K.; Aoki, Y.; Osaki, S.; Hu, C. Development of new oxygenator with cyclosiloxane coated polypropylene hollow fiber. *Jpn. J. Artif. Organs* **1996**, *25*, 329–332. (In Japanese)
20. Brickey, K.P.; Zydney, A.L.; Gomez, E.D. FIB-SEM tomography reveals the nanoscale 3D morphology of virus. *J. Membr. Sci.* **2021**, *640*, 119766. [CrossRef]
21. Roberge, H.; Moreau, P.; Couallier, E.; Abella, P. Determination of the key structural factors affecting permeability and selectivity of PAN and PES polymeric filtration membranes using 3D FIB/ SEM. *J. Membr. Sci.* **2022**, *653*, 120530. [CrossRef]
22. Carman, P.C. Chapter 3 Slip flow, free molecule flow, and diffusional flow. In *Flow of Gases Through Porous Media*; Carman, P.C., Ed.; Butterworths Scientific Publications: London, UK, 1956; pp. 62–80.

membranes

MDPI

Review

Development and Investigation of a New Polysulfone Dialyzer with Increased Membrane Hydrophilicity

Adam M. Zawada [1], Bettina Griesshaber [2], Bertram Ottillinger [3], Ansgar Erlenkötter [4], Nathan Crook [5], Skyler Boyington [5], Manuela Stauss-Grabo [2], James P. Kennedy [6] and Thomas Lang [2,*]

1 Product Development, Fresenius Medical Care Deutschland GmbH, 66606 Sankt Wendel, Germany; adam.zawada@freseniusmedicalcare.com
2 Clinical Research, Global Medical Office, Fresenius Medical Care Deutschland GmbH, 61352 Bad Homburg, Germany; bettina.griesshaber@freseniusmedicalcare.com (B.G.); manuela.stauss-grabo1@freseniusmedicalcare.com (M.S.-G.)
3 Ottillinger Life Sciences, 85649 Brunnthal, Germany; bertram.ottillinger@vicron.com
4 Biosciences, Fresenius Medical Care Deutschland GmbH, 66606 Sankt Wendel, Germany; ansgar.erlenkoetter@freseniusmedicalcare.com
5 Applications Laboratory, Fresenius Medical Care, Ogden, UT 84404, USA; nathan.crook1@freseniusmedicalcare.com (N.C.); skyler.boyington@freseniusmedicalcare.com (S.B.)
6 Product Development, Fresenius Medical Care, Ogden, UT 84404, USA; james.kennedy@freseniusmedicalcare.com
* Correspondence: thomas.lang@freseniusmedicalcare.com

Abstract: Innovation in dialysis care is fundamental to improve well-being and outcomes of patients with end-stage kidney disease. The dialyzer is the core element of dialysis treatments, as it largely defines which substances are removed from the patient's body. Moreover, its large surface size is the major place of interaction of the patient's blood with artificial surfaces and thus may lead to undesired effects such as inflammation or coagulation. In the present article we summarize the development path for a new dialyzer, including in vitro and clinical evidence generation. We use the example of the novel FX CorAL dialyzer, which has recently entered European and US markets, to show which steps are needed to develop and characterize a new dialyzer. The FX CorAL dialyzer includes a new hydrophilic membrane, which features reduced protein adsorption, sustained performance, and an improved hemocompatibility profile, characterized in numerous in vitro and clinical studies. Safety evaluations revealed a favorable profile, with low incidences of adverse device effects. Insights gained from both in vitro and clinical studies contribute to the advancement of dialyzer development, ultimately leading to improved patient care.

Keywords: dialyzer; membrane; hemocompatibility; performance; hydrophilicity; end stage kidney disease

1. Introduction

Hemodialysis is a life-saving treatment for millions of patients with end-stage kidney disease (ESKD) [1,2]. The core element of such treatments is the dialyzer, which replaces the function of the kidney during a ~4 h session, generally three times a week. In contrast to a continuously working kidney of healthy persons, the dialyzer has to eliminate accumulated uremic toxins in a very limited time frame from the patient's body. Thus, the performance of a dialyzer must approach the selective cleaning characteristics of a healthy human kidney, i.e., removal of small and middle molecules, while preserving essential proteins such as albumin. Moreover, during dialysis treatment, patient's blood contacts artificial surfaces, such as the capillary fibers, which potentially leads to an activation of different blood components, such as immune cells, platelets, or the complement system, contributing to undesirable outcomes such as inflammation or coagulation/clotting.

Thus, during the development of a new dialyzer, different aspects of performance and hemocompatibility must be considered to achieve optimal blood-purification properties and the reduction of short- and long-term adverse events in ESKD patients. Therefore, thoughtful design, together with excessive in vitro and clinical testing, are necessary to achieve a high-performing and safe product.

Taking the example of the novel FX CorAL dialyzer (Fresenius Medical Care, Bad Homburg, Germany), this article summarizes the development path of a new polysulfone dialyzer, including the scientific evidence-generation stream alongside the development to characterize performance, hemocompatibility, and safety of this new dialyzer.

2. Development of a New Dialyzer

A dialyzer consists of many parts, including the housing, flange, tubing connectors, potting, and the core element—the membrane, which mainly defines the dialyzer's performance and hemocompatibility features. While the housing, flange, and tubing connectors mainly represent the outer dialyzer design, the membrane and potting are located within the dialyzer.

2.1. Outer Dialyzer Design

With regards to the outer dialyzer design, two main designs are currently available (Figure 1). In the first variant, the blood enters and leaves the flanges in a vertical direction. Moreover, the dialysate tubing connections are located within the housing of the dialyzer. In the second design, both the blood and dialysate tubing connectors are located within the flanges, and the connections are in a horizontal direction. This design with lateral blood-inlet ports leads to a spiral and homogenous blood-flow path within the flange and may increase the performance of the dialyzer [3–6]. Moreover, in contrast to the first design, the dialysate distribution may be more homogeneous as the dialysate does not only enter the fiber compartment from one side, but the additional pinnacle structure of the housing supports the dialysate to enter the fiber compartment from all sides at a similar flow rate [4–6]. Moreover, the pinnacle structure prevents the detachment of the potting from the housing, which may happen with dialyzers of design 1, especially when exposed to higher temperatures. The FX CorAL dialyzer uses the second dialyzer design.

Moreover, different materials can be utilized for the dialyzer outer parts. The two main materials are polycarbonate and polypropylene. One advantage of polypropylene is the lower weight, which has positive environmental aspects during, e.g., transportation and disposal. While dialyzers of design 1 are available with polycarbonate or polypropylene housings and flanges, the second dialyzer design exclusively uses polypropylene housings and flanges.

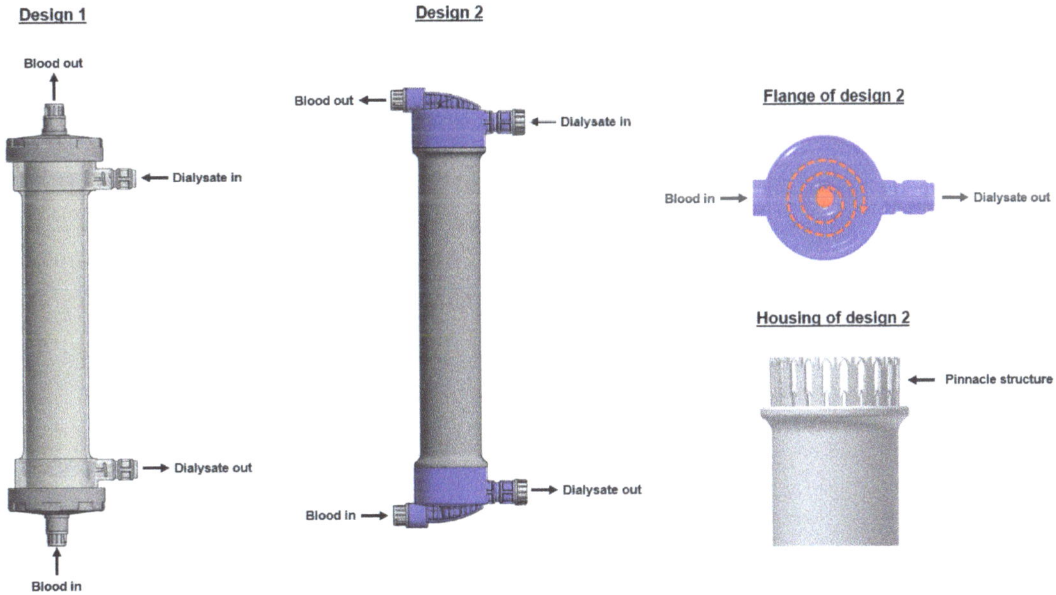

Figure 1. Outer dialyzer designs. Design 1 with blood inlet and outlet in vertical direction; Design 2 with inlet and outlet in horizontal direction, and dialysate ports within the flanges. The flange of design 2 leads to a spiral blood flow while the pinnacle structure within the housing allows dialysate entering from all sides and prevents detachments of the potting from the housing.

2.2. Dialyzer Membrane

Within the dialyzer, the fiber bundle, containing ~5.000–20.000 single capillary fibers, is potted at the two ends to allow separation of the blood and dialysate compartment. Based on the number of fibers, as well as the fiber length and geometry, the effective surface size can be calculated for each dialyzer. For the FX CorAL, nine different models are available, with surface sizes ranging from 0.6 m^2 to 2.5 m^2. Moreover, different fiber geometries are available for the FX CorAL with an inner diameter of 185 μm and wall thickness of 35 μm for the models FX CorAL 40 (0.6 m^2), FX CorAL 50 (1.0 m^2), FX CorAL 60 (1.4 m^2), FX CorAL 80 (1.8 m^2), FX CorAL 100 (2.2 m^2) and FX CorAL 120 (2.5 m^2), as well as an inner diameter of 210 μm and wall thickness of 40 μm for the models FX CorAL 600 (1.6 m^2), FX CorAL 800 (2.0 m^2) and FX CorAL 1000 (2.3 m^2), the latter three models optimized for hemodiafiltration (HDF) treatments. Due to the different surface sizes and geometry, the models also differ in their performance characteristics such as clearance values or ultrafiltration coefficients (e.g., ultrafiltration coefficient FX CorAL 40: 28 mL/h/mmHg, FX CorAL 120: 94 mL/h/mmHg, FX CorAL 600: 60 mL/h/mmHg, FX CorAL 1000: 82 mL/h/mmHg).

The hollow-fiber membranes may consist of different materials based on cellulose or synthetic polymers such as polysulfone, poly(aryl)ethersulfone, polyphenylene, acrylonitrile, polyester-polymer alloys, or polymethylmethacrylate [6–9]. Synthetic dialyzer membranes are most widely used in clinical practice. It is important to note that, although different dialyzers with the same base-material polysulfone exist on the market, continuous development of the polysulfone membrane during the last decades led to significant differences in the currently available polysulfone-based dialyzers. This development includes many different aspects during manufacturing and sterilization, leading to differences in, e.g., fiber geometry (e.g., inner diameter, wall thickness), pore size distribution, or hydrophilicity. One of the most important, and yet intransparent, attributes of polysulfone membranes is the concentration and distribution of polyvinylpyrrolidone (PVP). PVP is added to the membrane spinning mass as a hydrophilic agent, given that polysulfone is hydrophobic and increased membrane hydrophilicity is associated with reduced membrane fouling, improved hemocompatibility, and sustained performance [10–21].

For the production of standard hemodialysis membranes, a spinning solution, consisting of a polymer (e.g., polysulfone), co-polymer (PVP), and solvent, is run through the outer part of the spinneret, while the precipitation fluid (non-solvent) passes the spinneret through the inner orifice leading to the formation of the hollow fiber, which is then solidified in a bath containing the non-solvent [22]. During the development of the FX CorAL membrane (Helixone *hydro*), the spinning process was refined in a way, in that PVP was additionally added to the precipitation fluid leading to a PVP-enriched blood-side surface. Moreover, small amounts of the antioxidant α-tocopherol were added to the spinning solution, to prevent oxidation and elution of PVP during shelf life [10,13]. The increased hydrophilicity of the blood-side surface is important given that the inner membrane surface is the major place of interaction of the patient's blood during the extracorporeal treatment. The repulsive hydration force of the hydrophilic membrane reduces the interaction and adsorption of blood components, such as proteins, with the membrane and thus improves hemocompatibility [10,13,14]. Proteins generally bind to such artificial surfaces and undergo structural changes that subsequently trigger immune system activation [23]. Besides the FX CorAL dialyzer, another dialyzer with hydrophilic membrane modification is available on the market. The Toraylight NV (Toray Medical, Tokyo, Japan) with the "Hydrolink membrane" uses a hydrophilic polymer to make its polysulfone membrane hydrophilic [6,24–28]. In line, also this hydrophilic modification has been associated with positive hemocompatibility effects, including antithrombotic activity [24,26,28].

PVP stability is another aspect that differs among different dialyzers used in clinical practice. PVP may undergo oxidation, chain breaks, and subsequent elution from the membrane, which then may lead to a reduction of the positive effects of this compound. Generation of elutable PVP may occur either during manufacturing by, e.g., using radiation as a sterilization process or during shelf life as a continuous slow oxidation process. During the FX CorAL development and production, both PVP oxidation risks were reduced: (1) The FX CorAL sterilization process uses INLINE steam sterilization where no radiation is applied, but hot steam and sterile water are continuously rinsed through the dialyzer. (2) To reduce oxidation of PVP over time, small amounts of the anti-oxidant α-tocopherol are added to the spinning mass, leading to a stabilization of PVP over the complete shelf life [10].

3. Investigation of a New Dialyzer

3.1. Membrane Characterization

3.1.1. Characterization of PVP Content and Elution

To investigate whether the addition of PVP into the precipitation fluid during spinning leads to an increased amount of PVP on the blood-side surface, the Helixone *hydro* membrane was tested with X-ray photoelectron spectroscopy (XPS), which allows quantification of the atomic composition just of the fiber surface [13]. When compared to the predecessor membrane Helixone *plus* of the FX CorDiax (Fresenius Medical Care, Bad Homburg, Germany), which does not have additional PVP in the precipitation fluid during spinning, the novel Helixone *hydro* membrane had significantly higher PVP content on the blood-side surface (FX CorAL: 26.3% mass, FX CorDiax: 23.0%, $p < 0.05$). This was also the case when comparing to other polysulfone- or polyethersulfone-based membranes (xevonta [B.Braun, Melsungen, Germany]: 24.7%, $p = 0.356$; ELISIO [Nipro, Osaka, Japan]: 24.4%, $p = 0.219$; Polyflux [Baxter, Deerfield, IL, USA]: 18.3%, $p < 0.001$; Theranova [Baxter,

Deerfield, IL, USA]: 15.6%, $p < 0.001$). Moreover, in recirculation experiments up to 10 h, no elutable PVP from the Helixone *hydro* membrane was detected (below quantification limit of 0.5 mg/1.6 m^2 dialyzer surface). In contrast, other polysulfone- and polyethersulfone-based membranes, sterilized with gamma (xevonta: 3.2 mg/1.6 m^2, $p < 0.001$; ELISIO: 2.1 mg/1.6 m^2, $p < 0.001$) or autoclave steam (Polyflux: 9.0 mg/1.6 m^2, $p < 0.001$; Theranova: 9.1 mg/1.6 m^2, $p < 0.001$), showed a strong elution of PVP already after 5 h [13]. In addition, the generation of elutable PVP was investigated also over the complete shelf-life of 3 years [10]. Also in these experiments, no eluted PVP was detected from the Helixone *hydro* membrane after 3 years, supporting the stabilizing effect of α-tocopherol [29].

These results were in line with the measurements of the surface charge of the dialyzer membranes [14]. Zeta potential was determined for the FX CorAL and other polysulfone-, polyethersulfone- and cellulose-based dialyzers. Here, the FX CorAL showed the most neutral surface charge (−2.38 mV) as compared to all other investigated dialyzers (FX CorDiax: −5.56 mV, $p < 0.001$; xevonta: −7.30 mV, $p < 0.001$; Polyflux: −7.15 mV, $p < 0.001$; ELISIO: −7.12 mV, $p < 0.001$; FDX [Nikkiso, Tokyo, Japan]: −25.74 mV, $p < 0.001$; Theranova: −9.39 mV, $p < 0.001$; Sureflux [Nipro, Osaka, Japan]: −6.28 mV, $p < 0.001$).

3.1.2. Characterization of Membrane Hydrophilicity

It is well-established that increased PVP content on the membrane surface leads to increased hydrophilicity [15,20]. This association was also confirmed for the FX CorAL with contact angle measurements, where fibers are placed in a dye-containing water reservoir and the capillary height is measured after a predefined time [10]. The contact angle is then determined based on the capillary height and additional information such as the membrane geometry. When two membranes with the same geometry but different hydrophilicity are compared, the more hydrophilic membrane will have a lower contact angle than the less hydrophilic membrane. In these experiments, the Helixone *hydro* membrane of the FX CorAL dialyzer had a significantly lower contact angle then the predecessor membrane Helixone *plus* of the FX CorDiax dialyzer (FX CorAL: 45.1°, FX CorDiax: 51.7°, $p < 0.001$), confirming higher hydrophilicity of this novel membrane [10].

These results were confirmed by atomic force microscopy (AFM) measurements of the hydrolayer thickness. For the measurement, a fiber is cut lengthwise and affixed on a slide with the inner lumen exposed. A water droplet is placed on the fiber and on the AFM probe mount window. The two droplets are then brought together to form a continuous water bridge. AFM uses a cantilever with a tip that taps or moves on the membrane surface (Figure 2A). The cantilever bends to a different degree when the tip comes into contact with the hydrolayer or the membrane surface, leading to a reflection of the laser at different angles. These laser signals are measured and translated into a force-distance curve (Figure 2B). The force-distance curve comprises three different sections: (1) the approach section without increasing force, corresponding to the first step where the tip has no contact with the hydrolayer or the membrane and thus the cantilever shows no deflection, (2) a non-linear section, corresponding to the contact of the tip with the PVP-water gel ("hydrolayer") and (3) a final linear section with increasing force, corresponding to the contact of the tip with the solid membrane surface and the strong deflection of the cantilever. Based on this curve, the thickness of the hydrolayer can be determined as displayed in Figure 2B. These results showed that the Helixone *hydro* membrane of the FX CorAL has a significantly thicker hydrolayer as compared to the Helixone *plus* membrane of the FX CorDiax (Figure 2C), in line with the increased PVP content on the blood-side surface and the lower contact angle, showing that the AFM determination of hydrolayer thickness is a good measure of membrane hydrophilicity [10,13]. AFM also allows generation of 3D surface images of dry and wet membranes, showing that the FX CorAL membrane becomes smoother than the FX CorDiax under wet conditions (FX CorAL: Ra = 5.3; FX CorDiax Ra = 11.8) (Figure 2D). Lower membrane roughness is associated with positive effects, such as lower protein adsorption and blood cell interaction [30,31].

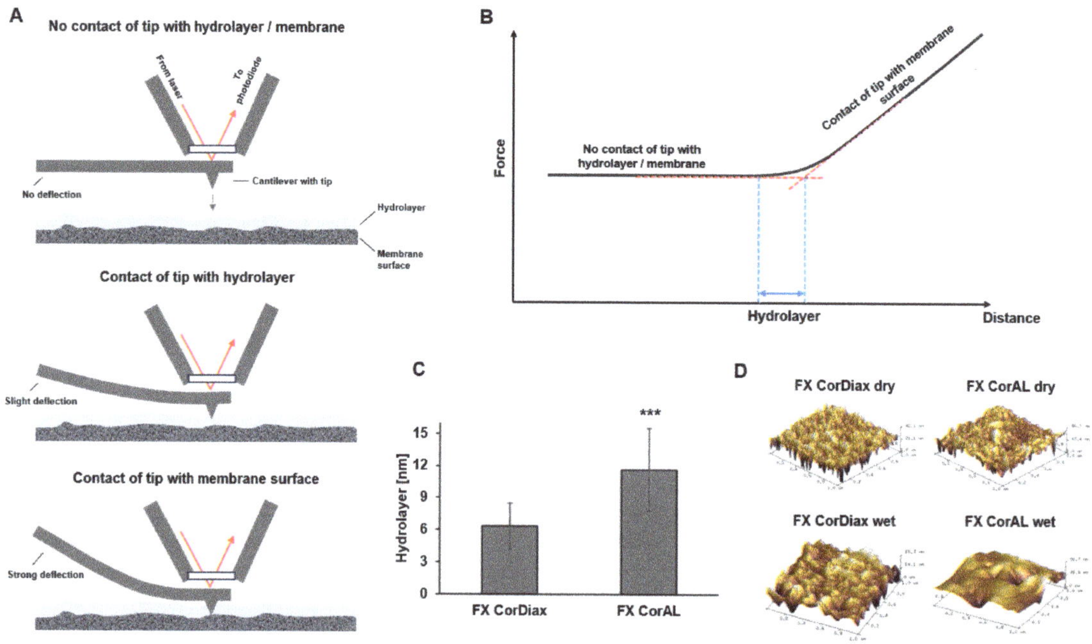

Figure 2. Measurement of hydrolayer. (**A**) Atomic force microscopy (AFM) determination of hydrolayer thickness. Membranes are cut open and exposed to water. Contact of the tip to the hydrolayer and membrane surface leads to a deflection of the cantilever. (**B**) For analysis, deflection of the cantilever is plotted in a force-distance diagram and the curved portion represents the thickness of the hydrolayer. (**C**) FX CorAL membrane shows a significantly thicker hydrolayer than the FX CorDiax ($n = 30$ each, *** $p < 0.001$). (**D**) AFM characterization of membrane surface in dry and wet conditions for FX CorDiax and FX CorAL.

3.1.3. Characterization of Membrane Fouling

Many previous studies demonstrated that increased membrane hydrophilicity, achieved via increased PVP content, reduces protein adsorption to the membrane via repulsive hydration force [15–20]. Different experimental approaches were applied to test whether also the Helixone *hydro* membrane of the FX CorAL dialyzer also shows lower secondary membrane formation than other membranes. In the first experiment, the sieving coefficient of albumin was investigated over time [14]. As proteins from the recirculated plasma adsorb to the membrane, the sieving coefficient decreases over time, especially in the first 15 min of plasma contact with the membrane. This is due to a reduction of the effective pore size by the generated protein layer [10]. These experiments found that the FX CorAL had the lowest reduction of the albumin sieving coefficient over time, pointing towards lowest secondary membrane formation. In the second experiment, the increase of the transmembrane pressure (TMP) at constant inlet and filtrate flows in a plasma recirculation experiment was investigated over time. This is also a good marker of secondary membrane formation, given that with an additional protein layer, more pressure is needed to achieve the same filtrate flow through the membrane. Also in this experiment, the FX CorAL showed the lowest increase of the TMP over time, especially in the first minutes of recirculation [12]. These results were also confirmed with other protein-containing solutions, showing generally low protein binding by the Helixone *hydro* membrane [10,11].

3.2. In Vitro and Clinical Investigation of Performance, Hemocompatibility and Safety

3.2.1. Overview of In Vitro and Clinical Studies with the New Polysulfone Dialyzer

Beyond the necessary performance and hemocompatibility testing, as requested by respective standards, such as ISO 8637 or ISO 10993, and the information which has to be presented in the instructions for use (IFU), more detailed investigations may allow a better understanding of the new product. For example, the clearance values presented in the IFU are measured in an aqueous solution. However, in clinical use the performance changes over treatment time due to protein adsorption to the dialyzer membrane [10,12,32–38]. Therefore, in vitro and clinical investigation of performance should take such considerations into account.

Such in vitro and clinical studies were also performed for the FX CorAL dialyzer. This chapter summarizes and analyzes these in vitro [10–14] and clinical studies [39–42] regarding the key dialyzer features. Moreover, Tables 1 and 2 give an overview of these in vitro and clinical studies, respectively.

Table 1. Overview of in vitro studies performed with the FX CorAL dialyzer.

Study Title	Comparators	Investigations	Study Findings for FX CorAL	Study Reference
Complement activation by dialysis membranes and its association with secondary membrane formation and surface charge	FX CorAL FX CorDiax xevonta Polyflux ELISIO FDX THERANOVA SUREFLUX	• Hemocompatibility (complement activation) • Surface charge • Protein adsorption (Albumin sieving coefficient decrease over time)	• Lowest C3a, C5a and sC5b-9 activation • Most neutral surface charge • Lowest albumin sieving coefficient decrease over time (i.e., protein adsorption)	Melchior et al. (2021) [14]
Polyvinylpyrrolidone in hemodialysis membranes: Impact on platelet loss during hemodialysis	FX CorAL FX CorDiax xevonta ELISIO Polyflux THERANOVA	• PVP content • PVP elution • Hemocompatibility (platelet loss)	• Highest PVP content • Lowest PVP elution (together with FX CorDiax) • Lowest platelet loss	Zawada et al. (2021) [13]
Time-resolving characterization of molecular weight retention changes among three synthetic high-flux dialyzers	FX CorAL ELISIO xevonta	• Sieving coefficients over time (β2-m, myoglobin, albumin) • Molecular weight retention curves over time • Effective pore radius over time • Predicted albumin loss	• Highest β2-m sieving coefficient and lowest decrease in albumin sieving coefficient over time • Lowest shift in molecular weight retention curves over time • Lowest reduction in effective pore radius over time • Low predicted albumin loss	Zawada et al. (2022) [12]
Impact of Hydrophilic Modification of Synthetic Dialysis Membranes on Hemocompatibility and Performance	FX CorAL FX CorDiax	• Contact angle (hydrophilicity) • PVP elution during shelf-life	• Lower contact angle (i.e., higher hydrophilicity) • No PVP elution during complete shelf-life (3 years)	Zawada et al. (2022) [10]

Table 1. *Cont.*

Study Title	Comparators	Investigations	Study Findings for FX CorAL	Study Reference
Hydrophilic Modification of Dialysis Membranes Sustains Middle Molecule Removal and Filtration Characteristics	FX CorAL FX CorDiax xevonta Diacap Pro HF ELISIO DORA Revaclear Cellentia	• Clearance over time (β2-m, inulin, creatinine, urea) • TMP/filtrate flow over time	• Highest β2-m clearance after protein fouling • Lowest TMP increase/filtrate flow decrease over time	Zawada et al. (2024) [11]

C3a: complement factor C3a; C5a: complement factor C5a; sC5b-9: complement factor sC5b-9; β2-m: β2-microglobulin; TMP: transmembrane pressure. Diacap Pro: B.Braun, Melsungen, Germany; HF: Wego, Weihai, China; DORA: Bain Medical, Guangzhou, China; Revaclear: Baxter, Deerfield, US; Cellentia: Nipro, Osaka, Japan.

Table 2. Overview of clinical studies performed with the FX CorAL dialyzer.

Study Acronym/ Study Title	Study Design	Therapy	Number of Patients *	Duration	Comparators	Primary Endpoint	Further Endpoints	Study Findings for FX CorAL	Study Reference
eMPORA I/ Safety and clinical performance of a dialyzer with a modified polysulfone membrane	Open, multicenter, randomized controlled trial (Germany)	Online-post dilution HDF/ without autosub plus	45	4 weeks	FX CorAL FX CorDiax FX class	β2-m removal rate (non-inferiority of FX CorAL)	• Removal rate and clearance of myoglobin and small molecules • Hemocompatibility makers • Safety	• Non-inferior in β2-m removal rate vs. both comparators • Lower complement activation vs. FX class • Comparable safety profile	Kempkes-Koch et al. (2021) [39]
eMPORA II/ Comparative safety and clinical performance of dialyzers applied during post-dilution online hemodiafiltration	Open, multicenter, randomized controlled trial (Germany)	Online-post dilution HDF/ without autosub plus	58	4 weeks	FX CorAL Polyflux SUREFLUX	β2-m removal rate (non-inferiority of FX CorAL)	• Removal rate and clearance of myoglobin and small molecules • Hemocompatibility makers • Safety	• Non-inferior in β2-m removal rate vs. both comparators • Lower complement and leukocyte activation • Comparable safety profile	Ehlerding et al. (2021) [40]

Table 2. *Cont.*

Study Acronym/ Study Title	Study Design	Therapy	Number of Patients *	Duration	Comparators	Primary Endpoint	Further Endpoints	Study Findings for FX CorAL	Study Reference
comPERFORM/ Comparative clinical performance of dialyzers applied during high volume online hemodiafiltration	Open, multicenter, randomized controlled trial (Germany)	Online-post dilution HDF/ with autosub plus	52	5 weeks	FX CorAL ELISIO xevonta	β2-m removal rate (non-inferiority/superiority of FX CorAL)	• Removal rate and clearance of myoglobin and small molecules • Albumin loss • Safety	• Superior in β2-m removal rate vs. both comparators • Low albumin loss • Comparable safety profile	Ehlerding et al. (2022) [41]
eMPORA III/ Comparison of clinical performance and hemocompatibility of dialyzers applied during postdilution online hemodiafiltration	Open, multi-center, randomized controlled trial (Germany, Czech Republic, Hungary)	Online-post dilution HDF/ without autosub plus	76	14 weeks	FX CorAL FX CorDiax xevonta	β2-m removal rate (non-inferiority/superiority of FX CorAL)	• Removal rate and clearance of myoglobin and small molecules • Hemocompatibility makers • PRO • Safety	• Non-inferior in β2-m removal rate vs. both comparators; superior vs. xevonta • Lower complement, leukocyte and platelet activation • Comparable PRO scores • Comparable safety profile	Ehlerding et al. (2024) [42]

* according to intention to treat population; β2-m: β2-microglobulin; HDF: hemodiafiltration; PRO: patient reported outcomes. FX class: Fresenius Medical Care, Bad Homburg, Germany.

3.2.2. Investigation of Performance

Importance of a High-Performing Dialyzer

The major function of the dialyzer is the removal of uremic toxins and excess water from patients with ESKD. These uremic toxins have a broad molecular weight range and have been associated with different comorbidities such as inflammation, malnutrition or cardiovascular disease [43–47]. This is especially the case for so-called middle molecules such as β2-microglobulin (~12 kDa), as higher levels have been linked to reduced survival among dialysis patients [46,48,49]. Therefore, efficient removal of such toxins during dialysis is fundamental for the well-being of patients with ESKD. Moreover, besides the efficient removal of a broad range of uremic toxins, the dialyzer shall have a sharp molecular weight cut-off to prevent significant loss of essential proteins, such as albumin. Reduced serum albumin concentration contributes to protein-energy wasting in dialysis patients and is a strong predictor of mortality [50–52].

Furthermore, while manufacturers present dialyzer performance values, such as clearances, in the respective IFUs, it is important to understand that dialyzer performance is not constant but may change during a dialysis session. This is mainly due to the formation of a secondary membrane as adsorption of plasma proteins to the dialyzer membrane reduce the effective pore size [11,12,34]. This effect of reduced performance is predominant for larger proteins than for smaller solutes such as urea, creatinine or vitamin B12 [11,12,34–37].

Therefore, in vitro and clinical testing during the development and investigation of a new dialyzer shall take all these considerations into account. The following two chapters summarize the in vitro and clinical studies regarding performance characterization of the FX CorAL dialyzer.

Investigation of Performance in In Vitro Studies

Beyond the standard performance testing for the IFU, including measurements of clearances at different flow rates (e.g., urea, creatinine, phosphate, vitamin B12, cytochrome C), sieving coefficients (e.g., β2-microglobulin, myoglobin, albumin), dialyzer mass transfer-area coefficient (K0A), ultrafiltration coefficients or pressure drop at the blood compartment, further in vitro testing was performed to characterize the impact of hydrophilic membrane modification of the FX CorAL with regard to performance stability over treatment time.

First, in contrast to the IFU sieving coefficient data, which are determined typically after 30 min of plasma recirculation, time-dependent characterization of the β2-microglobulin, myoglobin, and albumin sieving coefficients up to 240 min was performed [12,14]. In these investigations, sieving coefficients for all three proteins decreased over time, with the strongest decrease in the first ~20 min of plasma recirculation and with a higher magnitude for larger (i.e., albumin, 68 kDa) than for smaller (i.e., β2-microglobulin, 12 kDa) proteins. In these experiments, the FX CorAL had the lowest reduction of the sieving coefficients as compared to other synthetic dialyzers, pointing towards lower secondary membrane formation due to increased membrane hydrophilicity. Based on these time-dependent sieving coefficient data, molecular weight retention curves were generated, which showed the lowest shift over time towards lower Stokes radius for the FX CorAL, as compared to the other investigated dialyzers (FX CorAL: −0.23 nm at sieving coefficient of 50%; ELISIO: −0.29 nm; xevonta: −0.50 nm) [12]. These curves also allowed to determine the molecular weight cutoff (MWCO, molecular weight or Stokes radius at sieving coefficient of 10%) and molecular weight retention onset (MWRO, molecular weight or Stokes radius at sieving coefficient of 90%) over time. For all investigated time points, the FX CorAL showed the steepest molecular weight retention curve, leading to the lowest difference between MWCO (2 min: 2.66 nm; 120 min: 2.51 nm) and MWRO (2 min: 1.96 nm; 120 min: 1.65 nm), as compared to the other investigated dialyzers in this study. This supports the selective permeability of the FX CorAL membrane, with high permeability for middle molecules while preventing significant loss of essential proteins, such as albumin, in parallel. In line with these data, the mean effective pore radius declined to a lower extent for the Helixone *hydro* membrane of the FX CorAL, as compared to the other investigated dialyzers [12]. When, for example, comparing the pore distribution of the FX CorAL and xevonta dialyzers before and after protein fouling in relation to the Stokes radius of β2-microglobulin, the percentage of pores with an effective radius greater than the Stokes radius of β2-microglobulin before protein fouling was much higher (2 min: FX CorAL: 98.7%; xevonta: 97.2%) than after protein fouling (120 min: FX CorAL: 92.4%; xevonta: 85.8%), with a lower decrease for the FX CorAL dialyzer [11]. Interestingly, the performance for the small molecules urea and creatinine remained stable throughout the investigated time (60 min) for all dialyzers, supporting previous studies, that secondary membrane formation especially impacts the elimination of larger molecules [11,34–37].

Second, in recirculation experiments with human plasma and other protein-containing solutions, the impact of secondary membrane formation on membrane filtration characteristics was investigated [11,12]. Typically, due to the additional protein layer, which is formed during treatment, more pressure is needed to achieve the same filtration flow through the dialyzer membrane [7,11,12]. In recirculation experiments with constant inlet and filtration flows, the FX CorAL showed the lowest TMP increase throughout the experiment, as compared to all other investigated dialyzers, pointing towards lowest secondary membrane formation [11,12]. This was especially the case in the first ~30 min, where the protein fouling process is most active. Vice versa, when investigating the reduction of filtration flow at constant TMP, the FX CorAL showed the most stable filtration rate, which may help to achieve high exchange volumes during hemodiafiltration treatments [11], recently shown to improve mortality among dialysis patients as compared to standard high-flux hemodialysis treatments [53].

Third, in an in vitro study the potential albumin loss by the FX CorAL over a 240-min treatment was compared to other dialyzers [12]. In line with the lowest initial albumin sieving coefficient of the FX CorAL, this dialyzer showed a low level of albumin loss throughout the total duration [12,14].

Investigation of Performance in Clinical Studies

Four clinical studies investigated the performance of FX CorAL in a randomized controlled setting (Table 2) [39–42]. Table 3 presents an overview of patient characteristics and treatment parameters. The main difference in the treatment parameters between the four clinical studies was the use of autosub plus in the comPERFORM study, leading to higher substitution and, accordingly, higher convection volumes. When comparing the treatment parameters to the recommendation of $\geq$23 L convection volume during a HDF treatment, which has recently been associated with improved patient survival [53], comPERFORM and eMPORA III reached this threshold.

Table 3. Patient characteristics and treatment parameters in clinical studies performed with the FX CorAL dialyzer.

	eMPORA I	eMPORA II	comPERFORM	eMPORA III
Patient characteristics *				
Age [years]	66.3 ± 13.6	67.8 ± 13.4	64.8 ± 14.5	67.0 ± 15.6
Gender [% male]	76	74	85	74
Mean BMI [kg/m^2]	27.7 ± 5.8	28.9 ± 6.6	28.2 ± 7.5	27.3 ± 5.1
Diabetes [%]	31	36	21	34
Median time on RRT [months]	42	50	56	40
Treatment parameter (online post-dilution HDF) **				
Blood flow rate [mL/min]	303 ± 29	306 ± 29	331 ± 12	345 ± 44
Dialysate flow rate [mL/min]	499 ± 5	500 ± 0.1	529 ± 19	577 ± 96
Substitution volume [L]	20.2 ± 5.7	17.6 ± 4.5	25.4 ± 2.8	21.0 ± 3.2
Ultrafiltration volume [L]	1.9 ± 1.1	2.2 ± 1.2	2.0 ± 0.2	2.3 ± 1.0
Effective treatment time [min]	269 ± 29	272 ± 29	273 ± 14	262 ± 21

BMI: body mass index; RRT: renal replacement therapy; * Patient characteristics are presented for the total population in the respective studies; ** Treatment parameters are presented for the treatments with the FX CorAL dialyzer. No significant differences were found between treatments with other dialyzers.

All four studies investigated β2-microglobulin removal as the primary endpoint. Moreover, removal rates and clearances of other uremic toxins of different molecular ranges were also investigated in all studies. Albumin loss into the dialysate was investigated in the comPERFORM study only (Table 2). The results for the comparisons between dialyzers are summarized in Table 4.

Table 4. Primary and secondary performance endpoints in clinical studies performed with the FX CorAL dialyzer.

Performance Parameter	eMPORA I		eMPORA II		comPERFORM		eMPORA III	
	FX CorAL vs.		FX CorAL vs.		FX CorAL vs.		FX CorAL vs.	
	FX CorDiax	FX class	SUREFLUX	Polyflux	xevonta	ELISIO	FX CorDiax	xevonta
β2-m RR [non-inferiority]	★	★	★	★	★	★	★	★
β2-m RR [superiority]	-	-	★	c	★	★	c	★
β2-m clearance	c	★	★	c	★	★	c	★
Myoglobin RR	c	★	#	#	★	★	★	★
Myoglobin clearance	c	★	#	c	★	c	★	★
Phosphate RR	#	#	c	c	c	c	c	c
Phosphate clearance	c	c	c	c	c	c	c	c
Creatinine RR	#	c	c	c	c	c	c	c
Creatinine clearance	c	c	#	c	c	c	c	c
Urea RR	c	c	c	c	c	c	c	c
Urea clearance	c	c	#	#	c	c	c	c
Albumin loss	-	-	-	-	c	c	-	-

β2-m: β2-microglobulin; RR: removal rate; ★: FX CorAL significantly higher vs. comparators; #: FX CorAL significantly lower vs. comparators; c: comparable; -: not determined.

Across all four clinical studies, the FX CorAL showed the highest β2-microglobulin removal rate as compared to all other investigated dialyzers (Figure 3). Statistical analysis of non-inferiority and superiority testing is given in Table 4, showing that the predefined endpoints were reached. While these differences appear numerically low, it has to be taken into account that these results were obtained from a single treatment and these small differences during the repeating dialysis sessions may have a meaningful clinical impact on the patients over time. Longer clinical studies are needed to investigate this performance improvement relative to patient morbidity and mortality rates.

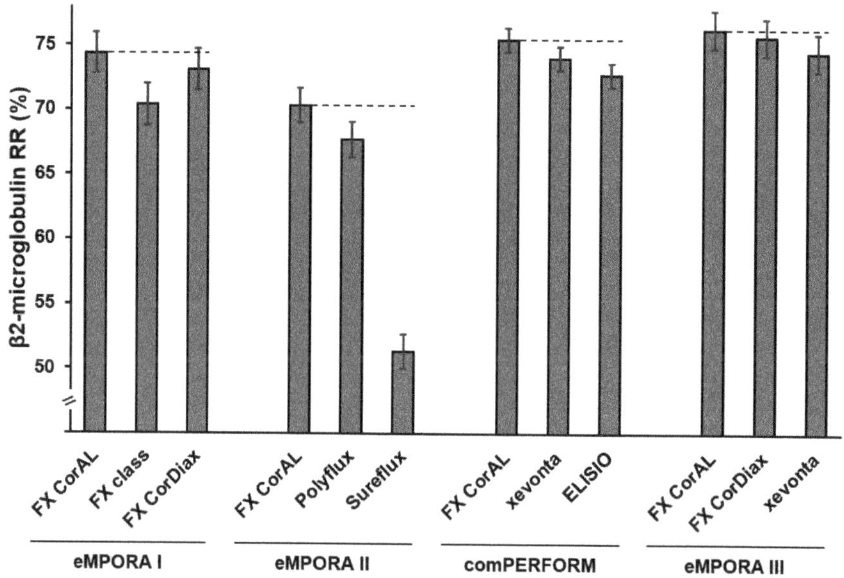

Figure 3. β2-microglobulin removal rates (least square mean ± standard error) of FX CorAL and comparators in the four randomized controlled trials eMPORA I, eMPORA II, comPERFORM, and eMPORA III [39–42].

When comparing the β2-microglobulin removal rates across the four clinical studies, eMPORA I and II achieved lower removal rates, as compared to eMPORA III and comPERFORM, which applied higher convection volumes.

Moreover, it is important to mention that these clinical results for the β2-microglobulin removal rate are in line with the in vitro investigations of the FX CorAL dialyzer [12]. Also here, the FX CorAL showed the highest β2-microglobulin sieving coefficient values throughout the experiment (2 min: 98.8%, 120 min: 91.6%) when compared to the ELISIO (2 min: 91.0%, 120 min: 84.0%) and xevonta (2 min: 96.3%, 120 min: 87.6%) dialyzers. To obtain high removal rates during treatment, it is crucial that the dialyzer membrane sustains the selective permeability throughout the treatment, which is achieved by the FX CorAL due to lower protein absorption to its membrane.

The results for the secondary endpoints in the four clinical studies regarding removal rates and clearances of middle molecules (β2-microglobulin, myoglobin) were in line with the primary endpoint, reflecting the optimization of the new Helixone *hydro* membrane of FX CorAL dialyzer for middle molecule removal. Regarding removal rates and clearances of smaller molecules, such as phosphate, creatinine, and urea, the results were mostly comparable between the different investigated dialyzers.

The comPERFORM study also investigated the albumin loss during the dialysis session by measuring the albumin concentration in the spent dialysate. Here, the FX CorAL demonstrated the lowest amount of albumin loss up to the 60-min time point. Albumin sieving of FX CorAL remained almost constant thereafter, whereas the other dialyzers' sieving rates showed a change of the sieving kinetics. These data are in line with in vitro examinations indicating lower secondary membrane formation by protein adsorption on the Helixone *hydro* membrane [14].

Besides the eMPORA and comPERFORM studies, two further clinical studies by other groups also investigated the performance of the FX CorAL dialyzer in high-volume post-dilution HDF treatments [27,54]. The first study was a prospective study with 19 patients, comparing different models of FX CorAL and FX CorDiax (surface size between 1.4 m^2 and 2.0 m^2) [54]. Beyond the performance markers investigated in the eMPORA and comPERFORM studies, this study additionally investigated the reduction ratios (RR) of the larger kappa free immunoglobulin light chains (22.5 kDa), prolactin (23 kDa), α1-microglobulin (33 kDa), α1-acid glycoprotein (41 kDa), lambda-free immunoglobulin light chains (45 kDa), as well as albumin RR and loss (68 kDa). No significant differences were found for the small, middle, and larger molecules when comparing the FX CorAL and FX CorDiax with the same surface size, except for the lambda-free immunoglobulin light chains, where FX CorDiax 80 had significantly higher RR than FX CorAL 80. Also, no differences were found in terms of albumin loss. These results are in line with the eMPORA I and eMPORA III studies, which also compared the FX CorAL and FX CorDiax dialyzers (FX CorAL 600 vs. FX CorDiax 600) and found slightly higher removal rates for the FX CorAL dialyzer (Figure 3) (ß2-microglobulin RR from Maduell et al. [54]: FX CorAL 600: 85.4%, FX CorDiax 600: 84.1%). As these differences are not significantly different, it has to be noted that the development of the FX CorAL aimed to improve hemocompatibility by reducing protein adsorption to the membrane. Indeed, reduced fouling is also associated with more stable performance during treatment, but also other parameters such as pore size distribution determine the overall dialyzer performance.

The second study was also a prospective study with 20 patients, comparing the performance of the Toraylight NV 2.1 dialyzer (Toray Medical, 2.1 m^2) with the performance of the FX CorAL 800 (2.0 m^2) dialyzer [27]. The study investigated the same markers as mentioned in the previous study and additionally measured the protein-bound uremic toxins p-cresyl sulfate (188 Da) and indoxyl sulfate (213 Da). No differences in the RR were found for the small molecules, as well as for β2-microglobulin, indoxyl-sulfate, and p-cresyl sulfate. For the larger molecules myoglobin, kappa free immunoglobulin light chains, prolactin, α1-microglobulin, α1-acid glycoprotein, and lambda-free immunoglobulin light chains, the Toraylight NV 2.1 dialyzer showed higher RR than the FX CorAL 800 dialyzer under the same convection conditions. However, under these conditions, a substantial albumin loss over 5 g per session in more than 50% of the patients was observed in treatments with the Toraylight NV 2.1 dialyzer, which was not the case in treatments with the FX CorAL dialyzer (FX CorAL 800: 1.3 g, Toraylight NV 2.1: 6.2 g).

3.2.3. Investigation of Hemocompatibility

Importance of a Hemocompatible Dialyzer

Hemocompatibility is an indispensable feature of dialyzer membranes and comprises many complex reactions between blood components and artificial surfaces, especially with the membrane, given its large surface size. These interactions trigger a cascade of immune, coagulative, and inflammatory responses, including complement activation, platelet activation, coagulation, as well as cell activation, inflammation, and oxidative stress [23,55,56].

1. Complement activation

Complement activation is one of the primary immune responses to foreign surfaces. After contact of blood to the dialyzer membrane, complement proteins, especially C3, bind to the surface and undergo activation. This initiates the alternative complement pathway, which is typically the primary pathway in this context [55,56]. However, the lectin pathway and classical pathway can also be involved depending on the membrane's surface properties [55–58]. Upon activation, C3 initiates a complement cascade activation, leading to the generation of different complement factors and finally the formation of the cytolytic membrane attack complex (C5b-9, MAC) [23,55,56,59]. Although full C5b-9 formation on cell surfaces in dialysis is rare, sublytic MAC fragments cause cellular activation and inflammation [23,55,56]. During the complement cascade activation and MAC formation, two fragments, C3a and C5a, also known as anaphylatoxins, are released [23,55,56,60]. These small peptides are potent pro-inflammatory molecules that attract and activate leukocytes, contributing to systemic inflammation. Chronic complement activation leads to a constant low-grade inflammatory state in dialysis patients, which contributes to cardiovascular complications, immune suppression, and poor clinical outcomes [23,55,61].

2. Platelet activation and coagulation

Platelet activation and coagulation is a further critical hemocompatibility challenge during the development of new dialyzers, given the thrombogenic potential of dialysis membranes [23,56]. Upon contact to non-endothelial surfaces like dialysis membranes, platelets adhere and become activated [23,56]. This activation is mediated by membrane receptors such as glycoprotein IIb/IIIa (GPIIb/IIIa), which undergo conformational changes that lead to platelet aggregation [21,23,56]. Additionally, complement fragments, especially C3a and C5a, interact with platelets and increase their activation. C5a can bind to platelet receptors, enhancing the release of pro-coagulant molecules and aggregation. Activated platelets release ADP, thromboxane A2, and other pro-thrombotic molecules, creating a positive feedback loop that encourages further platelet activation and thrombus formation on the dialysis membrane [23,56,62]. Beyond platelet activation, hemodialysis induces activation of the plasmatic coagulation system. This is evidenced by increasing levels of thrombin-antithrombin (TAT) complexes, indicating elevated thrombin generation during dialysis treatment [63–65]. Increased activation of platelets and the plasmatic coagulation system raise the risk of clot formation in the dialysis circuit, necessitating anticoagulation therapy. However, this also elevates the bleeding risk for patients, creating a complex clinical management challenge.

3. Cell activation, inflammation and oxidative stress

Cell activation during dialysis affects different cell types beyond platelets such as leukocytes or endothelial cells [23,55,56,66,67]. Activated leukocytes release pro-inflammatory cytokines such as TNF-α, IL-1β, and IL-6, as well as reactive oxygen species (ROS), contributing to systemic inflammation [23,55,56]. Increased ROS leads to oxidative stress, which exacerbates inflammation and may damage red and white blood cells, increasing cell fragility and disrupting normal cellular functions [23,55]. Besides ROS, also the anaphylatoxins C3a and C5a may activate different cells in the patient, such as endothelial cells [23,55]. Endothelial activation results in the upregulation of adhesion molecules like ICAM-1 and VCAM-1, facilitating leukocyte adherence and transmigration [23,68]. This response can impair vascular function and contribute to atherosclerosis in chronic dialysis patients [23]. Chronic cell activation in dialysis patients leads to a persistent inflammatory state and oxidative stress, which are linked to complications such as atherosclerosis, cardiovascular disease, anemia, and reduced immune function [23,61].

Investigation of Hemocompatibility in In Vitro Studies

During the development of the FX CorAL dialyzer, in vitro hemocompatibility studies were performed to characterize the activation of complement factors C3a, C5a, and sC5b-9 in a recirculation setup with human blood [14]. Here, the FX CorAL showed the lowest complement activation for all factors as compared to seven polysulfone-, polyethersulfone-

and cellulose-based dialyzers. While most of the investigated dialyzers showed higher complement activation than the reference dialyzer (FX class, Fresenius Medical Care), which was run in parallel for standardization, the FX CorAL showed a −39.4% [C3a], −57.5% [C5a], and −58.9% [sC5b-9] lower complement activation than the reference dialyzer [14].

In the same recirculation setup with human blood, also platelet loss, as a marker of platelet activation, was investigated [13]. Also here, the FX CorAL showed lower platelet loss than all other investigated dialyzers. When comparing platelet loss to the reference dialyzer, the FX CorAL showed a −225.2% lower platelet loss. This indicates lower activation of platelets by the hydrophilic membrane of the FX CorAL dialyzer.

The results of these in vitro hemocompatibility investigations strongly correlated with the slope of the albumin sieving coefficient decrease over time, supporting previous evidence that protein adsorption to the membrane may trigger complement activation, coagulation, and inflammation [10,13,14,23,55,56].

Investigation of Hemocompatibility in Clinical Studies

Hemocompatibility was also investigated in clinical studies. In total, three studies compared the hemocompatibility profile of the FX CorAL to other dialyzers on the market (Table 2) [39,40,42]. The results for these studies are summarized in Table 5.

Table 5. Overview of hemocompatibility markers in clinical studies performed with the FX CorAL dialyzer.

	Performance Parameter	**eMPORA I**		**eMPORA II**		**eMPORA III**	
		FX CorAL vs. FX CorDiax	FX class	FX CorAL vs. SUREFLUX	Polyflux	FX CorAL vs. FX CorDiax	xevonta
Complement activation	C3a	c^{15}	c^{15}	★^{15}	c^{15}	★^{15}	★^{15}
	C5a	c^{15}	★^{15}	★^{15}	c^{15}	-	-
	sC5b-9	c^{15}	★^{15}	★^{60}	★^{60}	c^{60}	★^{60}
Cell activation/ inflammation	Leukocyte counts	c^{15}	c^{15}	★^{15}	★^{15}	★^{15}	c^{15}
	Monocyte counts	c^{240}	c^{240}	c^{240}	c^{240}	c^{15}	★^{15}
	Neutrophile counts	c^{240}	c^{240}	c^{240}	c^{240}	★^{15}	c^{15}
	PMN elastase	c^{240}	c^{240}	c^{240}	★^{240}	★^{60}	★^{60}
	LTB4	-	-	-	-	★^{15}	c^{15}
	IL-6	-	-	-	-	c^{AUC}	c^{AUC}
	IL-8	-	-	-	-	c^{AUC}	c^{AUC}
	sICAM-1	-	-	-	-	c^{240}	c^{240}
	hsCRP	c^{240}	c^{240}	-	-	c^{0}	c^{0}
Platelet activation/ coagulation	Platelet counts	c^{240}	c^{240}	c^{240}	★^{240}	c^{15}	c^{15}
	β-TG	-	-	c^{240}	c^{240}	★^{60}	★^{60}
	TxB2	-	-	-	-	c^{15}	#^{15}
	TAT	c^{240}	c^{240}	c^{240}	c^{240}	-	-
Oxidative stress	MDA	-	-	-	-	c^{AUC}	c^{AUC}
	GSH-Px activity	-	-	-	-	c^{AUC}	c^{AUC}
Contact phase activation	Kallikrein	c^{240}	c^{240}	c^{240}	c^{240}	-	-

PMN elastase: Polymorphonuclear elastase; LTB4: Leukotriene B4; IL-6: Interleukin 6; IL-8: Interleukin 8; sICAM-1: Soluble intercellular adhesion molecule-1; hsCRP: high-sensitivity C-reactive protein; β-TG: β-Thromboglobulin; TxB2: Thromboxane B2; TAT: Thrombin-antithrombin; MDA: Malondialdehyde; GSH-Px: glutathione peroxidase; ★: FX Coral significantly better vs. comparator; #: comparator significantly better vs. FX CorAL; c: comparable; -: not determined; superscript numbers indicate the timepoint of measurement reported in the respective study: 0: pre-treatment, 15: 15 min after treatment start, 60: 60 min after treatment start; 240: 240 min after treatment start; AUC: area under the curve over 0–240 min of treatment.

All three studies investigated complement activation, platelet activation, as well as immune cell activation during the treatment. eMPORA III also investigated interdialytic changes of hemocompatibility markers [42].

Regarding complement activation, the studies investigated the course of complement factors C5a (eMPORA I and II [39,40]), C3a, and sC5b-9 (all three studies [39,40,42]) during the dialysis sessions at time points 15 min, 60 min, and 240 min as well as before the dialysis treatment. For all dialyzers, the complement factors showed a typical course over time, with a peak value at 15 min for C3a and C5a and at 60 min for sC5b-9. The FX CorAL dialyzer showed the lowest C3a and sC5b-9 peak levels across all three studies. Figure 4 shows the relative sC5b-9 peak levels at 60 min compared to the FX CorAL dialyzer. The increases of the complement factor levels were in most cases significantly lower for the FX CorAL dialyzer as compared to the other investigated dialyzers (Table 5).

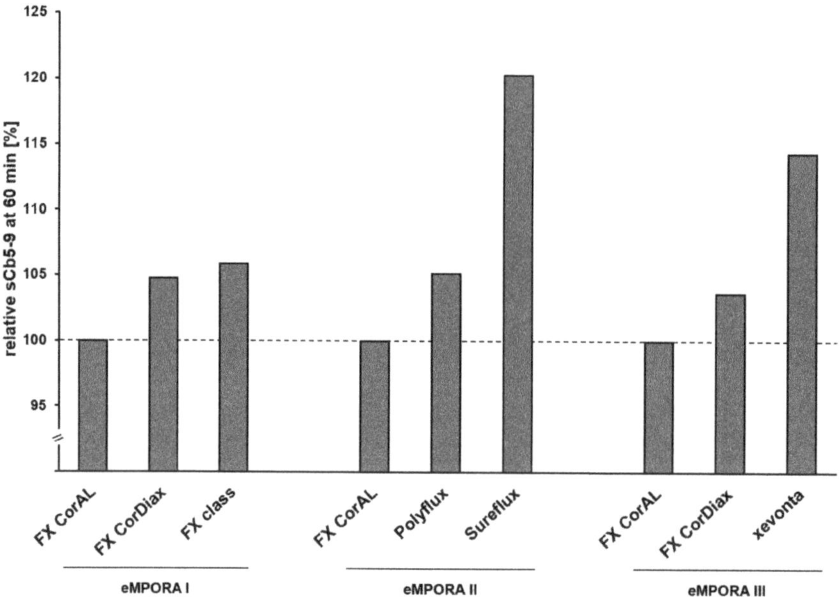

Figure 4. Relative sCb5-9 levels during treatment with of FX CorAL and comparators in the three randomized controlled trials eMPORA I, eMPORA II, and eMPORA III [39,40,42]. FX CorAL is used as reference.

Regarding platelet activation, all three studies investigated platelet counts during the dialysis sessions at time points 15 min, 60 min, and 240 min as well as before the dialysis treatment. For all dialyzers and all studies, platelet counts dropped after treatment start, as typically observed during dialysis treatments [69]. While the platelet drop with the FX CorAL was numerically lower in most comparisons with other polysulfone- or polyethersulfone-based dialyzers, statistical significance was only reached vs. Polyflux at 240 min (Table 5). However, when investigating β-Thromboglobulin (β-TG), as a sensitive marker of platelet activation, eMPORAIII showed a clear and significant differentiation between the FX CorAL and the other comparators, with the FX CorAL treatments having the lowest β-TG levels throughout the session. In contrast, Thromboxane B2 (TxB2), which is also a marker of platelet activation showed inconsistent findings, which could be caused by its strong elimination during treatment due to the low molecular weight (371 Da). Regarding plasmatic coagulation activation, eMPORA I and eMPORA II investigated

Thrombin-antithrombin III (TAT) but could not find any significant differences between the different dialyzers.

Regarding cell activation, inflammation, and oxidative stress, also different markers were investigated in the three clinical studies during the treatment sessions (Table 5). Leukocytes typically show a drop in their cell counts after treatment start [70], which was also observed in the eMPORA studies, and which is a marker of cell activation during dialysis treatments. The FX CorAL generally induced a lower leukocyte drop 15 min after treatment start, pointing towards lower immune cell activation. In line, FX CorAL also showed lower levels of Polymorphonuclear (PMN) elastase and Leukotriene B4 (LTB4), two markers of leukocyte activation, in eMPORA II and eMPORA III studies. All other investigated markers of cell activation, inflammation and oxidative stress did not differ between the different dialyzers (Table 5).

3.2.4. Investigation of Safety

Investigation of Safety in Clinical Studies

An essential requirement for each novel dialyzer is its safe use during dialysis treatments. Therefore, investigation of safety in clinical studies is a critical part and has also been part of all studies performed with the FX CorAL dialyzer.

Adverse events (AEs), which are evaluated in the context of such clinical studies, can be categorized based on their seriousness (serious vs. non-serious), relation to the medical device (i.e., dialyzer) and/or medical procedure (i.e., dialysis treatment), and whether they are expected or unexpected.

Serious adverse events (SAEs) are distinguished from non-serious adverse events (nsAEs) by their life-threatening nature, requirement for hospitalization, or significant medical consequences. AEs directly linked to the dialyzer and/or medical procedure are classified as adverse device effects (ADEs) and may include issues arising solely from the device or medical procedure or both. Expected ADEs are those anticipated based on prior clinical knowledge or device specifications, while unexpected events represent unanticipated reactions.

Table S1 gives an overview of AEs including the respective numbers of AEs in the four clinical studies [39–42], summarized for all investigated dialyzers.

To determine the safety profile of the FX CorAL dialyzer, particular focus was placed on collecting and analyzing (S)ADEs related to dialyzers alone or in combination with medical procedure reported across all four studies. Figure S1 illustrates known (S)ADEs in dialysis along with their clinical manifestations (medical conditions and symptoms).

ADE types occur with varying frequency and severity and significantly impact patient outcomes [71–77].

Among the most common ADEs are fluid- and electrolyte imbalances, which result from improper ultrafiltration or ionic clearance during dialysis sessions [78–80]. These issues can manifest as hypovolemia or hypervolemia, leading to symptoms such as dizziness, fatigue, headache, edema, or even cardiac arrhythmias in severe cases [79,81,82]. Intradialytic hypotension, occurring in up to 10–12% of dialysis sessions, is typically caused by excessive fluid removal, resulting in reduced blood pressure leading to symptoms like dizziness, fainting, nausea, vomiting, or shock in severe cases [83–85]. Electrolyte disturbances, including hypokalemia or hyperkalemia, are also frequent and may cause muscle cramps, weakness, headache, or life-threatening arrhythmias [86–88].

Hemocompatibility issues, including clotting within the dialyzer, are another prevalent complication [89,90]. Insufficient anticoagulation or patient-specific factors often contribute to clotting, reducing treatment efficiency and potentially causing anemia, fatigue, or thrombogenicity [91]. Hemolysis, though less common, arises from mechanical stress or improper dialysis settings [92,93]. It presents with symptoms such as dark urine, jaundice, or, in severe cases, circulatory shock [94].

Biocompatibility issues related to the dialyzer materials include allergic reactions and endotoxin-related pyrogenic reactions. Allergic reactions, which are relatively rare, represent a subset of hypersensitivity reactions and are caused by an IgE-mediated immune response to the dialyzer membrane or sterilization agents that begin a few minutes after starting dialysis [95,96]. Symptoms can range from mild itching and rash to severe anaphylaxis [97]. Regarding dialyzer materials, some previous studies speculated that PVP could be a possible cause for the rarely occurring hypersensitivity reactions during treatments with synthetic dialyzers [98–101]. Typically, these patients are then switched to cellulose-based or polymethylmethacrylate (PMMA) dialyzers for the next treatments. Whether dialyzers, which do not elute PVP into patient's blood, such as the FX CorAL, may be an adequate alternative for these patients, needs to be evaluated in the future. Regarding endotoxin retention, pyrogenic reactions can result from inadequate water treatment or filter contamination [102–104]. These reactions manifest as fever and chills and may exacerbate preexisting inflammation or lead to sepsis in vulnerable patients [77].

Vascular access complications, including infections or thrombosis are another significant concern [105–108]. Infections at the access site, such as arteriovenous fistulas or central venous catheters, can lead to systemic sepsis, with fever, chills, and tachycardia being common indicators [109]. Thrombosis or stenosis may cause access failure, pain, or swelling, requiring surgical intervention [107,108]. Hematomas and air embolisms, though rarer, can also arise, leading to discomfort or critical symptoms like syncope and dyspnea [75,110].

Device-related mechanical failures, while less common, include blood loss due to circuit disconnections, hemolytic reactions mechanically induced by kinked dialysis lines, pump malfunctions causing insufficient blood flow, and improper temperature of the dialysate or blood circuit, or sensor failures affecting ultrafiltration and blood pressure monitoring [92,111,112]. Such events can compromise treatment and result in feeling cold or in severe symptoms like anemia or intradialytic hypotension.

Dialysis disequilibrium syndrome is one of the rarest but most severe complications, arising from rapid solute removal that leads to osmotic shifts and cerebral edema [75,113,114]. Symptoms include headache, nausea, vomiting, seizures, and in extreme cases, coma.

Investigation of Safety in FX CorAL Studies

1. Safety population and methodology

The analysis presented in this article is a summary of safety information reported in the four clinical studies with the FX CorAL dialyzer [39–42]. The total number of study patients and treatment sessions for this safety analysis across all investigated dialyzers was 249 and 4149, respectively. The number of patients and treatments for the investigational device FX CorAL and the 6 comparator dialyzers are given in Table S2.

In all studies, adverse events were classified according to the MedDRA Preferred Terms (PT) and assigned to broader Level 1 categorization terms based on clinical context, seriousness, severity, relationship to the dialyzer and/or medical procedure, expectedness, and patient outcomes, as described before. The frequencies of adverse events were calculated based on the number of study patients and the number of treatments administered.

2. ADE Types and Frequency

Among the total safety population of 249 study patients and 4149 treatment sessions (Table S2), a total of 446 adverse events were documented, categorized as either serious (SAEs, n = 27) or non-serious (nsAEs, n = 419) (Table S1). Of these 446 adverse events,

94 were linked to the dialyzer and/or medical procedure (ADE: 93/SADE: 1). A total of 39 study patients (15.7% of the total population) experienced at least one ADE.

Most ADEs were solely related to the medical procedure, and only a lower frequency was related to the dialyzer or to both (Table S1). Of 94 (S)ADEs across the four clinicals studies, 46 AEs were linked solely to the medical procedure, while 14 AEs were related exclusively to the dialyzer, and an additional 34 AEs were related to the medical procedure and the dialyzer. These results are in line with the literature, which identified dialysis procedure per se as the most common cause for ADEs [73–77,85,115]. It is important to note that ADEs in dialysis treatments are often multifactorial, involving both the dialyzer and the procedure (Figure S1).

The respective adverse events according to MedDRA Preferred Terms (PT) are listed in Table 6. The most frequently observed Level 1 adverse event category was electrolyte imbalance and include events such as muscle spasms, myalgia, vomiting, malaise, pain, asthenia, palpitations, hyperphosphatemia, or hyperkalemia. Muscle spasms were the most reported adverse event in this category, observed in 7.2% of patients (n = 18) and in 0.4% of treatments. Across all studies, electrolyte imbalance occurred 41 times, leading to a frequency of 16.5% among all study patients and of 1.0% among all study treatments. The second most common observed Level 1 adverse event category was fluid imbalance with 27 occurrences, equating to 10.8% of patients and 0.7% of treatments. Intradialytic hypotension was the most common adverse event in the fluid imbalance category, affecting 3.6% of patients (n = 9) and occurring in 0.2% of treatments, a frequency notably lower than the 10–12% reported in the literature [84]. The other findings are consistent with the types and frequency categories of AEs typically associated with dialysis treatments, where large fluid shifts and electrolyte corrections are inherent to the procedure [115]. However, some symptoms, like ear pain, flatulence, and defecation urgency, also might be due to other underlying medical conditions. The other observed Level 1 adverse event categories had a much lower frequency, ranging from 0.4% to 2.4% of study patients and 0.1% or lower of study treatments. The absence of pyrogenic reactions, microbial contaminations and allergic reactions was a positive deviation from literature data [77,102–104], reflecting effective sterilization and infection control measures.

Importantly, none of the ADEs with relationship to the dialyzers were classified as serious (Table S1), indicating that no events led to significant deterioration in health or required major medical interventions. One ADE with relationship to the medical procedure was classified as serious, linked to the level 1 term electrolyte imbalance (MedDRA PT: hyperkalemia), and occurred during the treatment sequence with the xevonta dialyzer. In general, the severity of the ADEs was predominantly mild, with a few instances of moderate severity, particularly related to electrolyte and fluid imbalances.

Moreover, in line with the previous literature, all identified ADEs in the present studies were expected (Table S1), meaning they were expected as potential outcomes of dialysis treatments.

The outcomes of all reported ADEs were favorable, with all events either resolving completely or being ongoing without serious or fatal outcomes. No cases resulted in severe long-term effects or sequelae, further supporting the safety profile of the HDF procedures and the dialyzers used in these studies.

Table 6. Frequency of (S)ADEs in clinical studies performed with the FX CorAL dialyzer.

Level 1 Term	Adverse Event (MedDRA PT)	Total Occurrences [Device Related/ Procedure Related/Both]	Frequency Per Patient [%] * [Device Related/ Procedure Related/ Both]	Frequency Per Treatment [%] ** [Device Related/ Procedure Related/ Both]	Occurrences Per Device [Device Related/Procedure Related/Both]	
					FX CorAL ***	Other Dialyzers ****
Electrolyte imbalance	• Muscle contractions involuntary • Palpitations • Vomiting • Hyperkalemia • Muscle spasms • Hyperphosphatemia • Blood phosphorous increased • Ear pain • Abdominal pain • Pain in extremity • Flatulence • Myalgia • Hyponatremia • Asthenia • Malaise • Defecation urgency	41 [3/27/11]	16.5 [1.2/10.8/4.4]	1.0 [0.1/0.7/0.3]	11 [2/9/0] Frequency per treatment: 0.8%	30 [1/18/11] Frequency per treatment: 1.1%
Fluid imbalance	• Dyspnea • Hypovolemia • Hypotension • Dialysis hypotension • Headache • Head discomfort • Fatigue • Circulatory collapse • Dizziness • Dehydration • Blood pressure decreased • Visual impairment	27 [8/9/10]	10.8 [3.2/3.6/4.0]	0.7 [0.2/0.2/0.2]	10 [5/2/3] Frequency per treatment: 0.7%	17 [3/7/7] Frequency per treatment: 0.6%
Pruritus, non-allergic	• Pruritus	6 [0/0/6]	2.4 [0/0/2.4]	0.1 [0/0/0.1]	1 [0/0/1] Frequency per treatment: 0.1%	5 [0/0/5] Frequency per treatment: 0.2%

Table 6. *Cont.*

Level 1 Term	Adverse Event (MedDRA PT)	Total Occurrences [Device Related/Procedure Related/Both]	Frequency Per Patient [%] * [Device Related/Procedure Related/Both]	Frequency Per Treatment [%] ** [Device Related/Procedure Related/Both]	Occurrences Per Device [Device Related/Procedure Related/Both]	
					FX CorAL ***	Other Dialyzers ****
Device malfunction (clotting, thrombosis)	• Thrombosis in device • Dialysis related complication	5 [0/2/3]	2.0 [0/0.8/1.2]	0.1 [0/<0.1/0.1]	3 [0/2/1] Frequency per treatment: 0.2%	2 [0/0/2] Frequency per treatment: 0.1%
Procedural complication (blood-side)	• Hemodialysis complication • Vessel puncture site hematoma	4 [0/4/0]	1.6 [0/1.6/0]	0.1 [0/0.1/0]	1 [0/1/0] Frequency per treatment: 0.1%	3 [0/3/0] Frequency per treatment: 0.1%
Peripheral vascular disorder	• Feeling hot	3 [2/0/1]	1.2 [0.8/0/0.4]	0.1 [<0.1/0/<0.1]	1 [1/0/0] Frequency per treatment: 0.1%	2 [1/0/1] Frequency per treatment: 0.1%
Vascular hypertensive disorder	• Hypertension	2 [0/2/0]	0.8 [0/0.8/0]	<0.1 [0/<0.1/0]	2 [0/2/0] Frequency per treatment: 0.1%	0 [0/0/0] Frequency per treatment: 0%
Device mechanical failure	• Device leakage	2 [1/0/1]	0.8 [0.4/0/0.4]	<0.1 [<0.1/0/<0.1]	0 [0/0/0] Frequency per treatment: 0%	2 [1/0/1] Frequency per treatment: 0.1%
Anemia	• Anemia	1 [0/0/1]	0.4 [0/0/0.4]	<0.1 [0/0/<0.1]	0 [0/0/0] Frequency per treat-ment: 0%	1 [0/0/1] Frequency per treatment: <0.1%
Vasoconstriction, vascular insufficiency	• Feeling cold	2 [0/2/0]	0.8 [0/0.8/0]	<0.1 [0/<0.1/0]	0 [0/0/0] Frequency per treat-ment: 0%	2 [0/2/0] Frequency per treat-ment: 0.1%

Table 6. *Cont.*

Level 1 Term	Adverse Event (MedDRA PT)	Total Occurrences [Device Related/ Procedure Related/Both]	Frequency Per Patient [%] * [Device Related/ Procedure Related/ Both]	Frequency Per Treatment [%] ** [Device Related/ Procedure Related/ Both]	Occurrences Per Device [Device Related/Procedure Related/Both]	
					FX CorAL ***	Other Dialyzers ****
Infection	• Influenza like illness	1 [0/0/1]	0.4 [0/0/0.4]	<0.1 [0/0/<0.1]	0 [0/0/0] Frequency per treat-ment: 0%	1 [0/0/1] Frequency per treat-ment: <0.1%
Total		94 [14/46/34]	37.8 [5.6/18.5/13.7]	2.3 [0.3/1.1/0.8]	29 [8/16/5] Frequency per treat-ment: 2.1%	65 [6/30/29] Frequency per treat-ment: 2.4%

* Frequency per patient calculation: Occurrences/249 (total study population) × 100%; ** Frequency per treatment calculation: Occurrences/4149 (total study treatments) × 100%; *** FX CorAL: 1389 study treatments; **** Other dialyzers: xevonta, FX CorDiax, SUREFLUX, Polyflux, ELISIO, FX class (2760 study treatments).

3. FX CorAL safety profile

When analyzing the safety profile specifically for the FX CorAL dialyzer among all 1389 treatments, eight ADEs were related to the dialyzer alone (0.6% of study treatments) (Table 6). All these ADEs were attributed to two patients among the total analysis population (n = 241). Five more ADEs were related to the dialyzer and medical procedure. The majority of these ADEs were classified as fluid imbalance (n = 8), followed by electrolyte imbalance (n = 2). No ADE was attributed to hemocompatibility-related complications, such as chronic inflammation or immune responses. The severity of the ADEs was uniformly mild to moderate. Importantly, there were no unexpected new risks or complications associated with the use of the FX CorAL dialyzer. The measures to treat these related AEs were well-established, and all patient outcomes were favorable, with ADEs resolving completely. When compared to the grouped comparator dialyzers (xevonta, FX CorDiax, SUREFLUX, Polyflux, ELISIO, FX class), the FX CorAL showed comparable safety outcomes (frequency per treatment: FX CorAL 2.1%; other dialyzers: 2.4%), with consistently low rates of device-related adverse events and no significant differences in electrolyte or fluid imbalance occurrences. In line, the FX CorAL dialyzer did not differ in patient-reported outcomes (PROs) such as sleep quality, fatigue, or pruritus, either positively or negatively, when compared to the other dialyzers included in the eMPORA III study [42]. Altogether the FX CorAL demonstrated a high level of safety and efficacy, the benefit/risk profile is highly favorable.

4. Conclusions and Future Directions

In the present article, we summarized the path which is needed to develop of a new dialyzer including in vitro and clinical evidence generation. Dialyzer innovation helps to improve the life of patients with ESKD, as further development in dialyzer technology—especially membrane technology—may lead to better removal of uremic toxins and to a lower rate of complications and side effects.

We used the example of the new FX CorAL dialyzer with its new Helixone *hydro* membrane, which features enhanced hydrophilicity of the blood-side surface, leading to reduced protein adsorption and subsequently to sustained performance and improved hemocompatibility profile as compared to commonly used dialyzers in clinical practice. The dialyzer was tested in multiple in vitro studies and randomized controlled trials across various patient populations, including those with diverse comorbid conditions such as diabetes and hypertension, validating its suitability for routine treatment of ESKD.

Advances in dialyzer performance and hemocompatibility may contribute to better patient outcomes by potentially reducing the long-term risks associated with dialysis, such as cardiovascular events. Comprehensive safety analyses of the four clinical studies underline the FX CorAL's robust safety profile, characterized by a low frequency of dialyzer-related adverse events and an absence of unexpected and severe complications reported across thousands of treatment sessions. As FX CorAL undergoes real-world post-market surveillance, continued evaluation of its safety profile confirms its suitability for broader patient groups from different geographical regions. These data will also help to evaluate whether the FX CorAL may be an adequate alternative for patients suffering from hypersensitivity reactions during treatments with synthetic dialyzers and which currently are switched to cellulose-based or PMMA dialyzers. Looking further ahead, longitudinal clinical data will be essential to investigate any long-term benefits in terms of patient survival and quality of life, especially in the context of hemodiafiltration treatments [53,116,117]. Innovation in dialysis care is essential to improve well-being of ESKD patients, bringing hope for more effective, safer, and personalized treatment options worldwide.

Supplementary Materials: The following supporting information can be downloaded at: https://www.mdpi.com/article/10.3390/membranes15050132/s1, Table S1. Categorization of AEs and overview of AE numbers in the four clinical studies with the FX CorAL dialyzer. AEs are summarized across all studies and dialyzers; Table S2. Number of study patients and treatment sessions for safety analysis stratified by study and dialyzer; Figure S1. Dialyzer and medical procedure related (serious) adverse events ((S)ADEs) associated with dialysis treatments. All listed complications can be linked either to the dialyzer or to the medical procedure or both.

Author Contributions: A.M.Z., B.G., B.O. and T.L.: writing—original draft preparation, A.E., N.C., S.B., M.S.-G. and J.P.K.: writing—review and editing. All authors have read and agreed to the published version of the manuscript.

Funding: This research received no external funding.

Institutional Review Board Statement: Not applicable.

Informed Consent Statement: Not applicable.

Data Availability Statement: The original contributions presented in this study are included in the article/supplementary material. Further inquiries can be directed to the corresponding author.

Acknowledgments: The support of Craig Kamerath in the setup of hydrophilicity experiments and Brian Van Devener for his expertise in the AFM method development and optimization are greatly appreciated. The authors also extend their sincere thanks to C. Wade Dummer and Darrell Hendricks for their significant efforts regarding the development of AFM force-distance curve analysis techniques and associated measurements. The AFM work made use of Nanofab EMSAL shared facilities of the Micron Technology Foundation Inc. Microscopy Suite sponsored by the John and Marcia Price College of Engineering, Health Sciences Center, Office of the Vice President for Research.

Conflicts of Interest: A.M.Z., B.G, A.E., S.B., N.C., M.S.-G., J.P.K. and T.L. are employees of Fresenius Medical Care and may hold stock in the company. B.O. reports consulting fees from Fresenius Medical Care.

Abbreviations

The following abbreviations are used in this manuscript:

ADE	Adverse device effects
AE	Adverse event
AFM	Atomic force microscopy
AUC	Area under the curve
BMI	Body mass index
β2-m	β2-microglobulin
β-TG	β-Thromboglobulin
C3a	Complement factor C3a
C5a	Complement factor C5a
ESKD	End-stage kidney disease
GSH-Px	Glutathione peroxidase
HDF	Hemodiafiltration
hsCRP	High-sensitivity C-reactive protein
IL-6	Interleukin 6
IL-8	Interleukin 8
LTB4	Leukotriene B4
MDA	Malondialdehyde
MWCO	Molecular weight cut-off
MWRO	Molecular weight retention onset
nsAE	Non-serious adverse event

PMMA	Polymethylmethacrylate
PMN elastase	Polymorphonuclear elastase
PRO	Patient reported outcomes
PVP	Polyvinylpyrrolidone
RR	Removal rate
RRT	Renal replacement therapy
SAE	Serious adverse event
sC5b-9	Complement factor sC5b-9
sICAM-1	Soluble intercellular adhesion molecule 1
TAT	Thrombin-antithrombin
TMP	Transmembrane pressure
TxB2	Thromboxane B2
XPS	X-ray photoelectron spectroscopy

References

1. Saran, R.; Robinson, B.; Abbott, K.C.; Bragg-Gresham, J.; Chen, X.; Gipson, D.; Gu, H.; Hirth, R.A.; Hutton, D.; Jin, Y.; et al. US Renal Data System 2019 Annual Data Report: Epidemiology of Kidney Disease in the United States. *Am. J. Kidney Dis.* **2020**, *75*, A6–A7. [CrossRef] [PubMed]
2. Thurlow, J.S.; Joshi, M.; Yan, G.; Norris, K.C.; Agodoa, L.Y.; Yuan, C.M.; Nee, R. Global Epidemiology of End-Stage Kidney Disease and Disparities in Kidney Replacement Therapy. *Am. J. Nephrol.* **2021**, *52*, 98–107. [CrossRef] [PubMed]
3. Kim, J.C.; Kim, J.H.; Sung, J.; Kim, H.-C.; Kang, E.; Lee, S.H.; Kim, J.K.; Kim, H.C.; Min, B.G.; Ronco, C. Effects of Arterial Port Design on Blood Flow Distribution in Hemodialyzers. *Blood Purif.* **2009**, *28*, 260–267. [CrossRef] [PubMed]
4. Kulz, M. In Vitro and in Vivo Evaluation of a New Dialyzer. *Nephrol. Dial. Transplant.* **2002**, *17*, 1475–1479. [CrossRef]
5. Ronco, C.; Bowry, S.K.; Brendolan, A.; Crepaldi, C.; Soffiati, G.; Fortunato, A.; Bordoni, V.; Granziero, A.; Torsello, G.; La Greca, G. Hemodialyzer: From Macro-Design to Membrane Nanostructure; the Case of the FX-Class of Hemodialyzers. *Kidney Int. Suppl.* **2002**, *61*, S126–S142. [CrossRef]
6. Said, N.; Lau, W.J.; Ho, Y.-C.; Lim, S.K.; Zainol Abidin, M.N.; Ismail, A.F. A Review of Commercial Developments and Recent Laboratory Research of Dialyzers and Membranes for Hemodialysis Application. *Membranes* **2021**, *11*, 767. [CrossRef]
7. Lang, T.; Zawada, A.M.; Theis, L.; Braun, J.; Ottillinger, B.; Kopperschmidt, P.; Gagel, A.; Kotanko, P.; Stauss-Grabo, M.; Kennedy, J.P.; et al. Hemodiafiltration: Technical and Medical Insights. *Bioengineering* **2023**, *10*, 145. [CrossRef]
8. Bowry, S.K. Dialysis Membranes Today. *Int. J. Artif. Organs* **2002**, *25*, 447–460. [CrossRef]
9. Ronco, C.; Clark, W.R. Haemodialysis Membranes. *Nat. Rev. Nephrol.* **2018**, *14*, 394–410. [CrossRef]
10. Zawada, A.M.; Lang, T.; Ottillinger, B.; Kircelli, F.; Stauss-Grabo, M.; Kennedy, J.P. Impact of Hydrophilic Modification of Synthetic Dialysis Membranes on Hemocompatibility and Performance. *Membranes* **2022**, *12*, 932. [CrossRef]
11. Zawada, A.M.; Emal, K.; Förster, E.; Saremi, S.; Delinski, D.; Theis, L.; Küng, F.; Xie, W.; Werner, J.; Stauss-Grabo, M.; et al. Hydrophilic Modification of Dialysis Membranes Sustains Middle Molecule Removal and Filtration Characteristics. *Membranes* **2024**, *14*, 83. [CrossRef]
12. Zawada, A.M.; Melchior, P.; Schall, C.; Erlenkötter, A.; Lang, T.; Keller, T.; Stauss-Grabo, M.; Kennedy, J.P. Time-resolving Characterization of Molecular Weight Retention Changes among Three Synthetic High-flux Dialyzers. *Artif. Organs* **2022**, *46*, 1318–1327. [CrossRef] [PubMed]
13. Zawada, A.M.; Melchior, P.; Erlenkötter, A.; Delinski, D.; Stauss-Grabo, M.; Kennedy, J.P. Polyvinylpyrrolidone in Hemodialysis Membranes: Impact on Platelet Loss during Hemodialysis. *Hemodial. Int.* **2021**, *25*, 498–506. [CrossRef]
14. Melchior, P.; Erlenkötter, A.; Zawada, A.M.; Delinski, D.; Schall, C.; Stauss-Grabo, M.; Kennedy, J.P. Complement Activation by Dialysis Membranes and Its Association with Secondary Membrane Formation and Surface Charge. *Artif. Organs* **2021**, *45*, 770–778. [CrossRef] [PubMed]
15. Hayama, M. Nanoscopic Behavior of Polyvinylpyrrolidone Particles on Polysulfone/Polyvinylpyrrolidone Film. *Biomaterials* **2004**, *25*, 1019–1028. [CrossRef]
16. Zhu, L.; Song, H.; Wang, J.; Xue, L. Polysulfone Hemodiafiltration Membranes with Enhanced Anti-Fouling and Hemocompatibility Modified by Poly(Vinyl Pyrrolidone) via in Situ Cross-Linked Polymerization. *Mater. Sci. Eng. C Mater. Biol. Appl.* **2017**, *74*, 159–166. [CrossRef] [PubMed]
17. Jiang, J.; Zhu, L.; Zhu, L.; Zhang, H.; Zhu, B.; Xu, Y. Antifouling and Antimicrobial Polymer Membranes Based on Bioinspired Polydopamine and Strong Hydrogen-Bonded Poly (*N*-Vinyl Pyrrolidone). *ACS Appl. Mater. Interfaces* **2013**, *5*, 12895–12904. [CrossRef]

18. Ran, F.; Nie, S.; Zhao, W.; Li, J.; Su, B.; Sun, S.; Zhao, C. Biocompatibility of Modified Polyethersulfone Membranes by Blending an Amphiphilic Triblock Co-Polymer of Poly(Vinyl Pyrrolidone)-b-Poly(Methyl Methacrylate)-b-Poly(Vinyl Pyrrolidone). *Acta Biomater.* **2011**, *7*, 3370–3381. [CrossRef]
19. Hayama, M.; Yamamoto, K.; Kohori, F.; Sakai, K. How Polysulfone Dialysis Membranes Containing Polyvinylpyrrolidone Achieve Excellent Biocompatibility? *J. Membr. Sci.* **2004**, *234*, 41–49. [CrossRef]
20. Wang, H.; Yu, T.; Zhao, C.; Du, Q. Improvement of Hydrophilicity and Blood Compatibility on Polyethersulfone Membrane by Adding Polyvinylpyrrolidone. *Fibers Polym.* **2009**, *10*, 1–5. [CrossRef]
21. Koga, Y.; Fujieda, H.; Meguro, H.; Ueno, Y.; Aoki, T.; Miwa, K.; Kainoh, M. Biocompatibility of Polysulfone Hemodialysis Membranes and Its Mechanisms: Involvement of Fibrinogen and Its Integrin Receptors in Activation of Platelets and Neutrophils. *Artif. Organs* **2018**, *42*, E246–E258. [CrossRef] [PubMed]
22. Bowry, S.K.; Chazot, C. The Scientific Principles and Technological Determinants of Haemodialysis Membranes. *Clin. Kidney J.* **2021**, *14*, i5–i16. [CrossRef] [PubMed]
23. Ekdahl, K.N.; Soveri, I.; Hilborn, J.; Fellström, B.; Nilsson, B. Cardiovascular Disease in Haemodialysis: Role of the Intravascular Innate Immune System. *Nat. Rev. Nephrol.* **2017**, *13*, 285–296. [CrossRef]
24. Ronco, C.; Brendolan, A.; Nalesso, F.; Zanella, M.; De Cal, M.; Corradi, V.; Virzì, G.M.; Ferrari, F.; Garzotto, F.; Lorenzin, A.; et al. Prospective, Randomized, Multicenter, Controlled Trial (TRIATHRON 1) on a New Antithrombogenic Hydrophilic Dialysis Membrane. *Int. J. Artif. Organs* **2017**, *40*, 234–239. [CrossRef]
25. Kislikova, M.; Vega, A.; Verde, E.; Abad, S.; Vaca, M.; Acosta, A.; González, A.; Bascuñana, A.; Mijailova, A.; Nava, C.; et al. Depurative Capacity toward Medium Molecules of the Dialyzer Toray NV-U® Hydrolink™: A New Hydrophilic Membrane to Perform Online Hemodiafiltration. *Int. J. Artif. Organs* **2024**, *47*, 795–801. [CrossRef]
26. Kakuta, T.; Komaba, H.; Takagi, N.; Takahashi, Y.; Suzuki, H.; Hyodo, T.; Nagaoka, M.; Tanaka, R.; Iwao, S.; Ishida, M.; et al. A Prospective Multicenter Randomized Controlled Study on Interleukin-6 Removal and Induction by a New Hemodialyzer With Improved Biocompatibility in Hemodialysis Patients: A Pilot Study. *Ther. Apher. Dial.* **2016**, *20*, 569–578. [CrossRef]
27. Maduell, F.; Escudero-Saiz, V.J.; Cuadrado-Payán, E.; Rodriguez-Garcia, M.; Gómez, M.; Rodas, L.M.; Fontseré, N.; Salgado, M.D.C.; Casals, G.; Rico, N.; et al. A Study on the Safety and Efficacy of an Innovative Hydrophilic Dialysis Membrane. *Membranes* **2025**, *15*, 30. [CrossRef] [PubMed]
28. Oshihara, W.; Ueno, Y.; Fujieda, H. A New Polysulfone Membrane Dialyzer, NV, with Low-Fouling and Antithrombotic Properties. *Contrib. Nephrol.* **2017**, *189*, 222–229. [CrossRef]
29. Murakami, J.; Kaneko, I.; Kimata, N.; Mineshima, M.; Akiba, T. Problems in the Evaluation of Polyvinylpyrrolidone (PVP) Elution from Polysulfone Membrane Dialyzers Sterilized by Gamma-Ray Irradiation. *Ren. Replace. Ther.* **2016**, *2*, 36. [CrossRef]
30. Lee, G.T.; Hong, Y.K. Manufacturing and Separation Characteristics of Hemodialysis Membranes to Improve Toxin Removal Rate. *Adv. Polym. Technol.* **2022**, *2022*, 2565010. [CrossRef]
31. Abdelrasoul, A.; Westphalen, H.; Saadati, S.; Shoker, A. Hemodialysis Biocompatibility Mathematical Models to Predict the Inflammatory Biomarkers Released in Dialysis Patients Based on Hemodialysis Membrane Characteristics and Clinical Practices. *Sci. Rep.* **2021**, *11*, 23080. [CrossRef] [PubMed]
32. Kiguchi, T.; Ito, H.; Yamashita, A.C. Effect of Membrane Surface Area on Solute Removal Performance of Dialyzers with Fouling. *Membranes* **2022**, *12*, 684. [CrossRef] [PubMed]
33. Boschetti-de-Fierro, A.; Voigt, M.; Storr, M.; Krause, B. MCO Membranes: Enhanced Selectivity in High-Flux Class. *Sci. Rep.* **2015**, *5*, 18448. [CrossRef] [PubMed]
34. Morti, S.M.; Zydney, A.L. Protein-Membrane Interactions during Hemodialysis: Effects on Solute Transport. *ASAIO J.* **1998**, *44*, 319–326. [CrossRef]
35. Langsdorf, L.J.; Zydney, A.L. Effect of Blood Contact on the Transport Properties of Hemodialysis Membranes: A Two-Layer Membrane Model. *Blood Purif.* **1994**, *12*, 292–307. [CrossRef]
36. Kim, T.R.; Hadidi, M.; Motevalian, S.P.; Sunohara, T.; Zydney, A.L. Effects of Plasma Proteins on the Transport and Surface Characteristics of Polysulfone/Polyethersulfone and Asymmetric Cellulose Triacetate High Flux Dialyzers. *Artif. Organs* **2018**, *42*, 1070–1077. [CrossRef]
37. Röckel, A.; Hertel, J.; Fiegel, P.; Abdelhamid, S.; Panitz, N.; Walb, D. Permeability and Secondary Membrane Formation of a High Flux Polysulfone Hemofilter. *Kidney Int.* **1986**, *30*, 429–432. [CrossRef]
38. Canaud, B. Recent Advances in Dialysis Membranes. *Curr. Opin. Nephrol. Hypertens.* **2021**, *30*, 613–622. [CrossRef]
39. Kempkes-Koch, M.; Stauss-Grabo, M.; Erlenkötter, A.; Rauber, L.; Kennedy, J.; Gauly, A.; Schmidt-Gürtler, H. MO387 Clinical performance, hemocompatibility and safety of a new dialyzer with a modified polysulfone membrane. *Nephrol. Dial. Transplant.* **2021**, *36*, gfab082.0041. [CrossRef]
40. Ehlerding, G.; Erlenkötter, A.; Gauly, A.; Griesshaber, B.; Kennedy, J.; Rauber, L.; Ries, W.; Schmidt-Gürtler, H.; Stauss-Grabo, M.; Wagner, S.; et al. Performance and Hemocompatibility of a Novel Polysulfone Dialyzer: A Randomized Controlled Trial. *Kidney360* **2021**, *2*, 935–945. [CrossRef]

41. Ehlerding, G.; Ries, W.; Kempkes-Koch, M.; Ziegler, E.; Erlenkötter, A.; Zawada, A.M.; Kennedy, J.P.; Ottillinger, B.; Stauss-Grabo, M.; Lang, T. Randomized Comparison of Three High-Flux Dialyzers during High-Volume Online Hemodiafiltration—The comPERFORM Study. *Clin. Kidney J.* **2022**, *15*, 672–680. [CrossRef] [PubMed]
42. Ehlerding, G.; Ries, W.; Kempkes-Koch, M.; Ziegler, E.; Ronová, P.; Krizsán, M.; Verešová, J.; Böke, M.; Erlenkötter, A.; Nitschel, R.; et al. Randomized Investigation of Increased Dialyzer Membrane Hydrophilicity on Hemocompatibility and Performance. *BMC Nephrol.* **2024**, *25*, 220. [CrossRef] [PubMed]
43. Kuragano, T.; Kida, A.; Furuta, M.; Nanami, M.; Otaki, Y.; Hasuike, Y.; Nonoguchi, H.; Nakanishi, T. The Impact of B2-Microglobulin Clearance on the Risk Factors of Cardiovascular Disease in Hemodialysis Patients. *ASAIO J.* **2010**, *56*, 326–332. [CrossRef]
44. Kaizu, Y.; Ohkawa, S.; Odamaki, M.; Ikegaya, N.; Hibi, I.; Miyaji, K.; Kumagai, H. Association between Inflammatory Mediators and Muscle Mass in Long-Term Hemodialysis Patients. *Am. J. Kidney Dis.* **2003**, *42*, 295–302. [CrossRef] [PubMed]
45. Neirynck, N.; Vanholder, R.; Schepers, E.; Eloot, S.; Pletinck, A.; Glorieux, G. An Update on Uremic Toxins. *Int. Urol. Nephrol.* **2013**, *45*, 139–150. [CrossRef]
46. Wolley, M.; Jardine, M.; Hutchison, C.A. Exploring the Clinical Relevance of Providing Increased Removal of Large Middle Molecules. *Clin. J. Am. Soc. Nephrol.* **2018**, *13*, 805–814. [CrossRef]
47. Canaud, B.; Blankestijn, P.J.; Grooteman, M.P.C.; Davenport, A. Why and How High Volume Hemodiafiltration May Reduce Cardiovascular Mortality in Stage 5 Chronic Kidney Disease Dialysis Patients? A Comprehensive Literature Review on Mechanisms Involved. *Semin. Dial.* **2022**, *35*, 117–128. [CrossRef]
48. Cheung, A.K.; Rocco, M.V.; Yan, G.; Leypoldt, J.K.; Levin, N.W.; Greene, T.; Agodoa, L.; Bailey, J.; Beck, G.J.; Clark, W.; et al. Serum β-2 Microglobulin Levels Predict Mortality in Dialysis Patients: Results of the HEMO Study. *J. Am. Soc. Nephol.* **2006**, *17*, 546–555. [CrossRef]
49. Okuno, S.; Ishimura, E.; Kohno, K.; Fujino-Katoh, Y.; Maeno, Y.; Yamakawa, T.; Inaba, M.; Nishizawa, Y. Serum 2-Microglobulin Level Is a Significant Predictor of Mortality in Maintenance Haemodialysis Patients. *Nephrol. Dial. Transplant.* **2008**, *24*, 571–577. [CrossRef]
50. Pifer, T.B.; Mccullough, K.P.; Port, F.K.; Goodkin, D.A.; Maroni, B.J.; Held, P.J.; Young, E.W. Mortality Risk in Hemodialysis Patients and Changes in Nutritional Indicators: DOPPS. *Kidney Int.* **2002**, *62*, 2238–2245. [CrossRef]
51. Canaud, B.; Ye, X.; Usvyat, L.; Kooman, J.; van der Sande, F.; Raimann, J.; Wang, Y.; Kotanko, P. Clinical and Predictive Value of Simplified Creatinine Index Used as Muscle Mass Surrogate in End-Stage Kidney Disease Haemodialysis Patients—Results from the International MONitoring Dialysis Outcome Initiative. *Nephrol. Dial. Transplant.* **2020**, *35*, 2161–2171. [CrossRef] [PubMed]
52. de Mutsert, R.; Grootendorst, D.C.; Axelsson, J.; Boeschoten, E.W.; Krediet, R.T.; Dekker, F.W. the NECOSAD Study Group. Excess Mortality Due to Interaction between Protein-Energy Wasting, Inflammation and Cardiovascular Disease in Chronic Dialysis Patients. *Nephrol. Dial. Transplant.* **2008**, *23*, 2957–2964. [CrossRef] [PubMed]
53. Blankestijn, P.J.; Vernooij, R.W.M.; Hockham, C.; Strippoli, G.F.M.; Canaud, B.; Hegbrant, J.; Barth, C.; Covic, A.; Cromm, K.; Cucui, A.; et al. Effect of Hemodiafiltration or Hemodialysis on Mortality in Kidney Failure. *N. Engl. J. Med.* **2023**, *389*, 700–709. [CrossRef] [PubMed]
54. Maduell, F.; Broseta, J.J.; Rodríguez-Espinosa, D.; Rodas, L.M.; Gómez, M.; Arias-Guillén, M.; Fontseré, N.; Vera, M.; Salgado, M.D.C.; Rico, N.; et al. Comparison of Efficacy and Safety of the New Generation Helixone Dialyzers. *Nefrología* **2024**, *44*, 354–361. [CrossRef]
55. Losappio, V.; Franzin, R.; Infante, B.; Godeas, G.; Gesualdo, L.; Fersini, A.; Castellano, G.; Stallone, G. Molecular Mechanisms of Premature Aging in Hemodialysis: The Complex Interplay between Innate and Adaptive Immune Dysfunction. *Int. J. Mol. Sci.* **2020**, *21*, 3422. [CrossRef]
56. Bowry, S.K.; Kircelli, F.; Himmele, R.; Nigwekar, S.U. Blood-Incompatibility in Haemodialysis: Alleviating Inflammation and Effects of Coagulation. *Clin. Kidney J.* **2021**, *14*, i59–i71. [CrossRef]
57. Poppelaars, F.; Faria, B.; Gaya da Costa, M.; Franssen, C.F.M.; van Son, W.J.; Berger, S.P.; Daha, M.R.; Seelen, M.A. The Complement System in Dialysis: A Forgotten Story? *Front. Immunol.* **2018**, *9*, 71. [CrossRef]
58. Mares, J.; Richtrova, P.; Hricinova, A.; Tuma, Z.; Moravec, J.; Lysak, D.; Matejovic, M. Proteomic Profiling of Blood-Dialyzer Interactome Reveals Involvement of Lectin Complement Pathway in Hemodialysis-Induced Inflammatory Response. *Proteom. Clin. Appl.* **2010**, *4*, 829–838. [CrossRef]
59. Nilsson, B.; Ekdahl, K.N.; Mollnes, T.E.; Lambris, J.D. The Role of Complement in Biomaterial-Induced Inflammation. *Mol. Immunol.* **2007**, *44*, 82–94. [CrossRef]
60. Chenoweth, D.E.; Cheung, A.K.; Henderson, L.W. Anaphylatoxin Formation during Hemodialysis: Effects of Different Dialyzer Membranes. *Kidney Int.* **1983**, *24*, 764–769. [CrossRef]
61. de Borst, M.H. The Complement System in Hemodialysis Patients: Getting to the Heart of the Matter. *Nephron* **2016**, *132*, 1–4. [CrossRef] [PubMed]

62. Kourtzelis, I.; Markiewski, M.M.; Doumas, M.; Rafail, S.; Kambas, K.; Mitroulis, I.; Panagoutsos, S.; Passadakis, P.; Vargemezis, V.; Magotti, P.; et al. Complement Anaphylatoxin C5a Contributes to Hemodialysis-Associated Thrombosis. *Blood* **2010**, *116*, 631–639. [CrossRef]
63. François, K.; Orlando, C.; Jochmans, K.; Cools, W.; De Meyer, V.; Tielemans, C.; Wissing, K.M. Hemodialysis Does Not Induce Detectable Activation of the Contact System of Coagulation. *Kidney Int. Rep.* **2020**, *5*, 831–838. [CrossRef]
64. Lucchi, L.; Ligabue, G.; Marietta, M.; Delnevo, A.; Malagoli, M.; Perrone, S.; Stipo, L.; Grandi, F.; Albertazzi, A. Activation of Coagulation During Hemodialysis: Effect of Blood Lines Alone and Whole Extracorporeal Circuit. *Artif. Organs* **2006**, *30*, 106–110. [CrossRef]
65. Milburn, J.A.; Ford, I.; Mutch, N.J.; Fluck, N.; Brittenden, J. Thrombin-Anti-Thrombin Levels and Patency of Arterio-Venous Fistula in Patients Undergoing Haemodialysis Compared to Healthy Volunteers: A Prospective Analysis. *PLoS ONE* **2013**, *8*, e67799. [CrossRef]
66. Hernández, M.R.; Galán, A.M.; Lozano, M.; Bozzo, J.; Cases, A.; Escolar, G.; Calls, J.; Ordinas, A. Platelet-Leukocyte Activation during Hemodialysis Detected with a Monoclonal Antibody to Leukocyte Integrin CD11b. *Nephron* **1998**, *80*, 197–203. [CrossRef] [PubMed]
67. Daikidou, D.-V.; Lioulios, G.; Sampani, E.; Xochelli, A.; Nikolaidou, V.; Moysidou, E.; Christodoulou, M.; Iosifidou, A.; Iosifidou, M.; Briza, D.; et al. Prospective Analysis of B Lymphocyte Subtypes, before and after Initiation of Dialysis, in Patients with End-Stage Renal Disease. *Life* **2023**, *13*, 860. [CrossRef] [PubMed]
68. Singh, V.; Kaur, R.; Kumari, P.; Pasricha, C.; Singh, R. ICAM-1 and VCAM-1: Gatekeepers in Various Inflammatory and Cardiovascular Disorders. *Clin. Chim. Acta* **2023**, *548*, 117487. [CrossRef]
69. Daugirdas, J.T.; Bernardo, A.A. Hemodialysis Effect on Platelet Count and Function and Hemodialysis-Associated Thrombocytopenia. *Kidney Int.* **2012**, *82*, 147–157. [CrossRef]
70. Rogacev, K.S.; Ziegelin, M.; Ulrich, C.; Seiler, S.; Girndt, M.; Fliser, D.; Heine, G.H. Haemodialysis-Induced Transient $CD16^+$ Monocytopenia and Cardiovascular Outcome. *Nephrol. Dial. Transplant.* **2009**, *24*, 3480–3486. [CrossRef]
71. Maher, J.F.; Schreiner, G.E. Hazards and Complications of Dialysis. *N. Engl. J. Med.* **1965**, *273*, 370–377. [CrossRef]
72. Salem, M.; Ivanovich, P.T.; Ing, T.S.; Daugirdas, J.T. Adverse Effects of Dialyzers Manifesting during the Dialysis Session. *Nephrol. Dial. Transplant.* **1994**, *9* (Suppl. S2), 127–137. [PubMed]
73. Himmelfarb, J. Hemodialysis Complications. *Am. J. Kidney Dis.* **2005**, *45*, 1122–1131. [CrossRef] [PubMed]
74. Wong, B.; Zimmerman, D.; Reintjes, F.; Courtney, M.; Klarenbach, S.; Dowling, G.; Pauly, R.P. Procedure-Related Serious Adverse Events Among Home Hemodialysis Patients: A Quality Assurance Perspective. *Am. J. Kidney Dis.* **2014**, *63*, 251–258. [CrossRef] [PubMed]
75. Saha, M.; Allon, M. Diagnosis, Treatment, and Prevention of Hemodialysis Emergencies. *Clin. J. Am. Soc. Nephrol.* **2017**, *12*, 357–369. [CrossRef]
76. Scherer, J.S.; Combs, S.A.; Brennan, F. Sleep Disorders, Restless Legs Syndrome, and Uremic Pruritus: Diagnosis and Treatment of Common Symptoms in Dialysis Patients. *Am. J. Kidney Dis.* **2017**, *69*, 117–128. [CrossRef]
77. Greenberg, K.I.; Choi, M.J. Hemodialysis Emergencies: Core Curriculum 2021. *Am. J. Kidney Dis.* **2021**, *77*, 796–809. [CrossRef]
78. Davenport, A.; Vervloet, M. Effects of Convective Dialysis Techniques on Electrolytes and Mineral Metabolism. In *Hemodiafiltration*; Nubé, M.J., Grooteman, M.P.C., Blankestijn, P.J., Eds.; Springer International Publishing: Cham, Switzerland, 2016; pp. 121–135, ISBN 978-3-319-23331-4.
79. Dekker, M.J.E.; Kooman, J.P. Fluid Status Assessment in Hemodialysis Patients and the Association with Outcome: Review of Recent Literature. *Curr. Opin. Nephrol. Hypertens.* **2018**, *27*, 188–193. [CrossRef]
80. Hecking, M.; Madero, M.; Port, F.K.; Schneditz, D.; Wabel, P.; Chazot, C. Fluid Volume Management in Hemodialysis: Never Give Up! *Kidney Int.* **2023**, *103*, 2–5. [CrossRef]
81. Milinkovic, M.; Zidverc-Trajkovic, J.; Sternic, N.; Trbojevic-Stankovic, J.; Maric, I.; Milic, M.; Andric, B.; Nikic, P.; Jovanovic, P.; Rosic, P.; et al. Hemodialysis Headache. *Clin. Nephrol.* **2009**, *71*, 158–163. [CrossRef]
82. Loutradis, C.; Sarafidis, P.A.; Ferro, C.J.; Zoccali, C. Volume Overload in Hemodialysis: Diagnosis, Cardiovascular Consequences, and Management. *Nephrol. Dial. Transplant.* **2021**, *36*, 2182–2193. [CrossRef] [PubMed]
83. Kanbay, M.; Ertuglu, L.A.; Afsar, B.; Ozdogan, E.; Siriopol, D.; Covic, A.; Basile, C.; Ortiz, A. An Update Review of Intradialytic Hypotension: Concept, Risk Factors, Clinical Implications and Management. *Clin. Kidney J.* **2020**, *13*, 981–993. [CrossRef] [PubMed]
84. Sars, B.; van der Sande, F.M.; Kooman, J.P. Intradialytic Hypotension: Mechanisms and Outcome. *Blood Purif.* **2020**, *49*, 158–167. [CrossRef] [PubMed]
85. Davenport, A. Why Is Intradialytic Hypotension the Commonest Complication of Outpatient Dialysis Treatments? *Kidney Int. Rep.* **2023**, *8*, 405–418. [CrossRef]
86. Canzanello, V.J.; Burkart, J.M. Hemodialysis-Associated Muscle Cramps. *Semin. Dial.* **1992**, *5*, 299–304. [CrossRef]
87. Sav, M.Y.; Sav, T.; Senocak, E.; Sav, N.M. Hemodialysis-related Headache. *Hemodial. Int.* **2014**, *18*, 725–729. [CrossRef]

88. Pun, P.H.; Middleton, J.P. Dialysate Potassium, Dialysate Magnesium, and Hemodialysis Risk. *J. Am. Soc. Nephrol.* **2017**, *28*, 3441–3451. [CrossRef]
89. Boyer, C.J.; Swartz, R.D. Severe Clotting During Extracorporeal Dialysis Procedures. *Semin. Dial.* **1991**, *4*, 69–71. [CrossRef]
90. Hornig, C.; Bowry, S.K.; Kircelli, F.; Kendzia, D.; Apel, C.; Canaud, B. Hemoincompatibility in Hemodialysis-Related Therapies and Their Health Economic Perspectives. *J. Clin. Med.* **2024**, *13*, 6165. [CrossRef]
91. Fadem, S.Z. Blood Clotting Complications in Dialysis. In *Complications in Dialysis*; Fadem, S.Z., Moura-Neto, J.A., Golper, T.A., Eds.; Springer International Publishing: Cham, Switzerland, 2023; pp. 203–218, ISBN 978-3-031-44556-9.
92. Polaschegg, H. Red Blood Cell Damage from Extracorporeal Circulation in Hemodialysis. *Semin. Dial.* **2009**, *22*, 524–531. [CrossRef]
93. Tharmaraj, D.; Kerr, P.G. Haemolysis in Haemodialysis. *Nephrology* **2017**, *22*, 838–847. [CrossRef] [PubMed]
94. Yoon, J.; Thapa, S.; Chow, R.D.; Jaar, B.G. Hemolysis as a Rare but Potentially Life-Threatening Complication of Hemodialysis: A Case Report. *BMC Res. Notes* **2014**, *7*, 475. [CrossRef]
95. Röckel, A.; Klinke, B.; Hertel, J.; Baur, X.; Thiel, C.; Abdelhamid, S.; Fiegel, P.; Walb, D. Allergy to Dialysis Materials. *Nephrol. Dial. Transplant.* **1989**, *4*, 646–652.
96. Mukaya, J.E.; Jacobson, M.S.; Esprit, D.; Ajayi, T. Allergic Reaction to Polysulphone Membrane Dialyser Masquerading as Infection. *BMJ Case Rep.* **2015**, bcr2014208591. [CrossRef] [PubMed]
97. Sayeed, K.; Murdakes, C.; Spec, A.; Gashti, C. Anaphylactic Shock at the Beginning of Hemodialysis. *Semin. Dial.* **2016**, *29*, 81–84. [CrossRef] [PubMed]
98. Shu, K.-H.; Kao, T.-W.; Chiang, W.-C.; Wu, V.-C. A Case of Anaphylactic Shock Induced by FX60 Polysulfone Hemodialyzer but Not F6-HPS Polysulfone Hemodialyzer: Polysulfone Hemodialyzer Anaphylaxis. *Hemodial. Int.* **2014**, *18*, 841–845. [CrossRef]
99. Bacelar Marques, I.D.; Pinheiro, K.F.; de Freitas do Carmo, L.P.; Costa, M.C.; Abensur, H. Anaphylactic Reaction Induced by a Polysulfone/Polyvinylpyrrolidone Membrane in the 10th Session of Hemodialysis with the Same Dialyzer. *Hemodial. Int.* **2011**, *15*, 399–403. [CrossRef]
100. Martin-Navarro, J.; Esteras, R.; Castillo, E.; Carriazo, S.; Fernández-Prado, R.; Gracia-Iguacel, C.; Mas Fontao, S.; Ortíz, A.; González-Parra, E. Reactions to Synthetic Membranes Dialyzers: Is There an Increase in Incidence? *Kidney Blood Press. Res.* **2019**, *44*, 907–914. [CrossRef]
101. Ohashi, N.; Yonemura, K.; Goto, T.; Suzuki, H.; Fujigaki, Y.; Yamamoto, T.; Hishida, A. A Case of Anaphylactoid Shock Induced by the BS Polysulfone Hemodialyzer but Not by the F8-HPS Polysulfone Hemodialyzer. *Clin. Nephrol.* **2003**, *60*, 214–217. [CrossRef]
102. Glorieux, G.; Hulko, M.; Speidel, R.; Brodbeck, K.; Krause, B.; Vanholder, R. Looking beyond Endotoxin: A Comparative Study of Pyrogen Retention by Ultrafilters Used for the Preparation of Sterile Dialyis Fluid. *Sci. Rep.* **2014**, *4*, 6390. [CrossRef]
103. Hindman, S.; Carson, L.; Favero, M.; Petersen, N.; Schonberger, L.; Solano, J. Pyrogenic reactions during haemodialysis caused by extramural endotoxin. *Lancet* **1975**, *306*, 732–734. [CrossRef] [PubMed]
104. Catapano, G.; Morrone, G.; Hu, L.; Fragomeni, G.; Buscaroli, A. Endotoxin-Retentive Filters for the Online Preparation of Ultrapure Dialysis Fluid and Non-Pyrogenic Substitution Fluid: A Critical Review and Reference Guide. *Membranes* **2025**, *15*, 51. [CrossRef]
105. Axley, B.; Speranza-Reid, J.; Williams, H. Venous Needle Dislodgement in Patients on Hemodialysis. *Nephrol. Nurs. J.* **2012**, *39*, 435–445; quiz 446. [PubMed]
106. Tennankore, K.K.; d'Gama, C.; Faratro, R.; Fung, S.; Wong, E.; Chan, C.T. Adverse Technical Events in Home Hemodialysis. *Am. J. Kidney Dis.* **2015**, *65*, 116–121. [CrossRef]
107. MacRae, J.M.; Dipchand, C.; Oliver, M.; Moist, L.; Lok, C.; Clark, E.; Hiremath, S.; Kappel, J.; Kiaii, M.; Luscombe, R.; et al. Arteriovenous Access Failure, Stenosis, and Thrombosis. *Can. J. Kidney Health Dis.* **2016**, *3*, 2054358116669126. [CrossRef] [PubMed]
108. Franchin, M.; Tadiello, M.; Guzzetti, L.; Gattuso, A.; Mauri, F.; Cervarolo, M.C.; D'Oria, M.G.G.; Tozzi, M. Acute Problems of Hemodialysis Access: Thrombosis, Aneurysms, Symptomatic High-Flow Fistulas, and Complications Related to Central Lines. *Semin. Vasc. Surg.* **2023**, *36*, 300–306. [CrossRef]
109. Kumbar, L.; Yee, J. Current Concepts in Hemodialysis Vascular Access Infections. *Adv. Chronic. Kidney Dis.* **2019**, *26*, 16–22. [CrossRef] [PubMed]
110. Vesely, T.M. Air Embolism during Insertion of Central Venous Catheters. *J. Vasc. Interv. Radiol.* **2001**, *12*, 1291–1295. [CrossRef]
111. Sweet, S.J.; McCarthy, S.; Steingart, R.; Callahan, T. Hemolytic Reactions Mechanically Induced by Kinked Hemodialysis Lines. *Am. J. Kidney Dis.* **1996**, *27*, 262–266. [CrossRef]
112. Malinauskas, R.A. Decreased Hemodialysis Circuit Pressures Indicating Postpump Tubing Kinks: A Retrospective Investigation of Hemolysis in Five Patients. *Hemodial. Int.* **2008**, *12*, 383–393. [CrossRef]
113. Patel, N.; Dalal, P.; Panesar, M. *Fellows' Forum* Fellows' Forum in Dialysis Edited by Mark A. Perazella: Dialysis Disequilibrium Syndrome: A Narrative Review. *Semin. Dial.* **2008**, *21*, 493–498. [CrossRef] [PubMed]

114. Raina, R.; Davenport, A.; Warady, B.; Vasistha, P.; Sethi, S.K.; Chakraborty, R.; Khooblall, P.; Agarwal, N.; Vij, M.; Schaefer, F.; et al. Dialysis Disequilibrium Syndrome (DDS) in Pediatric Patients on Dialysis: Systematic Review and Clinical Practice Recommendations. *Pediatr. Nephrol.* **2022**, *37*, 263–274. [CrossRef] [PubMed]
115. Stuard, S.; Ridel, C.; Cioffi, M.; Trost-Rupnik, A.; Gurevich, K.; Bojic, M.; Karibayev, Y.; Mohebbi, N.; Marcinkowski, W.; Kupres, V.; et al. Hemodialysis Procedures for Stable Incident and Prevalent Patients Optimize Hemodynamic Stability, Dialysis Dose, Electrolytes, and Fluid Balance. *J. Clin. Med.* **2024**, *13*, 3211. [CrossRef] [PubMed]
116. Pedreros-Rosales, C.; Jara, A.; Lorca, E.; Mezzano, S.; Pecoits-Filho, R.; Herrera, P. Unveiling the Clinical Benefits of High-Volume Hemodiafiltration: Optimizing the Removal of Medium-Weight Uremic Toxins and Beyond. *Toxins* **2023**, *15*, 531. [CrossRef]
117. Rose, M.; Fischer, F.H.; Liegl, G.; Strippoli, G.F.M.; Hockham, C.; Vernooij, R.W.M.; Barth, C.; Canaud, B.; Covic, A.; Cromm, K.; et al. The CONVINCE Randomized Trial Found Positive Effects on Quality of Life for Patients with Chronic Kidney Disease Treated with Hemodiafiltration. *Kidney Int.* **2024**, *106*, 961–971. [CrossRef]

membranes

MDPI

Article

Quaternized Polysulfone as a Solid Polymer Electrolyte Membrane with High Ionic Conductivity for All-Solid-State Zn-Air Batteries

Luis Javier Salazar-Gastélum [1], Alejandro Arredondo-Espínola [2], Sergio Pérez-Sicairos [3], Lorena Álvarez-Contreras [4], Noé Arjona [2,*] and Minerva Guerra-Balcázar [1,*]

1 Facultad de Ingeniería, División de Investigación y Posgrado, Universidad Autónoma de Querétaro, Querétaro 76010, Santiago de Querétaro, Mexico; luis.javier.salazar@uaq.mx
2 Centro de Investigación y Desarrollo Tecnológico en Electroquímica S. C., Sanfandila 76703, Pedro Escobedo, Mexico; jarredondo@cideteq.mx
3 Centro de Graduados e Investigación en Química, Instituto Tecnológico de Tijuana, Tecnológico Nacional de México, Blvd. Alberto Limón Padilla, S/N Col. Otay Tecnológico, Tijuana 22510, Baja California, Mexico; sperez@tectijuana.mx
4 Centro de Investigación en Materiales Avanzados S. C., Complejo Industrial Chihuahua, Chihuahua 31136, Chihuahua, Mexico; lorena.alvarez@cimav.edu.mx
* Correspondence: wvelazquez@cideteq.mx (N.A.); minerva.guerra@uaq.mx (M.G.-B.)

Abstract: Solid polymer electrolytes (SPEs) are gaining attention as viable alternatives to traditional aqueous electrolytes in zinc–air batteries (ZABs), owing to their enhanced performance and stability. In this study, anion-exchange solid polymer electrolytes (A-SPEs) were synthesized via electrophilic aromatic substitution and substitution reactions. Thin films were prepared using the solvent casting method and characterized using proton nuclear magnetic resonance (^{1}H-NMR), Fourier-transform infrared spectroscopy (FT-IR), and thermogravimetric analysis (TGA). The ion-exchange capacity (IEC), KOH uptake, ionic conductivity, and battery performance were also obtained by varying the degree of functionalization of the A-SPEs (30 and 120%, denoted as PSf30/PSf120, respectively). The IEC analysis revealed that PSf120 exhibited a higher quantity of functional groups, enhancing its hydroxide conductivity, which reached a value of 22.19 mS cm^{-1}. In addition, PSf120 demonstrated a higher power density (70 vs. 50 mW cm^{-2}) and rechargeability than benchmarked Fumapem FAA-3-50 A-SPE. Postmortem analysis further confirmed the lower formation of ZnO for PSf120, indicating the improved stability and reduced passivation of the zinc electrode. Therefore, this type of A-SPE could improve the performance and rechargeability of all-solid-state ZABs.

Keywords: Zn-air batteries; separator; solid polymer electrolyte; all-solid-state

1. Introduction

The need for renewable energy sources such as solar and wind energy is critical because of the rapid development of the global economy [1,2]. Efficient utilization of these clean energy sources has driven significant interest in various electrochemical energy storage and conversion technologies designed to meet diverse energy needs [3]. Among these technologies, zinc–air batteries (ZABs) have emerged as a particularly promising option in the field of energy storage owing to their high theoretical energy density, low cost, and inherent safety [4,5].

The electrolyte, which has traditionally been in liquid form (6 M KOH), is a critical component for efficient and safe performance of Zn–air batteries. However, liquid electrolytes present significant challenges such as volatility, leakage, and limited long-term stability [6]. These issues compromise the safety and environmental sustainability of batteries and limit their application in portable devices. To address these challenges, transition to solid-state electrolytes has gained attention as a viable alternative [7]. Solid-state electrolytes not only eliminate issues associated with liquid electrolyte leakage and evaporation but also enhance the overall safety and environmental sustainability of batteries [8,9]. Moreover, solid-state electrolytes can function as both ionic conductors and physical separators, preventing internal short circuits and simplifying the design and fabrication of batteries [10].

Anion-exchange membranes (AEMs) have been extensively explored for diverse electrochemical applications because of their versatility. For instance, AEMs based on quaternary ammonium-functionalized polymers are pivotal in fuel cells because of their hydroxide ion conductivity under alkaline conditions [11]. In electrodialysis desalination, tailored AEMs enable efficient ion separation [12], whereas composite AEMs with enhanced mechanical strength are critical for water electrolyzers [13]. A recent study by Wang et al. demonstrated nanostructured AEMs with hierarchical porosity for high-performance energy devices, emphasizing the role of the membrane architecture in ion transport optimization [14]. Similarly, hybrid AEMs integrated with catalytic layers have improved oxygen reduction kinetics in aluminum–air batteries [15], underscoring the importance of material design across applications. Recent advances in anion-exchange membranes (AEMs) have highlighted the effectiveness of cross-linking strategies in enhancing performance. Wei et al. [16] reported a GA cross-linked AEM with a conductivity of 0.024 S cm^{-1} and low swelling, reinforcing the potential of chemical cross-linking to optimize both the conductivity and mechanical integrity in solid electrolytes. These studies highlight the critical role of tailored membrane designs in advancing high-performance AEMs for electrochemical applications.

In this context, A-SPEs have emerged as a promising solution [17]. Among the various available polymer materials, quaternized polysulfone (QPSf) is a particularly interesting candidate for use in solid polymer electrolytes. Polysulfones (PSf) are known for their excellent thermal and chemical stability, as well as their good mechanical properties [18]. The functionalization of PSf through quaternization introduces anionic groups that can facilitate ion conduction, making it suitable for application in anion-exchange batteries. Furthermore, these solid electrolytes can prevent the migration of zinc ions and formation of carbonates during battery operation, thus improving battery durability. However, the water-retaining properties of traditional membranes are relatively poor and directly limit the discharge of flexible, rechargeable ZABs [19,20].

Despite these advantages, the use of QPSfs as solid polymer electrolytes in zinc–air batteries has rarely been studied. Recent studies have predominantly focused on other types of polymers and electrolyte systems, leaving a significant gap in the literature regarding the potential of QPSfs for this application. These types of polymers have been used in other electrochemical technologies, such as fuel cells, electrodialysis desalination, water electrolyzers, and aluminum–air batteries [21–25]. Regarding the few studies found for ZABs, Zhang et al. reported the development of a solid-state polyelectrolyte for use in flexible Zn–air batteries. The solid electrolyte was synthesized through chemical functionalization of cellulose fibers and graphene oxide, followed by layer-by-layer filtration, crosslinking, and ion exchange processes. This solid electrolyte was evaluated in a rechargeable Zn–air battery and compared to the anion exchange membrane A201 from Tokuyama Corporation [20]. The nanocomposite solid polyelectrolyte exhibited superior performance in terms of the current density compared to the A201 membrane [26]. Similarly, Xu et al. developed a system using cellulose nanofiber matrices confined with ionic liquids to create flexible electrolytes with high ionic conductivity. This system achieved high hydroxide ion conductivity owing to the high hydrophilicity of the cellulose nanofibers and ionic liquids, which provided a solid polyelectrolyte with numerous conductive channels [27]. Alkaline membranes play a dual role as charge carriers and electrode separators in electrochemical systems, critically influencing the performance and durability. Their importance has surged in energy technologies, particularly in fuel cells, for which extensive research has been conducted. However, their widespread adoption remains limited by two key challenges: conductivity and stability. Strategies such as engineering microphase-separated structures (to accelerate ion transport) and optimizing micropore networks (to reduce resistance) have proven to be effective. These innovations have driven conductivity from <10 mS cm^{-1} to >100 mS cm^{-1}, marking transformative progress in the field [28].

The introduction of QPSFs into the design of solid polymer electrolytes could be beneficial for zinc–air batteries by providing safer, more efficient, and more durable all-solid-state ZABs. This study aims to explore the performance of QPSF-based solid polymer electrolytes in all-solid-state ZABs. For this purpose, different degrees of functionalization of the QPSFs were investigated to evaluate their impact on ionic conductivity and electrochemical stability. The battery performance (power density and charge/discharge behavior) of the resulting A-SPEs was compared with that of the benchmark Fumapem FAA-3-50 alkaline membrane (FuelCellStore ®), selected because of its status as one of the most extensively studied anion-exchange membranes for alkaline systems and its structural similarity to QPSF-based electrolytes (both utilize quaternary ammonium cations as ion-conductive groups).

2. Materials and Methods

2.1. Reagents

Polysulfone (Mw = 35,000), chloroform (≥99.5%), triethylamine (≥99%), *N*-methyl-2-pyrrrolidone (≥99%), paraformaldehyde (95.0%), chlorotrimethylsilane (≥98.0%), tin chloride (98.0%), ethyl acetate (≥99.5%), diethylether (≥99.7), dimethylsofoxide-d6 (99.9% D), chloroform-d1 (99.8% D) were purchased from Sigma-Aldrich (Saint Louis, MO, USA). Potassium hydroxide (85.0%) was purchased from Merck (Darmstadt, Germany). Hydrochloric acid, ethanol (95.0%) was purchased from J.T. Baker (Mexico City, Mexico). Bromothymol blue (95.0%) was purchased from Meyer (Mexico City, Mexico). All chemicals were used without further purification.

2.2. Chloromethylation and Quaternization of Polysulfone

To obtain chloromethylated polysulfone (PSf−Cl), an electrophilic substance was used following a previously reported methodology to avoid the use of chloromethyl methyl ether, which is dangerous for human health [29]. In a round-bottom flask, PSf pellets (2 g) were dissolved in chloroform under vigorous magnetic stirring. Following the addition of 44.96 mmol of paraformaldehyde, 0.44 mmol of tin chloride (IV) was added to the suspension dropwise. Finally, 44.96 mmol of chlorotrimethylsilane was added dropwise using the procedure described by Avram et al. [30]. Different functionalization degrees were achieved during the chloromethylation reaction by adjusting the reaction duration. After the reaction was complete, the solution was stirred and precipitated using cold methanol. After filtration and multiple methanol washes, the precipitated PSf functionalized with CH_2Cl groups (PSf−Cl) was dried in a vacuum oven at 50 °C. The quaternization reaction involved using those precursors with different degrees of functionalization; 1.5 g of PSf−Cl was dissolved into N-methyl-pyrrolidone (NMP) using magnetic stirring in a round-bottom flask, followed by the addition of 1 mL of triethylamine (TEA), then the mixture was heated up for 24 h at 80 °C to replace Cl with amine groups. The mixture was then precipitated with diethyl ether and washed with 99.5% ethyl acetate. Brown powder (QPSf) was obtained and kept at 60 °C in a vacuum oven [31].

2.3. Preparation of Alkaline Solid Polymer Electrolyte

A-SPEs were prepared using the solvent casting method, as described below. A certain amount of polymer powder was dissolved in NMP to obtain a polymer solution with a concentration of 16 wt. %. The polymer solution was cast onto a glass petri dish. The solvent was removed at 50 °C under ambient pressure for 48 h, and an A-SPE was formed, which was subsequently peeled off from the petri dish via immersion in DI water. A-SPE was washed with DI water several times to remove residual NMP. Subsequently, A-SPE in Cl^- form was obtained and stored in DI water before use. To obtain the OH^- form, A-SPE in the Cl^- form was immersed in 1 M KOH at room temperature for 48 h.

2.4. Characterization

^{1}H NMR spectra of the synthesized polymers were recorded at 400 MHz with a Bruker Advance III HD spectrometer at 30 °C. Chemical shifts were referenced to the peak of tetramethyl silane (TMS). The polymers were dissolved in chloroform-d for chloromethylated polysulfone and DMSO-d6 for quaternized polysulfone. FT-IR spectra of the A-SPEs were recorded using a Perkin Elmer Spectrum Gx with attenuated reflectance (ATR) in a scanning range of 600–4000 cm^{-1}, with a resolution of 2 cm^{-1} and 16 scans. Thermogravimetric analysis (TGA) was performed under nitrogen atmosphere using a TQ500 instrument. Thermogravimetric curves were recorded in the temperature range of 30 °C to 800 °C at a heating rate of 20 °C min^{-1}. To study membrane morphology, SEM images were acquired using a microscope (JEOL, model JSM-6610LV, Akishima, Japan) operated at an accelerating voltage of 15 kV.

The KOH uptake for the A-SPEs in the OH^- form was calculated based on the changes in the mass of the swollen and dried states. Each A-SPE was kept in a 1 M KOH aqueous solution for 24 h, excess of KOH solution was removed with tissue paper, and the weight of the wet membrane was recorded. The A-SPEs were dried in an oven at 60 °C until a constant weight was obtained. The KOU was calculated using Equation (1):

$$KOH\ uptake(\%) = \frac{m_{wet} - m_{dry}}{m_{dry}} * 100 \quad (1)$$

where m_{wet} and m_{dry} are the weights of the swollen and dried membranes, respectively.

The ion exchange capacity (IEC) was measured using an acid–base titration method as previously reported [32]. The A-SPEs were cut into pieces of 2 cm × 2 cm, then they were immersed in 15 mL of 0.01 M HCl and kept for 24 h to ensure the consumption of OH^- with H^+. Then, the remaining acidic solution was transferred to a flask and titrated with 0.01 M KOH using bromothymol blue (1%) as an indicator. The IEC was recorded in meq of H^+ g^{-1} and was determined based on the difference in the number of milliequivalents of HCl solution before (meq of HCl)$_0$ and after (meq of HCl)$_f$ the membrane being neutralized, according to Equation (2):

$$IEC\left(mmol\ g^{-1}\right) = \frac{(V_b - V_s) * C_{HCl} * 1000}{m_{dry}} \quad (2)$$

where V_b and V_s are the consumed volumes (L) of the KOH solution for the blank sample and A-SPE sample, respectively, C_{HCl} is the concentration (mol L^{-1}) of the HCl solution, and m_{dry} is the mass of the dry A-SPE.

The ionic conductivities (σ) of the A-SPEs were measured using a four-probe method with a VSP-3 Potentiostat/Galvanostat over a frequency range of 0.1 to 500 kHz in a Teflon conductivity cell (BT-112, Scribner, Southern Pines, NC, USA). A-SPE samples in the OH^- form were cut into 1 × 2.5 cm rectangular strips. The SPE strip was placed between the inner platinum wires to ensure contact with both the outer platinum wires and platinum mesh. A clamp was then applied to secure the membrane surfaces and was tightened using screws. The conductivity cell was then placed in a water bath at room temperature (25 °C). The resistance of the A-SPE was determined using Nyquist plots. The hydroxide conductivity was calculated using Equation (3):

$$\sigma\left(mS\ cm^{-1}\right) = \frac{L}{RWT} \tag{3}$$

where σ represents the hydroxide conductivity of the SPE, L is the distance between the inner platinum wires (5.5 cm), R is the resistance of the SPE (obtained from the high-frequency intercept on the real impedance axis), and W and T are the width and thickness of the SPE in its wet form, respectively [33,34].

2.5. Zn–Air Battery Test

For the battery tests, a design consisting of two acrylic slotted plates that were mechanically sealed with screws was used. This design allowed tests to be conducted using small amounts of material, and the contact area between the cathode and anode was 0.06 cm^2. A Sigracet 33 B (FuelCellStore ®) cathode was used as the current collector and coated with $Pt\text{-}IrO_2/C$. The metallic charge used in the air electrode was 1 mg cm^{-2}, and a piece of polished metallic zinc (99.9% purity) was used as the zinc electrode. In all cases, the membranes were activated in 1 M KOH for 24 h and then immersed in 6 M KOH before the cell assembly. Electrochemical Impedance Spectroscopy from 100 KHz to 100 Hz was used to evaluate the correct assembly of the ZAB. Linear voltammetry was used to evaluate the battery performance of different A-SPE materials. Charge and discharge cycles were evaluated at −0.5 mA cm^{-2} at discharge and 0.5 mA cm^{-2} at charge. Scheme 1 illustrates the battery assembly process.

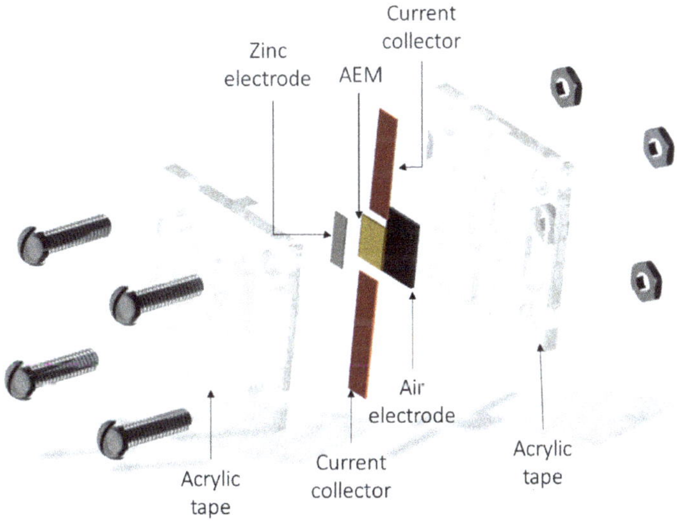

Scheme 1. Scheme of the all-solid-state Zn–air battery operated with an alkaline solid polymer electrolyte.

3. Results and Discussion

Figure 1 shows the ^{1}H NMR spectra of PSf (a) and PSf−Cl (b). The PSf spectrum showed peaks corresponding to isopropyl groups of the main chain and phenyl groups at 1.69 ppm and 7–8 ppm, respectively, where hydrogens close to the sulfone group appeared at 7.9 ppm. The ^{1}H NMR spectrum of PSf−Cl shows a new peak at 4.56 ppm (H_f), corresponding to hydrogens arising from the incorporation of the chloromethyl group, which add a new chemical environment to the aromatic hydrogen in the ortho position of the isopropyl group. The amount of CH_2Cl incorporated into the backbone of PSf was calculated from the ^{1}H NMR spectra of PSf−Cl according to Pérez-Prior [35] (Equation (4)):

$$DC = \frac{\int \frac{H_f}{2}}{\int \frac{H_e}{6}} * 100 \tag{4}$$

where $\int H_f$ and $\int H_e$ are the integral area of the signal at 4.56 ppm and 1.69 ppm, respectively. The DC value indicates the degree of chloromethylation.

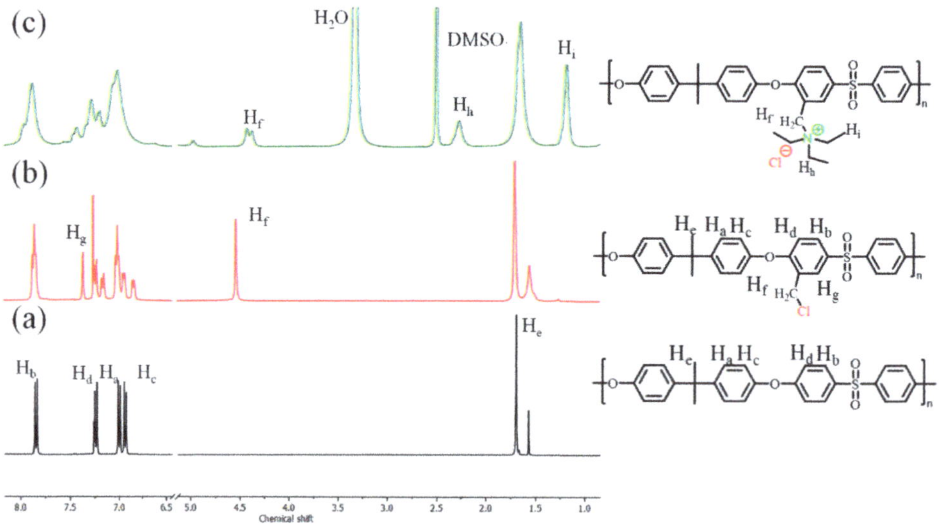

Figure 1. ^{1}H-NMR spectra of (**a**) PSf, (**b**) PSf−Cl, and (**c**) quaternized polysulfone polymers.

Figure 1c shows the ^{1}H NMR spectrum of quaternized PSf, where the signal corresponding to the methylene of the chloromethyl group shifts to lower ppm values. It was also observed that the signal split into two peaks because the reaction did not occur completely, indicating the presence of residual chloromethyl groups in the quaternized polymer. Two quaternized polymers were prepared: PSf30, obtained from PSf−Cl with a degree of functionalization (DC) of 30, and PSf120, obtained from PSf−Cl with a DC of 120%. It is important to point out that at 3.2 ppm, a new signal appeared that corresponds to methylene in triethylamine, which was linked due to the substitution reaction. In addition, the signal of the methyl groups of the amino compound appeared at a chemical shift of 1 ppm, and the spectrum of the quaternized polymer was obtained in DMSO-d6. This explains the peak at 3.4 ppm, attributed to the solvent and the peak at 2.54 ppm, related to the residual DMSO [36,37].

Figure 2 shows the FT-IR spectra of PSf and PSf−Cl, where the peak at 2968 cm^{-1} corresponds to the aliphatic stretching of C–H from the isopropyl groups of the backbone of PSf, and the peak at 1150 cm^{-1} corresponds to the symmetric stretching of the sulfone group.

Asymmetric stretching of the sulfone group was observed at 1320 cm^{-1} and 1296 cm^{-1} [38]. Finally, other signals were observed at 1238 and 1012 cm^{-1}, which were attributed to the asymmetric stretching of the ether species (C–O–C) [39,40]. The change in the shape of the band located at 660 cm^{-1} for the PSf−Cl spectra compared to PSf was attributed to C–Cl, as reported by Yang et al [41]. This is consistent with the ^{1}H NMR results, indicating that PSf was modified.

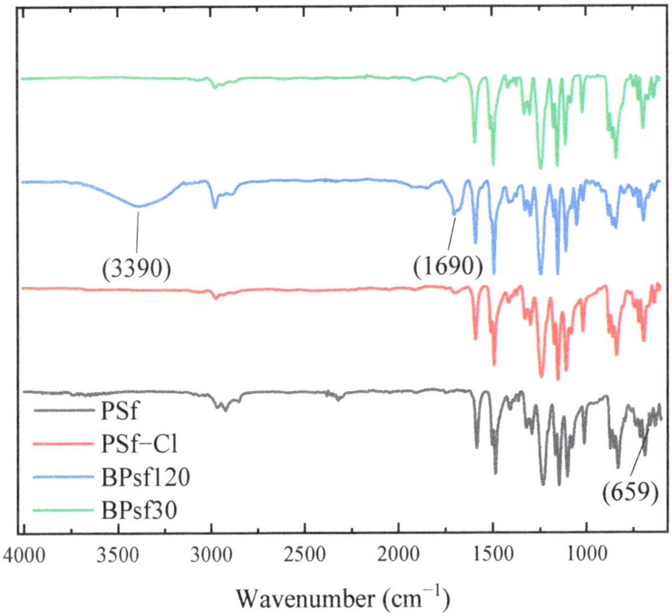

Figure 2. FT-IR spectra for PSf, PSf−Cl, and quaternized polymers.

For PSf120 and PSf30, the presence of ammonium quaternary cations was confirmed using FT-IR based on the appearance of three bands: the first two bands, one at 3390 cm^{-1}, attributed to the absorption of water from the atmosphere, corresponding to the hydrophilicity of the polymer; one at 1690 cm^{-1} for the moiety of the ammonium cations; and a third at 1010 cm^{-1} due to C–N [42].

TGA was performed to study the thermal stability and confirm the modification of pristine PSf (Figure 3). The figure shows that PSf shows only one stage of decomposition at 550 °C, which is due to the decomposition of the skeleton chain of the polymer. On the other hand, PSf−Cl shows two stages of decomposition: (i) at 350 °C, due to the loss of mass of the pedant groups of CH_2Cl for the chloromethylation reaction, and (ii) at 505 °C, due to decomposition of the main skeleton of PSf. This value shifted to lower temperatures as the pristine polymer became thermally unstable. For the ammonium-functionalized PSf, the first stage of decomposition occurred at 198 °C for PSf120 and 208 °C for PSf30. This behavior is usually observed for ammonium cations because the temperature of decomposition is reported to be between 180 and 220 °C [35,43]. Both polymers became unstable compared to pristine PSf. Furthermore, the degree of functionalization corresponds to the first stage of decomposition of ammonium-functionalized A-SPEs because PSf120 exhibits a higher mass loss than PSf30.

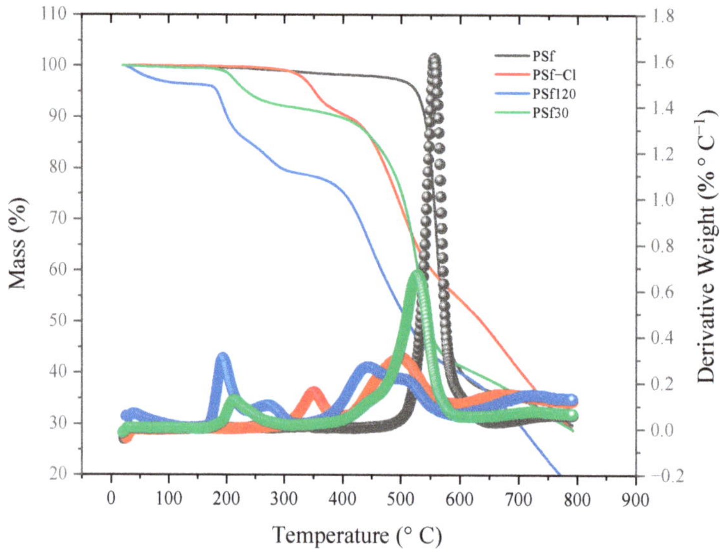

Figure 3. TG analysis of PSf, PSf–Cl, and quaternized polymers under a nitrogen atmosphere at 20 °C min^{-1}.

Table 1 lists the values obtained for IEC and KOH uptake, where the degree of functionalization plays an important role. On one hand, the IEC should be higher when the degree of functionalization increases; that is, when the availability of ammonium cations is higher. This agrees with the IR results, where it can be observed that the band at 1690 cm^{-1} is higher for PSf120. However, the KOH uptake value was higher for PSf120, which was also expected because its OH absorption band was higher. Under similar IEC experimental conditions, A-SPE PSf120 exhibited values similar to those of commercial materials. For hydroxide conductivity values, A-SPE PSf30 showed similar performance to that of Fumapem, indicating that both A-SPEs show a similar mechanism for hydroxide ion mobility in solid polyelectrolytes.

Table 1. Results of IEC, KOH uptake, and hydroxide conductivity for the A-SPEs.

A-SPE	Functionalization Degree in Cl form (%)	IEC (mmol/g)	KOH Uptake (%) Including Water Uptake	σ (mS cm^{-1})
PSf120	122	1.7124	56.05	22.19
PSf30	30	1.3393	6.17	8.82
Fumapem	NA	2.0200	37.56	8.71

Although commercial A-SPE exhibits a higher IEC value in terms of ionic conductivity, A-SPE PSf30 shows better performance than commercial A-SPE, which can be attributed to the better dispersion of ammonium moieties on the surface. Moreover, A-SPE PSf120 was the separator with the highest σ value, which may be due to the large quantity of ammonium cation groups compared to PSf30. On the other hand, comparing PSf12 and commercial A-SPE, PSf120 showed a greater capability to attach KOH molecules than commercial A-SPE, indicating that A-SPE is more hydrophilic. These results reveal that this type of polysulfone A-SPE is a promising candidate for use as a membrane and separator in Zn–air batteries.

Optical microscopy images were obtained to analyze the structure of the A-SPEs after activation in 6M KOH (Figure 4). The results showed that both Fumapem and PSF120 exhibited microporous structural characteristics of the AEMs. However, PSF130 displayed a significant reduction in porosity, and its surface primarily featured morphological defects.

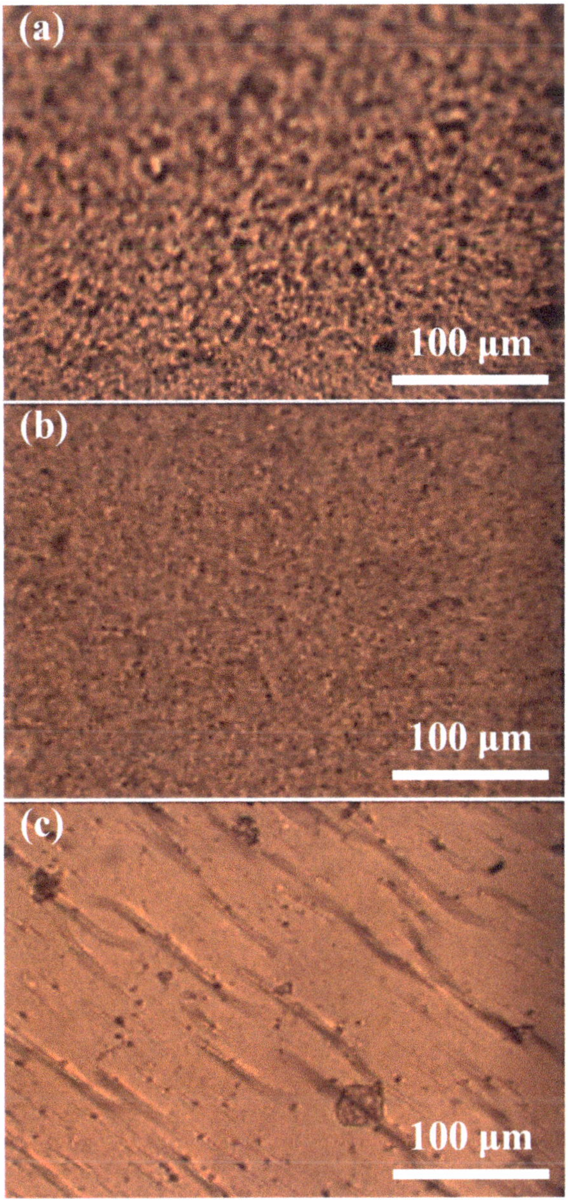

Figure 4. Optical micrographs of the A-SPEs after activation in KOH: (**a**) Fumapem, (**b**) PSf120, and (**c**) PSf30.

Nyquist plots (Figure 5a) revealed an increase in internal resistance in the ZABs when using the developed A-SPEs, primarily attributed to challenges in ion transverse transport across the membrane. Despite minor shifts toward higher resistances for the A-SPEs developed in this study, the measured battery resistance values indicated favorable interfacial interactions between the electrolyte and electrodes, which is a critical factor for functional device performance. Furthermore, the stable resistance values suggest successful cell assembly, achieved without precision tools, such as a torque wrench, highlighting the practicality of the proposed fabrication method. Notably, PSf120, with its high degree of functionalization, demonstrated a resistance value comparable to that of commercial A-SPE, underscoring its potential for scalable applications. This was further confirmed through cathodic linear polarization tests (Figure 5b), which showed poor performance for PSf30 and issues related to mixed control zones (between the activation and ohmic regions) for PSf120. Due to the higher ionic conductivity and KOH uptake values of PSf120, a current density of 74.58 mA cm^{-2} was achieved, representing a 42.43% increase compared to the commercial A-SPE.

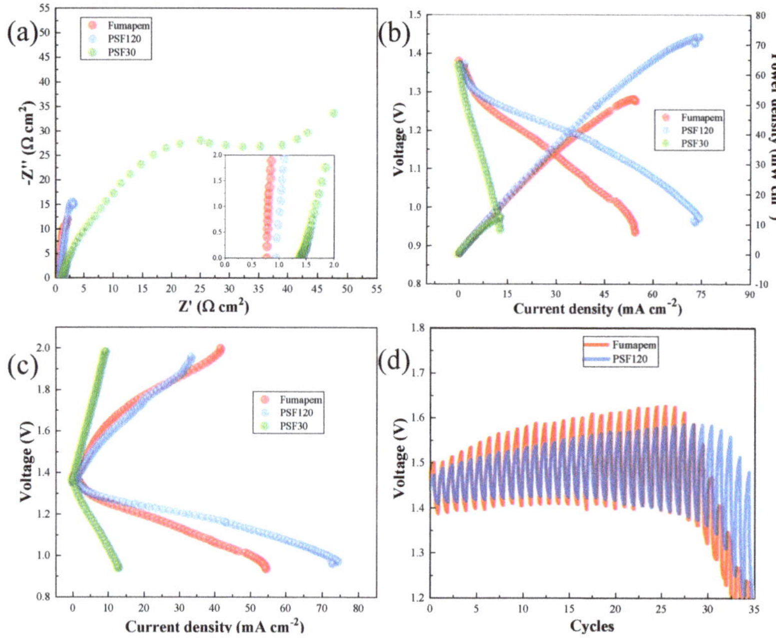

Figure 5. ZAB tests. (**a**) Nyquist plots for A-SPE, (**b**) curves of power density, (**c**) lineal polarizations, and (**d**) cycles in charge/discharge at 0.5 mA cm^{-2}.

Anodic linear polarization tests (Figure 5c) demonstrated the strong performance of PSF120, achieving a current density of 10 mA cm^{-2} at an overpotential of 205.5 mV and 248.5 mV for PSF120 and FUMAPEM, respectively, highlighting its excellent capability for zinc nucleation and redeposition processes at lower overpotentials. Charge/discharge cycles (Figure 5d) showed initial round-trip energy efficiencies of 95.24% and 91.39% for PSf120 and the commercial A-SPE, respectively. By the end of the test, both A-SPEs showed a decrease in round-trip efficiency, reaching 87.38% for PSf120 and 85.72% for the commercial A-SPE. This, combined with a 14% increase in cyclability, demonstrates that PSf120 exhibits greater stability and superior performance compared to the commercial Fumapem A-SPE.

The postmortem analysis of the air electrodes (Figure 6) revealed that both commercial A-SPE (Figure 6a) and PSf120 (Figure 6b) exhibit the formation of micrometric deposits. Through EDS analysis and elemental mapping of the commercial A-SPE (Figure 6c), it was confirmed that the composition of these deposits was associated with carbonate formation (due to the interaction of CO_2 with adsorbed KOH) and zinc migration [44]. Although similar effects were observed for PSf120, the migration of zinc was less pronounced, likely due to the smaller pore size of this A-SPE. Additionally, a greater formation of carbonates was identified, which was attributed to higher electrolyte absorption [45].

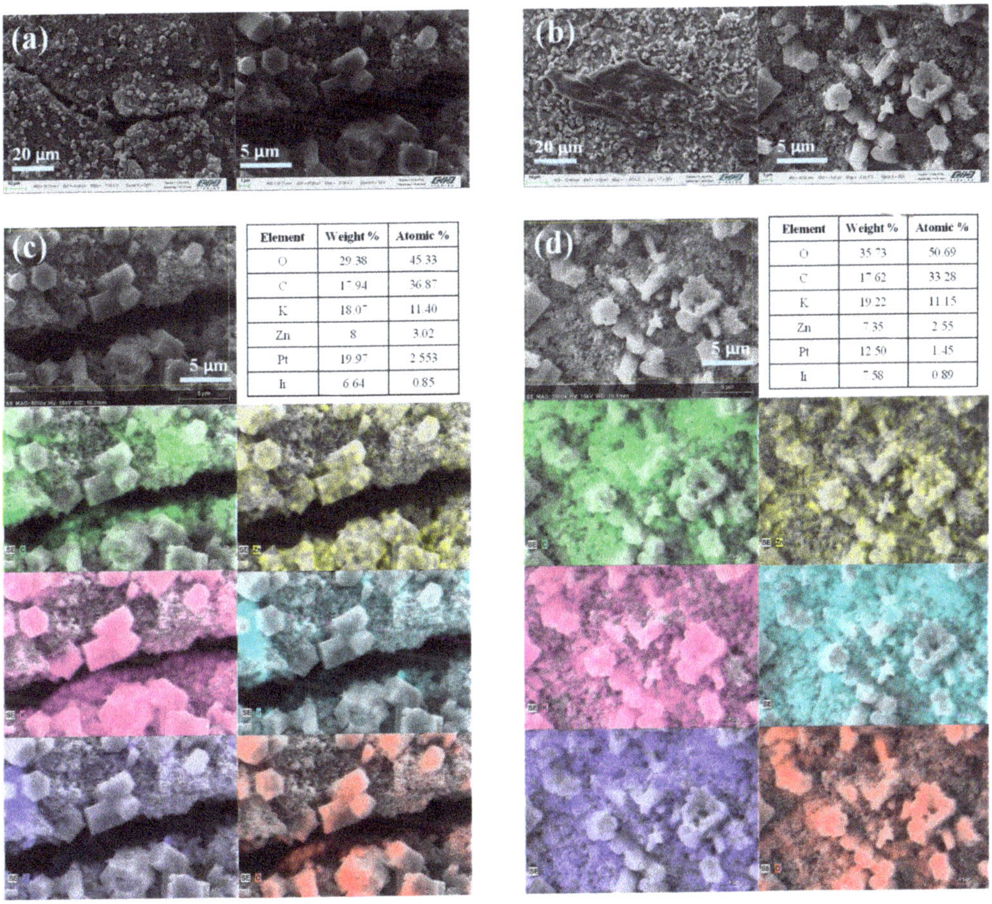

Element	Weight %	Atomic %
O	29.38	45.33
C	17.94	36.87
K	18.07	11.40
Zn	8	3.02
Pt	19.97	2.553
Ir	6.64	0.85

Element	Weight %	Atomic %
O	35.73	50.69
C	17.62	33.28
K	19.22	11.15
Zn	7.35	2.55
Pt	12.50	1.45
Ir	7.58	0.89

Figure 6. Postmortem analysis of cathodes. SEM micrographs of ZABs employed: (**a**) Fumapem and (**b**) PSF120. EDS and mapping analysis for ZAB with (**c**) Fumapem and (**d**) PSF120, each color represents a different element: C (green), Zn (yellow), K (pink), Ir (cyan), Pt (blue) and Cl (red).

As a result of using porous separators, dendritic growth was observed on the zinc electrodes in both commercial A-SPE (Figure 7a) and PSf120 (Figure 7b) [46]. Owing to the higher cyclability of PSF120, it exhibited larger dendrites, which did not pose a significant risk of sudden battery failure. Elemental analysis of the commercial A-SPE (Figure 7c) revealed the formation of carbonates and ZnO, with zinc electrode passivation identified as the primary cause of the system failure. In contrast, the use of PSf120 (Figure 7d) resulted in a higher presence of metallic zinc on the electrode surface and the formation of carbonates. This suggests that PSf120 enhances the electrochemical reversibility of zinc compared to commercial A-SPE, likely due to a higher distribution of atomic oxygen, indicating that this oxygen is associated with ZnO [47].

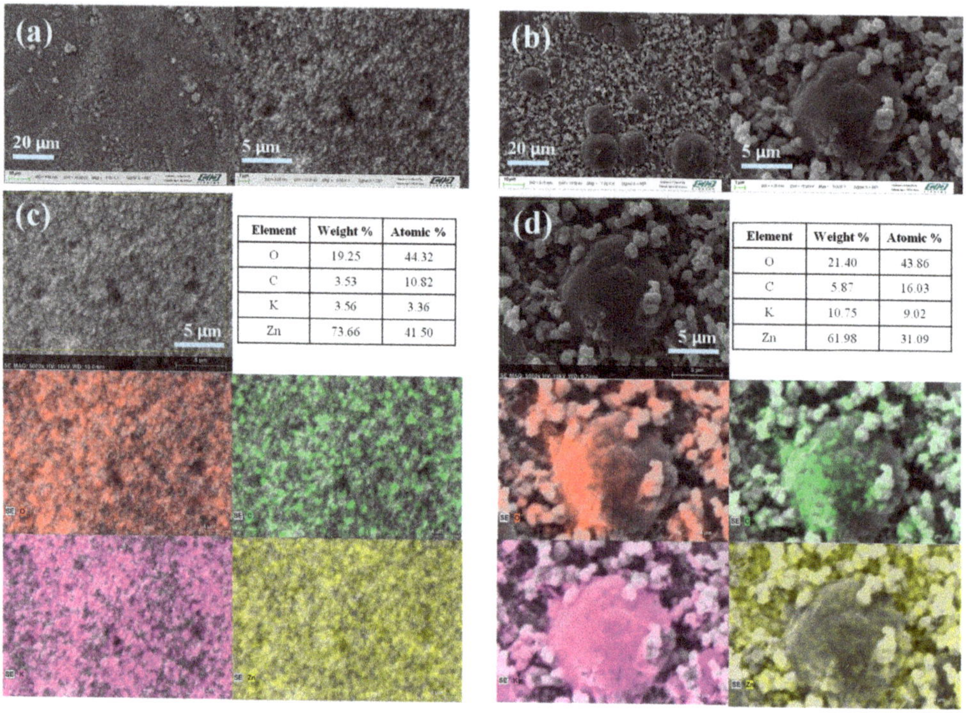

Element	Weight %	Atomic %
O	19.25	44.32
C	3.53	10.82
K	3.56	3.36
Zn	73.66	41.50

Element	Weight %	Atomic %
O	21.40	43.86
C	5.87	16.03
K	10.75	9.02
Zn	61.98	31.09

Figure 7. Postmortem analysis of Zinc electrodes. SEM micrographs of ZABs employed: (**a**) Fumapem and (**b**) PSF120. EDS and mapping analysis for ZAB with (**c**) Fumapem and (**d**) PSF120, each color represents a different element: O (red), C (green), K (pink) and Zn (yellow).

4. Conclusions

Anion-exchange solid polymer electrolytes were obtained by modifying polysulfone with ammonium cations, using triethylamine as a quaternizing agent. This was achieved using chloromethylation to avoid hazardous conditions. 1H-NMR confirmed the presence of an ammonium moiety in the main polysulfone skeleton. Two A-SPEs with different numbers of cations were obtained, as confirmed by the IEC values, which were similar to those of a commercial anion exchange membrane. In contrast, the KOU value for PSf120 was higher than that of PSf30 and the commercial membrane, indicating that PSf120 exhibits better KOH retention. The developed PSF120 A-SPE demonstrated superior performance and stability compared to commercial A-SPEs in zinc–air batteries (ZABs). Despite the increased internal resistance challenges due to ion transverse transport, the high functionalization of PSF120 enables impedance values comparable to those of commercial A-SPEs. Linear polarization tests demonstrated that PSF120 A-SPE reach higher values of current density (74.22 mA cm^{-2}); this was due to PSF120 exhibiting better hydroxide conductivity compared to the A-SPE commercial, demonstrating that hydroxide conductivity could be a limitation for the ZAB operation. In summary, postmortem analysis confirmed the presence of dendritic structures on the zinc electrode; nonetheless, PSf120 A-SPE exhibited an improvement in the reversibility of zinc owing to the higher presence of metallic zinc on the electrode surface. This may be due to the smooth surface of A-SPE PSf120, which does not allow Zn^{2+} to attach to the surface of A-SPE. This is in agreement with the EDS analysis of the air electrode, suggesting that the commercial A-SPE permeates a higher quantity of Zn ions through the separator. Although the formation of carbonates remains a shared challenge, PSF120's higher electrolyte absorption contributes to its enhanced overall performance. These findings establish PSF120 as a promising material for advancing ZAB technology, offering better efficiency, stability, and reversibility than commercial alternatives.

Author Contributions: Conceptualization, L.J.S.-G., M.G.-B. and N.A.; methodology, N.A., S.P.-S., A.A.-E. and L.J.S.-G.; software, L.J.S.-G. and A.A.-E.; validation, L.Á.-C., M.G.-B. and N.A.; formal analysis, L.J.S.-G., A.A.-E. and N.A.; investigation, L.J.S.-G. and A.A.-E.; resources, L.Á.-C., M.G.-B. and N.A.; data curation, M.G.-B. and N.A.; writing—original draft preparation, L.J.S.-G.; writing—review and editing, L.J.S.-G., M.G.-B. and N.A.; visualization, L.J.S.-G., N.A. and M.G.-B.; supervision, N.A. and M.G.-B.; project administration, N.A. and M.G.-B.; funding acquisition, L.Á.-C., M.G.-B. and N.A. All authors have read and agreed to the published version of the manuscript.

Funding: This research was funded by The Mexican Council of Humanities, Science and Technology (CONAHCyT) under the project "Ciencia de Frontera", grant number 39569.

Data Availability Statement: The raw data supporting the conclusions of this article will be made available by the authors on request.

Acknowledgments: L. J. Salazar-Gastélum and A. Arredondo-Espínola acknowledge the support of the Mexican Council of Humanities, Science, and Technology (CONAHCyT) for their postdoctoral and postgraduate scholarships, respectively.

Conflicts of Interest: The authors declare no conflicts of interest.

Abbreviations

The following abbreviations are used in this manuscript:

A-SPE	Anion Exchange Solid Polymer Electrolyte
PSf	Polysulfone
QPSf	Quaternized Polysulfone
PSf−Cl	Chlorometylated Polysulfone
NMP	N-methyl-pyrrolidone
TEA	Triethylamine
DI	Deionized Water
^{1}H NMR	Proton Nuclear Magnetic Resonance Spectroscopy
TMS	Tetramethylsilane
FT-IR	Fourier Transform–Infrared Spectroscopy
ATR	Attenuated Reflectance
TGA	Thermogravimetric Analysis
SEM	Scanning Electronic Microscope
KOU	Potassium Hydroxide Uptake
IEC	Ion Exchange Capacity
DC	Degree of Chloromethylation
DMSO	Dimethylsulfoxide

References

1. Abdalla, A.N.; Nazir, M.S.; Tao, H.; Cao, S.; Ji, R.; Jiang, M.; Yao, L. Integration of Energy Storage System and Renewable Energy Sources Based on Artificial Intelligence: An Overview. *J. Energy Storage* **2021**, *40*, 102811. [CrossRef]
2. Farghali, M.; Osman, A.I.; Chen, Z.; Abdelhaleem, A.; Ihara, I.; Mohamed, I.M.A.; Yap, P.-S.; Rooney, D.W. Social, Environmental, and Economic Consequences of Integrating Renewable Energies in the Electricity Sector: A Review. *Environ. Chem. Lett.* **2023**, *21*, 1381–1418. [CrossRef]
3. Amir, M.; Deshmukh, R.G.; Khalid, H.M.; Said, Z.; Raza, A.; Muyeen, S.M.; Nizami, A.-S.; Elavarasan, R.M.; Saidur, R.; Sopian, K. Energy Storage Technologies: An Integrated Survey of Developments, Global Economical/Environmental Effects, Optimal Scheduling Model, and Sustainable Adaption Policies. *J. Energy Storage* **2023**, *72*, 108694. [CrossRef]
4. Iqbal, A.; El-Kadri, O.M.; Hamdan, N.M. Insights into Rechargeable Zn-Air Batteries for Future Advancements in Energy Storing Technology. *J. Energy Storage* **2023**, *62*, 106926. [CrossRef]

5. Hosseini, S.; Masoudi Soltani, S.; Li, Y.-Y. Current Status and Technical Challenges of Electrolytes in Zinc–Air Batteries: An in-Depth Review. *Chem. Eng. J.* **2021**, *408*, 127241. [CrossRef]
6. Fu, J.; Cano, Z.P.; Park, M.G.; Yu, A.; Fowler, M.; Chen, Z. Electrically Rechargeable Zinc–Air Batteries: Progress, Challenges, and Perspectives. *Adv. Mater.* **2017**, *29*, 1604685. [CrossRef]
7. Sankaralingam, R.K.; Seshadri, S.; Sunarso, J.; Bhatt, A.I.; Kapoor, A. Effect of Electrolyte Parameters on the Discharge Characteristics of Planar Zinc-Air Flow Battery with Polymer Gel Electrolyte as Separator. *Energy Storage* **2022**, *4*, e304. [CrossRef]
8. Wang, Z.; Li, H.; Tang, Z.; Liu, Z.; Ruan, Z.; Ma, L.; Yang, Q.; Wang, D.; Zhi, C. Hydrogel Electrolytes for Flexible Aqueous Energy Storage Devices. *Adv. Funct. Mater.* **2018**, *28*, 1804560. [CrossRef]
9. Wang, M.; Xu, N.; Fu, J.; Liu, Y.; Qiao, J. High-Performance Binary Cross-Linked Alkaline Anion Polymer Electrolyte Membranes for All-Solid-State Supercapacitors and Flexible Rechargeable Zinc–Air Batteries. *J. Mater. Chem. A* **2019**, *7*, 11257–11264. [CrossRef]
10. Huang, S.; Zhu, J.; Tian, J.; Niu, Z. Recent Progress in the Electrolytes of Aqueous Zinc-Ion Batteries. *Chem.–A Eur. J.* **2019**, *25*, 14480–14494. [CrossRef]
11. Das, G.; Choi, J.-H.; Nguyen, P.K.T.; Kim, D.-J.; Yoon, Y.S. Anion Exchange Membranes for Fuel Cell Application: A Review. *Polymers* **2022**, *14*, 1197. [CrossRef]
12. Dong, H.; German, M.; Tian, L.; SenGupta, A.K. Multifunctional Ion Exchange Pretreatment Driven by Carbon Dioxide for Enhancing Reverse Osmosis Recovery during Impaired Water Reuse. *Desalination* **2020**, *485*, 114459. [CrossRef]
13. Yang, Y.; Li, P.; Zheng, X.; Sun, W.; Dou, S.X.; Ma, T.; Pan, H. Anion-Exchange Membrane Water Electrolyzers and Fuel Cells. *Chem. Soc. Rev.* **2022**, *51*, 9620–9693. [CrossRef] [PubMed]
14. Wang, Q.; Wei, T.; Peng, Z.; Zhao, Y.; Jannasch, P.; Yang, J. High-Performance Anion Exchange Membranes Based on Poly(Oxindole Benzofuran Dibenzo-18-Crown-6)s Functionalized with Hydroxyl and Quaternary Ammonium Groups for Alkaline Water Electrolysis. *J. Colloid Interface Sci.* **2025**, *686*, 304–317. [CrossRef] [PubMed]
15. Fujiwara, N.; Yao, M.; Siroma, Z.; Senoh, H.; Ioroi, T.; Yasuda, K. Reversible Air Electrodes Integrated with an Anion-Exchange Membrane for Secondary Air Batteries. *J. Power Sources* **2011**, *196*, 808–813. [CrossRef]
16. Wei, Y.; Wang, M.; Xu, N.; Peng, L.; Mao, J.; Gong, Q.; Qiao, J. Alkaline Exchange Polymer Membrane Electrolyte for High Performance of All-Solid-State Electrochemical Devices. *ACS Appl. Mater. Interfaces* **2018**, *10*, 29593–29598. [CrossRef]
17. Fu, J.; Lee, D.U.; Hassan, F.M.; Yang, L.; Bai, Z.; Park, M.G.; Chen, Z. Flexible High-Energy Polymer-Electrolyte-Based Rechargeable Zinc–Air Batteries. *Adv. Mater.* **2015**, *27*, 5617–5622. [CrossRef]
18. Teresa Pérez-Prior, M.; Ureña, N.; Tannenberg, M.; del Río, C.; Levenfeld, B. DABCO-Functionalized Polysulfones as Anion-Exchange Membranes for Fuel Cell Applications: Effect of Crosslinking. *J. Polym. Sci. Part B Polym. Phys.* **2017**, *55*, 1326–1336. [CrossRef]
19. Song, F.; Fu, Y.; Gao, Y.; Li, J.; Qiao, J.; Zhou, X.-D.; Liu, Y. Novel Alkaline Anion-Exchange Membranes Based on Chitosan/Ethenylmethylimidazoliumchloride Polymer with Ethenylpyrrolidone Composites for Low Temperature Polymer Electrolyte Fuel Cells. *Electrochim. Acta* **2015**, *177*, 137–144. [CrossRef]
20. Xu, N.; Zhang, Y.; Wang, M.; Fan, X.; Zhang, T.; Peng, L.; Zhou, X.-D.; Qiao, J. High-Performing Rechargeable/Flexible Zinc-Air Batteries by Coordinated Hierarchical Bi-Metallic Electrocatalyst and Heterostructure Anion Exchange Membrane. *Nano Energy* **2019**, *65*, 104021. [CrossRef]
21. Liu, Y.; Wang, J. Preparation of Anion Exchange Membrane by Efficient Functionalization of Polysulfone for Electrodialysis. *J. Memb. Sci.* **2020**, *596*, 117591. [CrossRef]
22. Tang, H.; Geng, K.; Hao, J.; Zhang, X.; Shao, Z.; Li, N. Properties and Stability of Quaternary Ammonium-Biphosphate Ion-Pair Poly(Sulfone)s High Temperature Proton Exchange Membranes for H_2/O_2 Fuel Cells. *J. Power Sources* **2020**, *475*, 228521. [CrossRef]
23. Lekoane, T.; Msomi, P.F. Quarternized Polysulfone/ZSM-5 Zeolite Composite Anion Exchange Membrane Separators for Aluminum-Air Battery. *J. Appl. Polym. Sci.* **2023**, *140*, e54006. [CrossRef]
24. Simari, C.; Caprì, A.; Ur Rehman, M.H.; Enotiadis, A.; Gatto, I.; Baglio, V.; Nicotera, I. Composite Anion Exchange Membranes Based on Polysulfone and Silica Nanoscale Ionic Materials for Water Electrolyzers. *Electrochim. Acta* **2023**, *462*, 142788. [CrossRef]
25. Dong, D.; Xiao, Y.; Zhang, M.; Yang, Z.; Wang, K.; Fan, M. Crosslinked Anion Exchange Membranes with Regional Intensive Ion Clusters Prepared from Quaternized Branched Polyethyleneimine/Quaternized Polysulfone. *Int. J. Hydrogen Energy* **2022**, *47*, 24991–25006. [CrossRef]
26. Zhang, J.; Fu, J.; Song, X.; Jiang, G.; Zarrin, H.; Xu, P.; Li, K.; Yu, A.; Chen, Z. Laminated Cross-Linked Nanocellulose/Graphene Oxide Electrolyte for Flexible Rechargeable Zinc–Air Batteries. *Adv. Energy Mater.* **2016**, *6*, 1600476. [CrossRef]
27. Xu, M.; Dou, H.; Zhang, Z.; Zheng, Y.; Ren, B.; Ma, Q.; Wen, G.; Luo, D.; Yu, A.; Zhang, L.; et al. Hierarchically Nanostructured Solid-State Electrolyte for Flexible Rechargeable Zinc–Air Batteries. *Angew. Chem. Int. Ed.* **2022**, *61*, e202117703. [CrossRef]
28. Song, W.; Zhang, X.; Yang, C.; Yang, Z.; Wu, L.; Ge, X.; Xu, T. Alkaline Membranes toward Electrochemical Energy Devices: Recent Development and Future Perspectives. *ACS Cent. Sci.* **2023**, *9*, 1538–1557. [CrossRef]

29. Dumbrava, O.; Filimon, A.; Marin, L. Tailoring Properties and Applications of Polysulfone Membranes by Chemical Modification: Structure-Properties-Applications Relationship. *Eur. Polym. J.* **2023**, *196*, 112316. [CrossRef]
30. Avram, E.; Butuc, E.; Luca, C.; Druta, I. Polymers with Pendant Functional Group. III. Polysulfones Containing Viologen Group. *J. Macromol. Sci.-Pure Appl. Chem.* **1997**, *34*, 1701–1714. [CrossRef]
31. Vinodh, R.; Purushothaman, M.; Sangeetha, D. Novel Quaternized Polysulfone/ZrO2 Composite Membranes for Solid Alkaline Fuel Cell Applications. *Int. J. Hydrogen Energy* **2011**, *36*, 7291–7302. [CrossRef]
32. Wan, Y.; Peppley, B.; Creber, K.A.M.; Bui, V.T. Anion-Exchange Membranes Composed of Quaternized-Chitosan Derivatives for Alkaline Fuel Cells. *J. Power Sources* **2010**, *195*, 3785–3793. [CrossRef]
33. Xu, S.; Sun, Z.; Sun, C.; Li, F.; Chen, K.; Zhang, Z.; Hou, G.; Cheng, H.; Li, F. Homogeneous and Fast Ion Conduction of PEO-Based Solid-State Electrolyte at Low Temperature. *Adv. Funct. Mater.* **2020**, *30*, 2007172. [CrossRef]
34. Chen, N.; Wang, H.H.; Kim, S.P.; Kim, H.M.; Lee, W.H.; Hu, C.; Bae, J.Y.; Sim, E.S.; Chung, Y.-C.; Jang, J.-H.; et al. Poly(Fluorenyl Aryl Piperidinium) Membranes and Ionomers for Anion Exchange Membrane Fuel Cells. *Nat. Commun.* **2021**, *12*, 2367. [CrossRef]
35. Swaby, S.; Ureña, N.; Pérez-Prior, M.T.; Várez, A.; Levenfeld, B. Synthesis and Characterization of Novel Anion Exchange Membranes Based on Semi-Interpenetrating Networks of Functionalized Polysulfone: Effect of Ionic Crosslinking. *Polymers* **2021**, *13*, 958. [CrossRef]
36. An, E.J.; Sim, G.H.; Yu, S.; Kim, H.G.; An, S.J.; Lee, C.; Kim, M.; Kim, J.H.; Lee, J.H.; Chi, W.S. Cross-Linked Polysulfone Membranes with Controllable Cross-Linkers for Anion-Exchange Membrane Water Electrolysis. *Eur. Polym. J.* **2024**, *221*, 113543. [CrossRef]
37. Celik, A.; Hasar, H. Effect of Same Chloromethylation and Sulfonation Process on the Ion Exchange Membranes in Terms of Polymer Types and Ionic Properties. *Ionics (Kiel)* **2021**, *27*, 1243–1254. [CrossRef]
38. Lin, S.W.; Martínez-Ayala, A.V.; Pérez-Sicairos, S.; Félix-Navarro, R.M. Preparation and Characterization of Low-Pressure and High $MgSO_4$ Rejection Thin-Film Composite NF Membranes via Interfacial Polymerization Process. *Polym. Bull.* **2019**, *76*, 5619–5632. [CrossRef]
39. Kumar, R.; Isloor, A.M.; Ismail, A.F.; Rashid, S.A.; Matsuura, T. Polysulfone–Chitosan Blend Ultrafiltration Membranes: Preparation, Characterization, Permeation and Antifouling Properties. *RSC Adv.* **2013**, *3*, 7855–7861. [CrossRef]
40. Singh, K.; Devi, S.; Bajaj, H.C.; Ingole, P.; Choudhari, J.; Bhrambhatt, H. Optical Resolution of Racemic Mixtures of Amino Acids through Nanofiltration Membrane Process. *Sep. Sci. Technol.* **2014**, *49*, 2630–2641. [CrossRef]
41. Yang, Z.; Zhang, M.; Zhao, Z.; Zhang, X.; Fan, M. Construction of Quaternized Polysulfone/Polyquaternium-10 Anion Exchange Membrane with Semi-Interpenetrating Network for Alkaline Fuel Cell. *Macromol. Mater. Eng.* **2022**, *307*, 2100539. [CrossRef]
42. Bhowmick, S.; Qureshi, M. Vanadium Oxide Nanosheet-Infused Functionalized Polysulfone Bipolar Membrane for an Efficient Water Dissociation Reaction. *ACS Appl. Mater. Interfaces* **2023**, *15*, 5466–5477. [CrossRef]
43. Qaisrani, N.A.; Ma, L.; Hussain, M.; Liu, J.; Li, L.; Zhou, R.; Jia, Y.; Zhang, F.; He, G. Hydrophilic Flexible Ether Containing, Cross-Linked Anion-Exchange Membrane Quaternized with DABCO. *ACS Appl. Mater. Interfaces* **2020**, *12*, 3510–3521. [CrossRef]
44. Audeves−Audeves, Y.; Arredondo−Espínola, A.; Nava, O.; Olivas, A.; Torres-González, J.; Álvarez–Contreras, L.; Guerra-Balcázar, M.; Arjona, N. High Activity of Cobalt-Atomically Dispersed Catalyst on Mesoporous Carbon for Rechargeable Zn-Air Batteries via Effective Removal of the Hard Template. *Microporous Mesoporous Mater.* **2025**, *381*, 113359. [CrossRef]
45. Dewi, E.L.; Oyaizu, K.; Nishide, H.; Tsuchida, E. Cationic Polysulfonium Membrane as Separator in Zinc–Air Cell. *J. Power Sources* **2003**, *115*, 149–152. [CrossRef]
46. Jabbari, V.; Foroozan, T.; Shahbazian-Yassar, R. Dendritic Zn Deposition in Zinc-Metal Batteries and Mitigation Strategies. *Adv. Energy Sustain. Res.* **2021**, *2*, 2000082. [CrossRef]
47. Contreras-Martínez, M.V.; Arredondo-Espínola, A.; Guerra-Balcázar, M.; Álvarez-Contreras, L.; España-Sánchez, B.L.; Arriaga, L.G.; Arjona, N. Development of Electrospun and Casted Poly(Vinyl Alcohol)/Poly(Acrylic Acid) Membranes and Their Effect on the Performance of Rechargeable Zinc-Air Batteries. *J. Mater. Sci.* **2025**, *60*, 4502–4518. [CrossRef]

membranes

MDPI

Review

Pressure-Driven Membrane Processes for Removing Microplastics

Priscila Edinger Pinto [1,*], Alexandre Giacobbo [1,*], Gabriel Maciel de Almeida [1], Marco Antônio Siqueira Rodrigues [2] and Andréa Moura Bernardes [1]

1 Post-Graduation Program in Mining, Metallurgical and Materials Engineering (PPGE3M), Federal University of Rio Grande do Sul (UFRGS), Av. Bento Gonçalves, n. 9500, Porto Alegre 91509-900, RS, Brazil; gabrielmacielrs@gmail.com (G.M.d.A.); amb@ufrgs.br (A.M.B.)

2 Post-Graduation Program in Materials Technology and Industrial Processes, Feevale University, Rodovia RS-239, n. 2755, Vila Nova, Novo Hamburgo 93525-075, RS, Brazil; marcor@feevale.br

* Correspondence: priscila.edingerp@gmail.com (P.E.P.); alexandre_giacobbo@yahoo.com.br (A.G.)

Abstract: The intense consumption of polymeric materials combined with poor waste management results in the dissemination of their fragments in the environment as micro- and nanoplastics. They are easily dispersed in stormwater, wastewater, and landfill leachate and carried towards rivers, lakes, and oceans, causing their contamination. In aqueous matrices, the use of membrane separation processes has stood out for the efficiency of removing these particulate contaminants, achieving removals of up to 100%. For this review article, we researched the removal of microplastics and nanoplastics by membrane processes whose driving force is the pressure gradient. The analysis focuses on the challenges found in the operation of microfiltration, ultrafiltration, nanofiltration, and reverse-osmosis systems, as well as on the innovations applied to the membranes, with comparisons of treatment systems and the peculiarities of each system and each aqueous matrix. We also point out weaknesses and opportunities for future studies so that these techniques, known to be capable of removing many other contaminants of emerging concern, can subsequently be widely applied in the removal of micro- and nanoplastics.

Keywords: microfiltration; microplastics; nanofiltration; nanoplastics; reverse osmosis; ultrafiltration

1. Introduction

Produced mainly from the refining of oil and natural gas, polymers, popularly known as plastics, are present in in a large variety of items surrounding the human population. Their production process and properties justify their constant presence in the daily human routine. They are light, have electrical and thermal insulating characteristics, can be flexible, and have good optical properties [1]. Due to their widespread consumption, in 2019 annual global production reached 460 million tons, representing an increase of almost 230 times since the 1950s, when the rapid growth in plastic production began [2]. Furthermore, according to Ritchie et al. [3], approximately 0.5% of the global plastic waste generated is transported to the oceans, totaling somewhere between 1 and 2 million tons, representing alarming data regarding water pollution. Water is one of the most important routes by which microplastics are transferred from one place to another [4], which makes the presence of micro- and nanoparticles of plastics an emerging concern.

According to the vast majority of scientists in the field, microplastics (MPs) are defined as any polymeric particle smaller than 5 mm. Their origin may stem from the degradation of larger materials disposed of improperly, fragmenting in the oceans, or from weathering, as well as from fibers released during the washing of clothes and the production of synthetic fabrics. In all these cases, the particles are termed secondary MPs. In addition, a recent study [5] indicated that particles emitted by tire tread wear and asphalt pavement fragmentation residues represent a significant portion of secondary MPs in urban stormwater. On the other hand, MPs can be particles intentionally produced in this format, classified as primary MPs, which are present in personal care products such as toothpaste, soaps, and exfoliating creams [6].

In fact, the classification of MPs by size is already established, covering polymeric particles ranging from 1 μm to 5 mm [7–9]. However, particles smaller than 1 μm have also been detected and classified as nanoplastics (NPs) [10]. Since studies on NPs are incipient, their classification is still under development. Thus, the scientific literature typically classifies polymeric particles smaller than 1000 nm as NPs [9–11]; however, some studies consider only those smaller than 100 nm as NPs [12,13] while others consider particles between 1 and 1000 nm to be NPs [7,14]. These definitions can influence and shape directions in the elaboration of laws and standards, but in fact, regarding the environment, it is known that both particle sizes have harmful effects on the ecosystem. Furthermore, studies indicate that microplastics can act as bioaccumulators and adsorb heavy metals or other contaminants of emerging concern [15–17] increasing their toxicological potential. They can cause serious problems for aquatic fauna, as the ingestion of MPs can lead to gastrointestinal blockages and a feeling of satiety in animals without providing the necessary nutrients [17].

Similarly, there is already evidence of MPs in human organs, tissues, and fluids, including the lungs, brain, and placenta, and even in breast milk [18]. Although the effects of MPs in humans are still being studied, their ingestion or inhalation has already been associated with adverse effects in several organs. It is known that the presence of MPs measuring 44 to 60 μm in the intestine can alter the microbiota, causing dysfunction of the intestinal barrier, inflammation, and even cell death [19]. In organs such as the lungs, liver, and heart, in addition to the risks of inflammation and cell death, there are also potentially more severe effects. In the heart, MPs can impair the development of heart valves and form blood clots, resulting in cardiovascular problems [19]. Another recent highlight is the discovery of NPs and MPs in deep layers of the human brain: this condition is capable of causing an imbalance in energy production, reduced cell viability, and impaired development of brain organoids [20].

The problem related to the occurrence of MPs in humans is so great that the ingestion of MPs—which can be found in milk, honey, beer, and other edibles and potables—is equivalent to weekly values of 0.1 to 5 g, something that could be represented as the consumption of approximately one "credit card" per week [21]. This further highlights the need to include MP/NP removal methods in water and wastewater treatment systems [18,21,22]. Due to the prospects of increasing MPs in the ocean and the limited knowledge about the adverse effects they can cause on the environment [23], MPs can be considered contaminants of emerging concern [24]. Therefore, it is necessary to investigate and implement technologies capable of efficiently removing them.

There are several studies proposing a variety of technologies for removing these contaminants from aqueous matrices. The most popular alternatives are physical treatment methods, such as adsorption, gravitational separation, conventional filtration, and membrane processes, as seen in Figure 1. Chemical (oxidation and coagulation, for example) and biological (activated sludge processes and anaerobic digestion) processes are also utilized, which can be combined in one or more treatment stages, such as primary, secondary, and tertiary [25,26]. Additionally, techniques such as dissolved air flotation [27] UV radiation, ozone [28] electrokinetic-assisted filtration [11] and even electrocoagulation, are being employed. Many of these methods, from the simplest to the most advanced, have the potential to remove MPs, but like membrane processes, they also present some drawbacks [28,29].

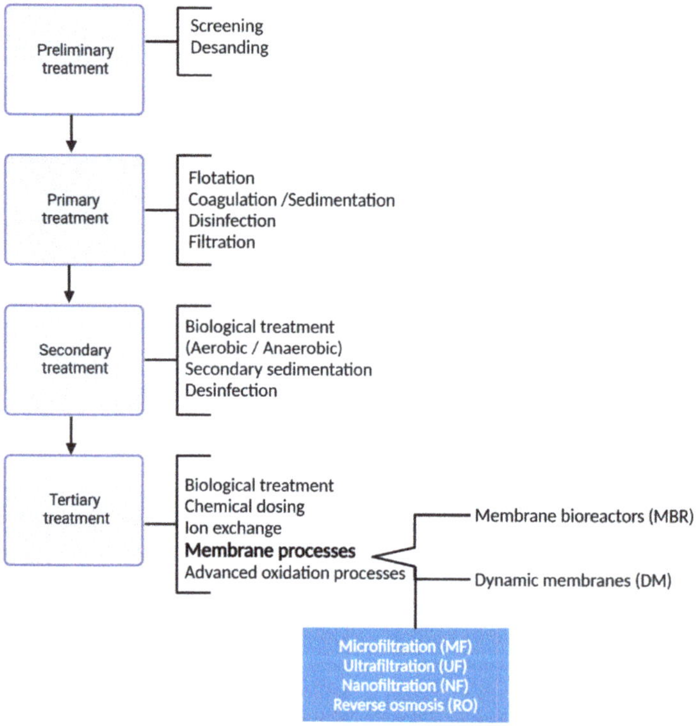

Figure 1. Stages and main processes used in wastewater treatment. Adapted from [25,30].

With this in mind, this review analyzes studies carried out on the removal of MPs using pressure-driven membrane processes, namely microfiltration (MF), ultrafiltration (UF), nanofiltration (NF), and reverse osmosis (RO). Several factors influence MP removal efficiency by membrane processes, such as pore size, membrane charge, dipole moment, and hydrophilicity [9]. Through the studies already published on the subject, these membrane processes will be assessed based on aspects such as difficulties, advantages, and opportunities, as well as suggestions for future work. Despite the distinction in the literature between MPs and NPs, throughout this paper we will use the generic term MPs, which encompasses plastic particles smaller than 5 mm. We will return to the term NPs only in specific cases where it is necessary to differentiate them from MPs.

2. Methodology

The data for this review were extracted from the Scopus database through searches performed on 23 December 2024 and 22 February 2025 using the following keywords: "microplastic" AND "removal" OR "catch" OR "rejection" OR "retention" AND "microfiltration" OR "ultrafiltration" OR "nanofiltration" OR "reverse osmosis". The search was limited to research articles published from 2017 to 31 December 2024. Among the 107 articles found, those whose applied methodologies included the four pressure-driven membrane processes (PDMPs) for removing MPs were selected for data processing. For this reason, 44 articles were removed, since some did not encompass both subjects, MPs and PDMPs, used another type of membrane separation process, or were focused on MP quantification studies and did not add important information to the present study. Another 31 references were added due to a screening in the references of the selected articles or included to ground some information, totaling 94 references in this review. It is worth mentioning that MPs are a current topic, which justifies the absence of articles before 2017 in the search with the keywords we used. The term microplastic itself was incorporated into the scientific literature only in 2004 by Thompson et al. [31] and since then, countless studies have been developed in several areas of knowledge.

3. Removal of MPs from Aqueous Matrices by Pressure-Driven Membrane Processes

Pressure-driven membrane processes (PDMPs) can segregate solid particles such as microplastics and nanoplastics from fluids depending on their size. Separation methods in which exclusion occurs by size, as represented in Figure 2, seem to be a good strategy, since the interactions of plastic particles with their surroundings encompass several phenomena, such as van der Waals forces, electrostatic repulsion, adsorption, steric interactions, hydrophobic interactions, size exclusion, and hydration forces [32,33]. For this reason, for adequate retention of MPs, it is necessary to pay attention to the properties of the solution, the hydrodynamic conditions, and other parameters that can influence the separation.

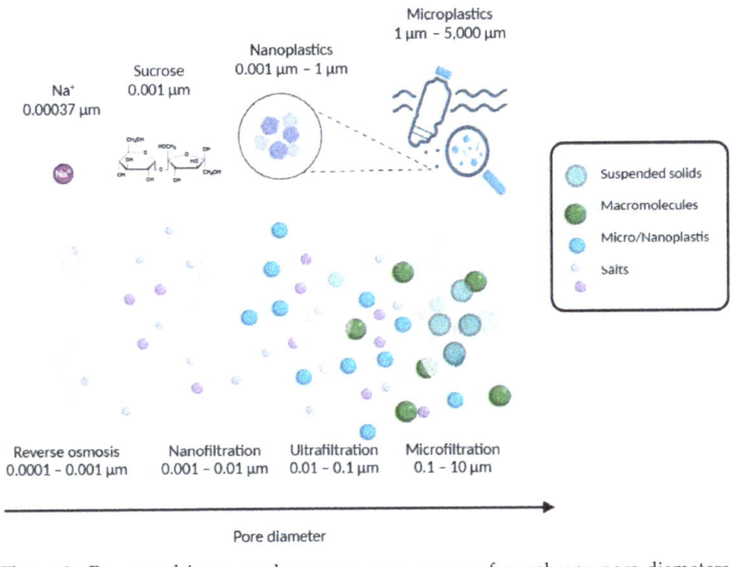

Figure 2. Pressure-driven membrane processes, range of membrane pore diameters, and some examples of contaminants capable of being removed.

In PDMPs, the membrane acts as a selective barrier, allowing certain substances, such as molecules, ions, or small particles, to permeate, while blocking others. PDMPs are categorized based on membrane pore size for the purposes of microfiltration, ultrafiltration, nanofiltration, and reverse osmosis [34] Membranes are typically produced with organic such polymers as polytetrafluoroethylene (PTFE), polyethylene terephthalate (PET), polyvinylidene fluoride (PVDF), polypropylene (PP), polyethylene (PE), polyamide (PA), polysulfone (PSU), and polyethersulfone (PES) [35] as well as with biodegradable materials such as polylactic acid [36]. The use of polymeric membranes in separation processes is considered advantageous due to their characteristics, such as low energy consumption, absence or minimization of chemicals, ease in scaling up, and high efficiency even under mild operating conditions [37,38]. These characteristics benefit applications in drinking water treatment systems, wastewater, or other purposes, such as in the separation stage in producing agroindustrial products or other products for human consumption [39]. Research relating to the usage of PDMPs for MP removal is very diverse, with the vast majority of PDMPs showing excellent removal performance, whether using commercial membranes, modified membranes, or membranes produced in research laboratories [40–42].

3.1. Microfiltration

MF uses membranes with pore sizes ranging from 0.1 to 10 μm and operating pressures from 0.01 to 2 bar. It is usually applied to remove bacteria and suspended solids and to treat water and wastewater [43]. MF is sometimes employed as a pretreatment for operations with tighter membranes such as NF or RO, contributing to the reduction in fouling in the membranes of downstream processes in water treatment plants (WTPs) or wastewater treatment plants (WWTPs), as reported by Barbier et al. [8], whose research evaluated a WTP in Paris that had MF as a pretreatment for nanofiltration [8,44]. Because it uses larger pores and lower working pressure, which reduce its operating costs compared to other PDMPs, MF can be used in supplementary drinking water filtration devices for end consumers.

In a study conducted by Cherian et al. [23], MP removal capacity was assessed by comparing three different types of domestic filters combining MF, activated carbon, and ion exchange, one without a membrane, one using microfiltration, and the other with a geotextile membrane. The best performance was achieved using the device equipped with MF (pore opening $\geq$ 0.2 μm), capable of retaining 93.6% $\pm$ 2.2% of the polyvinyl chloride (PVC), polyethylene terephthalate (PET), and nylon microfiber particles. It should be noted that due to the impossibility of performing a backwash, the consumer must change the filters according to the manufacturer's instructions in order to maintain their efficiency in removing microplastics and other contaminants [23].

In a study by Pizzichetti et al. [45], microfiltration membranes were used for polishing treated water. In the three membranes tested, polycarbonate (PC), cellulose acetate (CA), and polytetrafluoroethylene (PTFE), whose pore size was 5 μm, the microparticles of PA and polystyrene (PS) were efficiently removed. Table 1 shows the average particle sizes, as well as details on the operation of the membrane systems. Although all membranes removed >94% of MPs, reaching very close to 100% in PA removal, the study highlights some factors that may influence the permeation of microplastics through the membranes, such as low mechanical resistance of some membranes, irregular shape of some MPs, and MP fracture when subjected to the filtration stage. Other important factors are the abrasion of MPs on the membrane and hydrophobicity characteristics. The authors concluded that the most recommended membrane among those tested was the CA membrane, due to its performance in terms of MP removal, operating pressure, and flow pattern [45].

Hyeon et al. [46] tested two ceramic MF membranes, one made of alumina and the other of silicon carbide, to remove MPs from laundry wastewater. Both membranes demonstrated good performance in removing MPs, particularly the alumina membrane, which was attributed to its narrower pore size distribution. However, the silicon carbide membrane exhibited better performance in terms of permeate flux, with a smaller decline throughout the tests.

Kook et al. [47] evaluated the behavior of alumina ceramic membranes in removing PE microplastics in the presence of hydrophobic compounds used as organic UV filters, contaminants of emerging concern commonly found in personal care products designed to protect the skin against UV radiation. They observed the adsorption of the organic UV filters onto the PE microplastics through hydrophobic interactions, which increased the removal efficiency of the organic UV filters from 34.2–37.8% (without MPs) to 82.2–97.9% (with MPs), as the MPs were nearly completely removed.

In another study, Kook et al. [48] compared three MF membranes with a nominal pore size of 0.1 μm, two of which were polymeric (PVDF) and one ceramic (alumina), in removing PE and PS microplastics with sizes of 0.1 and 1 μm. The authors observed a sharp decline in flux during the first 15–20 minutes of experiments with 0.1 μm PS microplastics for all three membranes. Thereafter, it remained relatively constant at values below 10% of the initial flux. This significant loss of productivity was attributed to fouling in all three membranes caused by standard or complete pore blocking, since these MPs were 0.1 μm in size, equal to the average pore size of the membranes. Particles with sizes very close to the pore size can penetrate and attach to the inner walls of the membrane pores, leading to severe membrane fouling. The authors also emphasize that both polymeric and ceramic membranes can remove a significant fraction of MPs from WWTP, noting that the ceramic membrane was particularly effective at removing PE microplastics but less effective in rejecting PS microplastics and was less susceptible to productivity loss in filtering PE microplastics compared to polymeric membranes.

Sriani et al. [49] investigated the production of membranes from waste-expanded polystyrene (WEPS) containing up to 8% polyimide (PI). The membranes were tested for removal of MPs and bovine serum albumin. All membranes showed hydrophilic characteristics, good antifouling properties, and MP rejections above 80%. However, the one with 8% polyimide showed the lowest porosity and pore size values, while improving the antifouling property by 67% compared to the pristine membrane prepared solely with WEPS. The research spotlight was the investigation of waste with great potential for generating secondary MPs as raw material for producing membranes that could serve as a solution for removing MPs from aqueous matrices.

In fact, MF is the most studied PDMP for the removal of MPs, with Table 1 showing the conditions under which the MF studies evaluated in this work were conducted.

Table 1. Articles found in the literature related to microfiltration (MF) for the removal of microplastics (MPs).

<table>
<tr><th colspan="6">Membrane</th><th rowspan="2">Operating Pressure (Bar)</th><th rowspan="2">Aqueous Matrix</th><th rowspan="2">Scale</th><th colspan="4">Microplastic</th><th rowspan="2">Reference</th></tr>
<tr><th>Name</th><th>Supplier</th><th>Type</th><th>Material</th><th>Pore Size (μm)</th><th>Hydrophilicity</th><th>Type</th><th>Concentration</th><th>Size (μm)</th><th>Removal (%)</th></tr>
<tr><td rowspan="3">VVLP</td><td rowspan="3">Millipore</td><td rowspan="3">Flat sheet</td><td rowspan="3">PVDF</td><td rowspan="3">0.10</td><td rowspan="3">Hydrophilic</td><td rowspan="3">0.7</td><td rowspan="3">Milli-Q water</td><td rowspan="3">Lab</td><td>PE</td><td>100 mg/L</td><td rowspan="3"><2</td><td rowspan="3">>90</td><td rowspan="3">[13]</td></tr>
<tr><td>PVC</td><td>100 mg/L</td></tr>
<tr><td>PES</td><td>100 mg/L</td></tr>
<tr><td colspan="4" rowspan="3">N.A. [a]</td><td rowspan="3">≥0.2</td><td rowspan="3">N.A.</td><td rowspan="3">N.A.</td><td rowspan="3">RO water</td><td rowspan="3">Lab</td><td>PVC</td><td>39 ± 9 particles/L</td><td>39–246</td><td rowspan="3">93.6 ± 2.2</td><td rowspan="3">[23]</td></tr>
<tr><td>PET</td><td>36 ± 7 particles/L</td><td>28–121</td></tr>
<tr><td>Nylon</td><td>64 ± 15 fibers/L</td><td>496–1862</td></tr>
<tr><td rowspan="3">N.A.</td><td rowspan="3">Filter-Lab</td><td rowspan="3">Flat sheet</td><td>PC</td><td rowspan="3">5.0</td><td>Hydrophilic</td><td rowspan="2">0.5</td><td rowspan="3">Milli-Q water</td><td rowspan="3">Lab</td><td>PA</td><td>100 mg/L</td><td>20–300</td><td rowspan="3">~94</td><td rowspan="3">[45]</td></tr>
<tr><td>CA</td><td>Hydrophilic</td><td rowspan="2">PS</td><td rowspan="2">100 mg/L</td><td rowspan="2">20–300</td></tr>
<tr><td>PTFE</td><td>Hydrophobic</td><td>1.5</td></tr>
<tr><td>N.A.</td><td>Cembrane</td><td>Flat sheet</td><td>Silicon carbide</td><td>0.10</td><td>Hydrophilic</td><td>0.25</td><td>Laundry wastewater</td><td>Lab</td><td>PET</td><td>3000–45,000 fibers/L</td><td>29–36 thickness and 220–550 length</td><td>87.7–98.9</td><td rowspan="2">[46]</td></tr>
<tr><td>Meiden</td><td>Meindensha</td><td>Flat sheet</td><td>Al_2O_3</td><td>0.10</td><td>Hydrophilic</td><td>0.25</td><td>Laundry wastewater</td><td>Lab</td><td>PET</td><td>3000–45,000 fibers/L</td><td>29–36 thickness and 220–550 length</td><td>90.3–97.7</td></tr>
<tr><td>T1-70, Membralox</td><td>Pall</td><td>Tubular</td><td>Al_2O_3</td><td>0.10</td><td>Hydrophilic</td><td>1.0</td><td>Synthetic wastewater (organic UV filter + water)</td><td>Lab</td><td>PE</td><td>100 mg/L</td><td>75–90; 300–355; 600–710</td><td>80.2–97.9</td><td>[47]</td></tr>
<tr><td>V0.1</td><td>Synder Filtration</td><td>Flat sheet</td><td>PVDF</td><td>0.10</td><td>Hydrophilic</td><td rowspan="3">1.0</td><td rowspan="3">DI water + surfactant</td><td rowspan="3">Lab</td><td rowspan="3">PE; PS</td><td rowspan="3">50 mg/L</td><td rowspan="3">0.1–1.0</td><td rowspan="3">>99</td><td rowspan="3">[48]</td></tr>
<tr><td>SteriLUX</td><td>Meissner</td><td>Flat sheet</td><td>PVDF</td><td>0.10</td><td>Hydrophilic</td></tr>
<tr><td>Anopore/ Anodisc</td><td>Whatman</td><td>Flat sheet</td><td>Al_2O_3</td><td>0.10</td><td>Hydrophilic</td></tr>
<tr><td>N.A.</td><td>Lab-made</td><td>Flat sheet</td><td>WEPS/PI</td><td>0.17–0.20</td><td>Hydrophilic</td><td>1.0</td><td>DI water + SDS</td><td>Lab</td><td>PTFE</td><td>N.A.</td><td>7.0</td><td>>80</td><td>[49]</td></tr>
</table>

Table 1. *Cont.*

Membrane						Operating Pressure (Bar)	Aqueous Matrix	Scale	Microplastic				Reference
Name	Supplier	Type	Material	Pore Size (µm)	Hydrophilicity				Type	Concentration	Size (µm)	Removal (%)	
N.A.	Lab-made	Hollow fiber	Sericin coated on PP	N.A.	Hydrophilic	0.2	50 L DI water + 1.75 kg rock salt or sea salts	Lab	PE PET PP PVC Nylon PU PS Undefined	287–417 particles/kg rock salt or 1434–2284 particles/kg sea salt	≥20–≤5000	99.30	[50]
N.A.	Pall [b]	Hollow fiber	PVDF	0.10	N.A.	N.A.	Wastewater	Full	HDPE, LDPE, PP	N.A.	>5	>94	[51]
Durapore	Millipore	Flat sheet	PVDF	0.10 0.22 0.45 5.0	Hydrophilic	0.5	DI water + ethanol (1:1)	Lab	PE PA	50 mg/L 50 mg/L	10–106 15–55	N.A.	[52]
N.A.	N.A.	Flat sheet	PVDF	0.22	Hydrophobic	1–3	Secondary wastewater	Lab	PE PVC	200 particles/L	150 250	100	[53]
N.A.	Mervilab	Flat sheet	CA	5.0	Hydrophilic	0.1–0.7	Milli-Q water	Lab	PA PS	1–20 mg/L	10–105 20–320	N.A.	[54]
N.A.	N.A.	Flat sheet	PES	0.22	N.A.	0.5	DI water	Lab	PE PET	10 mg/L 10 mg/L	40–48 300	100	[55]
N.A.	ADVANTEC	Flat sheet	MCE	0.80	Hydrophilic	1.0	DI water	Lab	PS	2.5 mg/L	1.0	99.9	[56]
Filtanium Filtanium TÜBITAK 1-4	TAMI TAMI TÜBITAK	Tubular	TiO_2	0.80 0.14 0.40	N.A.	0.1–1.0	Secondary wastewater	Full	N.A.	10–100 particles/L	17–427	96	[57]
N.A.	Lab-made	Flat sheet	Modified PVC	0.86–3.44	Hydrophilic	1.0	DI water	Lab	PS	2.5 mg/L	1.0	~100	[58]
N.A.	Lab-made	Flat sheet	Modified PVDF	0.34–1.36	Amphiphilic	1.0	DI water + 0.1% Triton X	Lab	PS	0.1%	0.1–0.5	90.41–97 2	[59]

Table 1. *Cont.*

Membrane						Operating Pressure (Bar)	Aqueous Matrix	Scale	Microplastic				Reference
Name	Supplier	Type	Material	Pore Size (μm)	Hydrophilicity				Type	Concentration	Size (μm)	Removal (%)	
N.A.	Lab-made	Flat sheet	Glass fiber membrane + chitosan-modified geopolymer sub-microparticles	0.058	Hydrophilic	0.9	Wastewater	Lab	PS	N.A.	0.05	92.04	[60]
N.A.	Lab-made	Flat sheet	SA/GO/CS	0.70	Hydrophilic	0.04	Synthetic wastewater	Lab	PS	N.A.	0.05	97.10 ± 1.17	[61]

Abbreviations: N.A.: not available; CA: cellulose acetate; CS: chitosan; DI: deionized; GO: graphene oxide; HDPE: high-density polyethylene; LDPE: low-density polyethylene;
MCE: mixed cellulose ester; PA: polyamide; PC: polycarbonate; PE: polyethylene; PES: polyethersulfone; PET: polyethylene terephthalate; PI: polyimide; PP: polypropylene;
PS: polystyrene; PTFE: polytetrafluoroethylene; PU: polyurethane; PVC: polyvinylchloride; PVDF: polyvinylidene chloride; SA: sodium alginate; SDS: sodium dodecyl sulfate;
WEPS: waste-expanded polystyrene. [a] Point-of-use device composed of microfiltration + granular activated carbon + ion exchange resins. [b] from ref. [59].

3.2. Ultrafiltration

UF membranes have pores ranging from 1 to 100 nm and generally operate under pressures from 0.5 bar to a maximum of 10 bar. They have the capacity to remove suspended solids, colloids, and bacteria like microfiltration membranes, and can also be used to remove viruses and high-molecular-weight organic compounds from water and wastewater [43]. In addition of being a great alternative in the treatment of drinking water by removing algae, cyanobacteria, and other organic substances [62], UF can be used as pretreatment for reverse osmosis [63]. There are a variety of studies using UF, either individually or combining one or more processes, such as biological processes, activated carbon adsorption, electrocoagulation, and RO, among others, obtaining excellent results [64,65].

Ma et al. [16] conducted studies using coagulation and UF to remove MPs from water. Among their conclusions, they realized that coagulation alone was not sufficient to remove MPs from polymers such as PE, especially because its density is very similar to that of water. The removal efficiency was below 15%, showing some improvement with higher concentrations of Fe-based coagulant accompanied by an anionic polyacrylamide flocculant. The combination of coagulation with UF was also evaluated. Good performance was observed, especially in aspects related to fouling. When the flocs were smaller, a more compact cake on the membrane was formed, resulting in a greater drop in ultrafiltration flux. When larger MPs were present, larger flocs were formed and the cake became less compact, which resulted in higher permeate fluxes [16]. Pretreatment with Al-based coagulants exerts a better performance than those based on Fe [16,66]. In the presence of humic acid, the cake becomes thicker and even less compact due to the larger flocs, improving UF performance [16,66].

Furthermore, regarding the assessment of fouling on an UF membrane, during the processing of aqueous matrices intentionally contaminated with PET microfibers, Hachemi et al. [10] observed that the microfibers, in addition to only marginally influencing permeability, did not interact with the other components of the wastewater. The cake formed on the PVDF membrane showed similar compaction regardless of the presence or absence of PET microfibers. In other words, the presence of MPs did not significantly influence the system's operation or its performance [10]. On the other hand, Ghasemi et al. [67] observed that the presence of MPs during the use of UF with hollow fiber membranes facilitated the removal of fouling caused by alginate. The membranes subjected to samples containing only alginate required a more thorough cleaning (hydraulic and chemical), whilst those subjected to samples containing alginate and MPs had less pore obstruction and the incrustation was released from the membrane with only hydraulic cleaning.

Yuan et al. [68] evaluated the performance of UF in two water treatment plants in Canada. The authors reported MP removals above 90%. Comparing different forms of filtration with various aqueous matrices, such as urban wastewater and wastewater, from a plastic recycling industry, González-Camejo et al. [69] highlighted the better performance of UF compared to gravitational filtration. However, differences were observed between the two operating modes regarding energy consumption, which in turn varied depending on the type of wastewater being treated [69]. This indicates the need to conduct pilot plant tests specifically for each type of aqueous matrix and desired objective. In addition, using pretreatments to reduce the MP load influent on the UF provides better membrane performance, mitigating the incidence of fouling. According to Xiong et al. [70], the presence of MPs favors the multiplication of bacteria inside the membrane; that is, they act as facilitators of biofouling.

In a study conducted in Thailand, UF achieved 96.97% efficiency in MP removal in a sewage treatment pilot plant [71]. Studies also underline the importance of MP removal methods that do not generate sludge, as the use of sludge from WWTPs or WTPs as a soil conditioner can be a major source of soil contamination due to the transfer of MPs from the liquid matrix (water or wastewater) to the solid matrix (sludge) [72].

Mitigating fouling is a crucial factor for the effective performance of membrane processes, which is why it has been the focus of many studies. In this regard, Enfrin et al. [12] performed UF tests with a commercial PSU membrane, observing significant fouling formation after 4 hours of ultrafiltration of samples containing micro- and nanoparticles. After 48 hours, the drop in flux throughout the tests became even more evident, along with the observation of an increasingly lower concentration of MPs in the retentate stream, confirming the suspicion that the MPs were adsorbed onto the membranes. Observations were also made using scanning electron microscopy (SEM) of the membrane before and after the tests, corroborating the findings regarding fouling. Focused on minimizing fouling issues, Enfrin et al. [73] conducted another study testing the intermittent injection of nitrogen gas into the feed stream during the UF of samples containing MPs and NPs to perform mechanical cleaning throughout the tests. Pristine PSU membranes and some modified with low-pressure plasma were tested. In all membranes, the gas injection together with the feed stream was able to reduce the adhesion of MPs on the membranes, resulting in better fluxes. Luogo et al. [74], aiming to remove micro- and nanoplastics from synthetic fabrics, tested two ceramic membranes: a silicon carbide (SiC) MF membrane and a zirconium oxide (ZrO_2) UF membrane. The tests were conducted initially with synthetic wastewater containing nylon fibers and later with real wastewater from washing PVC tarpaulins. It was observed that the MF membrane displayed a drop in flux during the test with synthetic wastewater containing MPs, while the UF membrane remained constant. Both membranes showed complete rejection of MPs in both tests. However, in the test with real wastewater, the UF membrane showed a slight drop in flux throughout the experiment, but remained lower than the drop reached with the MF. In tests of 4 days of continuous operation, the MF membrane had a 95% decrease in permeate flux, while the UF exhibited only a 37% drop.

While some studies evaluated the modification of MF membranes for the removal of MPs, others focused on the modification of UF membranes, which demonstrated good anti-fouling performance, facilitating the cleaning and reuse of the membranes. This was observed by Fryczkowska and Przywara [75], who evaluated polyacrylonitrile (PAN) membranes with the addition of reduced graphene oxide. Mohana et al. [9] in turn prepared nanocomposite membranes based on a double-charged MOF (metal–organic framework), and the hydraulic permeability was superior to that of pristine membranes. In summary, UF has also been extensively studied for MP removal, as illustrated in Table 2, which outlines the main conditions under which the assessed studies were performed.

Table 2. Articles found in the literature related to ultrafiltration (UF) for the removal of microplastics (MPs).

Membrane						Operating Pressure (Bar)	Aqueous Matrix	Scale	Microplastic				Reference
Name	Supplier	Type	Material	Pore Size/MWCO	Hydrophilicity				Type	Concentration	Size (μm)	Removal (%)	
BY [a]	Synder Filtration	Flat sheet	PVDF	100 kDa	Hydrophobic	0.5	DI water and synthetic wastewater	Lab	PET fibers	1 mg/L	0.142	>99	[10]
N.A.	Pureach Beijing	Flat sheet	PSU	30 kDa	Hydrophilic	1.0	Milli-Q water	Lab	PE (nano/micro-plastics from a facial scrub)	10 mg/L	0.013–0.69	N.A.	[12]
N.A.	Ande	Flat sheet	PVDF	100 kDa	N.A.	1.0	DI water + $FeCl_3 \cdot 6H_2O$ + 0.1 M $NaHCO_3$	Lab	PE	N.A.	500–5000	100	[16]
N.A.	N.A.	N.A.	N.A.	N.A.	N.A.	N.A.	Landfill leachate	Full	PE, PES, PP, PA, EPM, PVAC	1.2 ± 0.57 particles/L	1000–5000 (41.7%); 500–1000 (16.6%); <500 (41.7%)	N.A.	[34]
N.A.	Yuling	Flat sheet	PES	100 kDa	Hydrophilic	0.8	DI water + humic acid (10 mg/L) + $AlCl_3 \cdot 6H_2O$ and $FeCl_3 \cdot 6H_2O$	Lab	PS	100 mg/L	50	91.2–92.7	[66]
SIP-1013 (SIP-1023	Asahi Kasei	Hollow fiber	PSU	6 kDa	N.A.	0.11–0.15	DI water + alginate	Lab	PE	10–100 ppm	125	N.A.	[67]
N.A.	Millipore	Flat sheet	PC	N.A.	N.A.	1.4	Drinking water	Lab	PET PVC PET Nylon fibers	>30 particles/L	74 61 15 16	>90	[68]
KMS Puron	Koch	Hollow fiber	PVDF	0.03 μm	N.A.	0.05–0.45	Wastewater from a PET recycling plant and urban wastewater	Pilot	HDPE PET Other MPs	5.8 ± 2.1 mg/L 12.3 ± 1.8 mg/L 0.17 ± 0.08 mg/L	25–500	100	[69]
N.A.	Ande	Flat sheet	PVDF	100 kDa	Hydrophilic	0.05–0.7	Surface water from Qinghe River	Lab	PE	1 mg/L	40–48	N.A.	[70]

Table 2. *Cont.*

Membrane						Operating Pressure (Bar)	Aqueous Matrix	Scale	Microplastic				Reference
Name	Supplier	Type	Material	Pore Size/MWCO	Hydrophilicity				Type	Concentration	Size (µm)	Removal (%)	
AAc	Lab-made	Flat sheet	PSU modified	N.A.	Hydrophilic	0.25–1.2	Milli-Q water	Lab	PE (nano/micro-plastics from a facial scrub)	10 mg/L	0.093	N.A.	[73]
CPAm					Hydrophilic								
HMDSO					Hydrophobic								
N.A.	LiqTech Ceramics	Tubular	ZrO_2	0.074 µm	Hydrophilic	0.1–1.4	DI water	Pilot	Nylon fibers	180 mg/L	80	100	[74]
							Laundry wastewater		PVC fibers	~22.6 mg/L	N.A.	99.2	
N.A.	Lab-made	Flat sheet	rGO/PAN	0.15 µm	Hydrophilic	1–2	Wastewater from a PET recycling plant	Lab	PET	N.A.	N.A.	>80	[75]
N.A.	Motimo	Flat sheet	PVDF	100 kDa	N.A.	1.0	DI water + $AlCl_3 \cdot 6H_2O$ + 0.1 M $NaHCO_3$	Lab	PE	N.A.	500–5000	100	[76]
UP150	Microdyn-Nadir [b]	Flat sheet [b]	PES [b]	150 kDa [b]	N.A.	1.0	Personal care product wastewater	Lab	N.A.	80 particles/L	N.A.	100	[77]

Abbreviations: N.A.: not available; DI: deionized; EPM: ethylene–propylene polymer; HDPE: high-density polyethylene; PAN: polyacrylonitrile; PC: polycarbonate; PE: polyethylene; PES: polyethersulfone; PET: polyethylene terephthalate; PP: polypropylene; PS: polystyrene; PSU: polysulfone; PVC: polyvinylchloride; PVDF: polyvinylidene chloride; rGO: reduced graphene oxide. [a] from ref. [78]. [b] from ref. [79].

3.3. Nanofiltration

NF is characterized by the usage of membranes with pores of 1 to 10 nm and operating conditions at pressures generally between 5 and 15 bar. In addition, it is capable of removing multivalent salts and organic molecules with molecular weights greater than 200 Da [43]. In many cases, NF is implemented as a polishing step after other treatments, usually aiming at water reuse.

In a study aimed at producing reused water from secondary effluent contaminated with microplastics, Lin et al. [80] assessed the influence of electrocoagulation as a pretreatment for nanofiltration. The combination of the two technologies improved the removal of MPs and reduced membrane fouling. However, the authors emphasized that attention must be paid to the applied electric current for electrocoagulation to be effective, ultimately contributing to better performance of the NF positioned downstream.

Akarsu et al. [77] also evaluated electrocoagulation followed by PDMPs for water reuse purposes, but for treating wastewater from personal care products containing microplastics. In electrocoagulation, various electrode materials, including iron, aluminum, stainless steel (SS), titanium, and graphite, were tested under different operating conditions. The effluent was subsequently treated with UF (UP150), NF (NF90 or NF270), or RO (BW30 or SW30) to produce reusable water. The SS electrode demonstrated the best performance in electrocoagulation. Under optimal conditions, it reduced the concentration of microplastics from 1030 particles/L to 80 particles/L and lowered chemical oxygen demand (COD) from 15,930 mg/L to 3200 mg/L (~80% removal). Although all PDMPs achieved complete removal of microplastics (see Table 3), only the SW30 RO membrane produced a permeate with a COD low enough for water reuse purposes.

When comparing the behavior of a polyamide NF membrane over 50 days of operation in treating two types of synthetic wastewater, Lin et al. [81] observed the influence of MPs on the incidence of fouling. The presence of MPs contributed to the formation of bacterial colonies and the accumulation of polysaccharides, proteins, and humic compounds in the NF membrane. It was also found that the wastewater containing MPs resulted in a significantly greater drop in permeate flux than that reached with the wastewater without MPs. However, the tested membrane performed well in rejecting MPs, achieving removal close to 100%. Furthermore, regarding the larger contaminants in the wastewater, such as polysaccharides, their rejection was not impaired, although there was greater fouling formation [81].

It is well known that landfills are also major sources of MP discharge and transport MPs and other contaminants through their leachate to watercourses [82]. When comparing the leachate treatment systems of two landfills in Istanbul, Kara, Sari Erkan and Onkal Engin [83] classified the MPs they found by size, color, and shape. Although the vast majority of MPs were fibrous in shape, which is already known from previous studies to hinder membrane performance, the UF and NF systems assessed achieved 96–99% MP removals, i.e., values much higher than those observed with conventional aerobic and anaerobic systems. In general, since most MPs were large, ranging in size from 0.5 to 2 mm, high removals were expected with the PDMPs assessed. Zhang et al. [34] investigated the distribution and removal characteristics of MPs in different stages of the leachate treatment plant of the Shanghai Laogang landfill (Shanghai, China), the largest in Asia. The leachate treatment plant uses pretreatment (homogenization), a biological system consisting of a membrane bioreactor (MBR) and a two-stage anoxic/oxic system, and advanced treatment (UF, NF, and RO). While the effluent from the plant (RO permeate) is discharged into the environment, the concentrates from the PDMPs and MBR are sent to the sludge treatment system and then disposed of in the landfill itself. According to the study, PE was the predominant type of MP found in the leachate, which correlated with the higher generation of waste from this material. Additionally, fibers represented the most abundant form, comprising 42.2% of the MPs. In terms of size, the largest particles were most common—58.3% measured between 0.5 and 5 mm—while the remaining 41.7% were smaller than 0.5 mm. Even though the leachate treatment plant removed a large portion, significant amounts of MPs were found in the NF and RO permeates. According to the authors, what may justify this finding is the possibility that the membrane itself releases MPs, since fractions of cellulose nitrate (CN) were found. Moreover, it is also possible that fibrous materials passed through the pores of the membranes due to their shape. Another important point mentioned by the authors is that the sludge formed during leachate treatment, as well as the concentrate from membrane processes, carried a high load of microplastics, which were returned to the landfill. Despite this practice being technically and environmentally appropriate, it does not definitively eliminate MPs, which remain in a cycle, being transferred from the solid phase (sludge or landfill soil) to the liquid phase (leachate), and are subject to migration into the environment through wind or leachate leaks.

Fortin et al. [84] focused their studies on evaluating MPs smaller than 20 μm and within the detection limit of Raman spectroscopy. Secondary post-treatment domestic sewage seeded with MPs was used as the aqueous matrix, which was then subjected to advanced treatments based on activated carbon and NF/RO followed by disinfection. The authors reported how difficult it was to work with MPs of those dimensions since contamination care needed to be redoubled throughout the process and Raman spectroscopy did not accurately detect particles smaller than 1 μm. Nevertheless, the treatments with NF and RO eliminated 100% of the MPs detectable by the analytical methodology. The main conditions under which the studies we found regarding the use of NF and RO for removing MPs are presented in Table 3.

Table 3. Articles found in the literature related to nanofiltration (NF) and reverse osmosis (RO) for the removal of microplastics (MPs).

<table>
<tr><th rowspan="2">Process</th><th colspan="6">Membrane</th><th rowspan="2">Operating Pressure (Bar)</th><th rowspan="2">Aqueous Matrix</th><th rowspan="2">Scale</th><th colspan="4">Microplastic</th><th rowspan="2">Reference</th></tr>
<tr><th>Name</th><th>Supplier</th><th>Type</th><th>Material</th><th>MWCO</th><th>Hydrophilicity</th><th>Type</th><th>Concentration</th><th>Size (μm)</th><th>Removal (%)</th></tr>
<tr><td>NF and RO</td><td>N.A.</td><td>N.A.</td><td>N.A.</td><td>N.A.</td><td>N.A.</td><td>N.A.</td><td>N.A.</td><td>Landfill leachate</td><td>Full</td><td>PE, PES, PP, PA, EPM, PVAC</td><td>1.2 ± 0.57 particles/L</td><td>1000–5000 (41.7%); 500–1000 (16.6%); <500 (41.7%)</td><td>N.A.</td><td>[34]</td></tr>
<tr><td rowspan="2">NF</td><td>NF270</td><td rowspan="4">Dow—FilmTec</td><td rowspan="4">Flat sheet</td><td rowspan="4">PA [b]</td><td>400 Da [a]</td><td rowspan="2">Hydrophilic [a]</td><td rowspan="4">20</td><td rowspan="4">Personal care products wastewater</td><td rowspan="4">Lab</td><td rowspan="4">N.A.</td><td rowspan="4">80 particles/L</td><td rowspan="4">N.A.</td><td rowspan="4">100</td><td rowspan="4">[77]</td></tr>
<tr><td>NF90</td><td>200 Da [a]</td></tr>
<tr><td rowspan="2">RO</td><td>SW30</td><td rowspan="2">99.4% NaCl rejection [b]</td><td rowspan="2">Hydrophilic [b]</td></tr>
<tr><td>BW30</td></tr>
<tr><td>NF</td><td>NF90</td><td>Dow—FilmTec</td><td>Flat sheet</td><td>PA</td><td>200 Da [a]</td><td>Hydrophilic [a]</td><td>4.0</td><td>Synthetic wastewater</td><td>Lab</td><td>PET</td><td>1 mg/L</td><td>100</td><td>~100</td><td>[81]</td></tr>
<tr><td>NF</td><td>NF90</td><td>Dow—FilmTec</td><td>Flat sheet</td><td>PA</td><td>200 Da [a]</td><td>Hydrophilic [a]</td><td>4.0</td><td>Synthetic wastewater</td><td>Lab</td><td>PET</td><td>1 mg/L</td><td>100</td><td>100</td><td>[80]</td></tr>
<tr><td>RO</td><td>N.A.</td><td>N.A.</td><td>Spiral wound</td><td>N.A.</td><td>N.A.</td><td>N.A.</td><td>N.A.</td><td>Surface water</td><td>Full</td><td>Various types</td><td>0.96 ± 0.46 particles/L</td><td>20–5000</td><td>93 ± 5</td><td>[85]</td></tr>
</table>

Abbreviations: N.A.: not available; EPM: Ethylene-propylene polymer; PA: polyamide; PE: polyethylene; PES: polyethersulfone; PET: polyethylene terephthalate; PP: polypropylene.
[a] from ref. [86]. [b] from ref. [87].

3.4. Reverse Osmosis

Reverse-osmosis membranes differ from other membranes due to their operating driving force. In addition to operating under high pressures (>20 bar), this means that transport through the membranes is based on the solution-diffusion mechanism. Their pores are smaller than 1 nm, and their applications range from obtaining high-purity water to the desalination of seawater, including for drinking water. Indeed, RO is one of the most important technologies for purifying water and wastewater [32,88]. Since it operates under higher-pressure conditions compared to other PDMPs, it incurs higher operating costs associated with its greater energy consumption [43]. Greater care must be taken when using RO to remove MPs, as the high pressure applied can fragment the MPs, reducing their size and generating nanoplastics [9]. This may be more relevant for plastics that have undergone weathering, which makes them more fragile and brittle.

Considering that many studies with PDMPs operating under milder pressures, such as MF, UF, and NF, have demonstrated success in removing MPs, few studies exist with RO for this purpose (see Table 3). Furthermore, thanks to RO mainly being used as a polishing step for other PDMPs, few studies have truly evaluated the influence of MPs on fouling formation in RO membranes, since MPs are generally removed in previous steps. Ziajahromi et al. [89] focused on developing sampling and quantification methods for MPs in wastewater and found MPs in different treatment stages of the evaluated WWTPs. In a complete WWTP, with primary, secondary, and tertiary treatment—including screening (3 mm mesh), sedimentation, biological treatment, flocculation, disinfection/dechlorination processes, UF, RO, and decarbonation—the authors found approximately 0.21 MPs per liter in the RO permeate. Considering the flow rate of the studied WWTP, this represents a discharge of 10 million MPs per day into the environment. The authors suggested the possibility of membrane imperfections, as well as internal sealing problems in the membrane modules or pipes [89]. Although a small fraction of MPs escaped from RO, it is worth noting that it is much more effective than conventional systems commonly used in WWTPs, making WWTPs sources of MPs released into water bodies [90]. The same can be said regarding the WTP tested by Dalmau-Soler et al. [85], in which advanced treatment methods, such as UF and RO, performed better for MP removal than conventional ones or those followed by ozonation and activated carbon. The authors emphasized that the complete treatment, including the stages of membrane processes, achieved a removal efficiency of 93% ± 5% of MPs, and evaluating the performance of RO alone, it managed to eliminate 54% ± 27% of the particles that remained from the other stages. Although microplastics cannot be completely eliminated, it is important to highlight the ability of RO to remove viruses, bacteria, and virtually all organic and inorganic contaminants, providing greater safety for the treatment of drinking water [85,91]. Another interesting contribution of Dalmau-Soler et al. [85] is that they investigated the migration of MPs from the hydraulic components of the WTP. They concluded that even with a low input, it was possible to find PVC fractions, for example, coming from the WTP's own pipes. Therefore, the contamination of MPs in the system should be considered non-negligible.

RO membranes have also been modified and evaluated for MP removal. Farahbakhsh et al. [92] developed modified high-performance thin-film composite reverse-osmosis membranes, evaluating their hydraulic permeability, salt rejection, wettability, and fouling resistance, among other characteristics. The membrane that performed best in nanoplastic removal was the double-modified one, classified as TFC-RO-DM-2. This membrane was prepared using a PSU membrane as a base that underwent the incorporation of titanium-based metallic organic structures and multiwalled carbon nanotubes in the support layers and the selective layer, respectively. This modified bilayer membrane was able to remove NPs, achieving high permeate fluxes and good fouling resistance, attributed to its hydrophilic character and negative charge. The authors reported that it was possible to clean it with simple hydraulic washing.

As can be seen in Table 3, studies on NF and RO have predominantly focused on evaluating MPs in the micrometer and millimeter size ranges involving experiments performed on a laboratory scale or monitoring large-scale treatment plants. In the latter case, information regarding the membranes and operational conditions of the membrane processes or the treatment plants themselves is in fact scarce. It may seem counterintuitive to investigate NF and RO membranes, which have pore sizes below 10 nm, to remove particles as large as micrometer-scale. However, it is important to note that this may be related to the simple fact that larger particles are more easily detected and quantified than smaller ones, such as nanoplastics. Nonetheless, studies assessing large-scale treatment plants indicated the passage of MPs/NPs through both NF and RO membranes. While this may seem illogical, it could be attributed to a wider membrane pore-size distribution, imperfections in the membranes or modules, or even membrane aging. In addition, these studies did not report how long the membranes had been in operation, which is a critical factor, especially in the context of membrane application in leachate treatment, as was the case in the study by Zhang et al. [34]. In short, these are some hypotheses that may explain why NF and RO are being assessed to remove micrometric and even millimetric particles and also shed some light on the strange detection of MPs in NF and RO permeates.

4. Conclusions and Future Perspectives

This review presents a compilation of studies focused on PDMPs, unlike others that compared other strategies for removing MPs, allowing a comparison among the four types of PDMP. It was observed that most studies were carried out with MF or UF. This is associated with the fact that these PDMPs use membranes with smaller pores than the vast majority of MPs studied, which provides high removal rates. MF and UF operate at lower pressures than NF and RO and therefore have lower energy consumption, which is a crucial factor to consider in full-scale plants.

Among the studies analyzed, most were conducted on a laboratory scale using synthetic suspensions of microplastics in deionized or Milli-Q water or seeding them in some real (waste)water. In turn, some studies assessed industrial or domestic wastewater, landfill leachate, surface water, or water from different treatment stages in WTP or WWTP. Furthermore, studies performed with aqueous matrices consisting of synthetic wastewater or suspensions prepared with deionized or Milli-Q water were prepared with virgin polymer microplastics. Considering that the literature states that weathered MPs may present different behavior than virgin MPs, further studies evaluating PDMPs and other treatment systems in removing weathered MPs are necessary.

There was a predominance of studies whose concern in the preparation or functionalization of membranes or optimization of operational conditions focused on mitigating fouling and maintaining a stable permeate flux throughout the working regime. Indeed, new materials for manufacturing or modifying/functionalizing membranes to improve performance are a hot topic. Improvements in materials, preparation of 2D membranes, membranes based on graphene oxide, reduced graphene oxide, or MOFs, or membranes prepared with biodegradable materials can be applied in removing MPs, opening up a wide range of space for innovation in the field of material selection for membrane preparation.

Fouling is in fact the main drawback of membrane processes, which generally results in performance loss and increased operating costs. Although there are several studies on the role of MPs in membrane fouling, the information is still controversial, sometimes indicating that the presence of MPs in an aqueous matrix improves PDMP productivity or reporting the opposite; therefore, further studies are needed to consolidate the kinetics and mechanisms of fouling occurrence in different PDMPs.

Another important issue that requires in-depth investigation is defining the role of MPs in the clogging of feed channels in spiral-wound modules, particularly those of NF and RO, where spacer thickness typically ranges from 0.66 mm (26 mil) to 0.86 mm (34 mil) [93,94]. As shown in Tables 1–3, several studies have assessed the usage of PDMPs to remove MPs as large as 5 mm, demonstrating the potential for MPs to clog feed channels in spiral-wound modules containing NF or RO membranes. Similarly, despite having spiral-wound MF and UF modules featuring spacers thicker than 2 mm [78], their feed channels are also prone to clogging by MPs, especially by the larger ones, MP clusters, or MP agglomerates with colloidal substances.

Although some studies on rubber MPs exist, they evaluated treatment methods other than PDMPs. While microplastics are a recent subject of study and there are adversities related to the quantification and detection analyses of rubber microparticles, it would be interesting to evaluate the performance of PDMPs in their removal. This fact becomes even more important as the high deformation rates of rubbers, even under low shear stresses, can allow rubber MPs to undergo large deformations due to the operating conditions used in PDMPs, mainly due to the effect of pressure, and to pass through the membrane.

Considering the literature data we assessed, only one study cited a concern regarding the sludge generated from the PDMP concentrates and the pollutants adsorbed on the membranes. It is known that sludge treatments can fragment MPs during sludge drying by thermal processes or by lime stabilization. In addition, the sludge has more significant concentrations of MPs, often being disposed of in the soil, contaminating it to the point that it is possible to detect MPs in the soil even many years after sludge disposal. Therefore, studies based on the disposal of sludge and of PDMP concentrates rich in microplastics in the soil are necessary, aiming to investigate the interaction of these contaminants with different plant crops and soil microbiota. Physicochemical processes such as coagulation–flocculation–sedimentation (or flotation) could be an alternative for removing MPs from PDMP concentrates. In fact, the ideal would be to degrade or eliminate MPs from the environment, which in turn can be achieved by the action of microorganisms or worms, or even to evaluate the possibilities of recycling MPs or membranes, which would require an assessment of financial feasibility. Although chemical or mechanical recycling methodologies may not be feasible, avoiding the reintroduction of MPs into the environment is imperative.

Linking academic studies to the UN Sustainable Development Goals (SDGs) is an interesting way to demonstrate that the research is aligned with global goals and that the knowledge generated is relevant and has social and environmental impacts. In addition, associating a study with the SDGs can spark the interest of funding agencies, government agencies, or non-governmental organizations in making its implementation viable, which can drive public policies, economic transitions, and the provision of technologies to address crises such as climate change, for example. None of the studies evaluated in this review associated the removal of MPs with the SDGs, though some studies involving the removal of contaminants of emerging concern using membranes already made this connection. Goal 6—clean water and sanitation, in its subitems 6.3 and 6.a, and goal 14—life below water, in its subitems 14.1, 14.2, and 14.c are just two examples of SDGs with which a connection can be established with the removal of MPs by PDMPs. Considering the issues related to the management of PDMP concentrates and sludge formed in other treatment systems, there is also a relationship with SDGs 12—responsible consumption and production and 15—life on land, since PDMP concentrates and sludge generated in leachate treatment contain a high load of MPs, classifying them as causes of soil contamination. Both SDGs have targets that aim to reverse soil degradation and adopt sustainable waste management strategies to minimize negative impacts on human health and the environment. Considering that some areas of study related to water and wastewater treatment already use this association successfully, future work associating MP removal with SDGs tends to achieve promising results.

Indeed, MPs are emerging as a major concern, particularly for human health, but NPs may pose an even greater problem, since they can be as small as a few nanometers. This size allows for easier absorption into the human body, leading to deeper penetration into organs and tissues than MPs. Furthermore, MF and UF, the most commonly used PDMPs for MP removal, may be ineffective in removing small NPs. Additionally, some of the smallest NP particles could pass through NF membranes and also RO membranes. This is because the membranes have a pore size distribution, which means that they may have a fraction of pores larger and another of pores smaller than the nominal size, as well as some imperfections. On the other hand, despite concerns about micro- and nanoplastic pollution being a recent issue, several studies have shown that PDMPs are effective in removing these contaminants from aqueous systems. In most cases, rejections of almost 100% of particles were achieved, providing greater safety to the treatment system concerning the removal of these contaminants. Considering this and also the fact that PDMPs are capable of retaining contaminants of emerging concern whose removal in conventional treatment systems is ineffective, an expansion of the usage of membrane technologies in WTPs and WWTPs is expected in the near future, making it crucial to expand knowledge about the phenomena involved in the removal of MPs and NPs from water and wastewater by PDMPs.

Author Contributions: Conceptualization, P.E.P., A.G. and A.M.B.; methodology, P.E.P.; software, P.E.P.; validation, P.E.P. and G.M.d.A.; formal analysis, P.E.P. and G.M.d.A.; investigation, P.E.P., A.G. and G.M.d.A.; resources, M.A.S.R. and A.M.B.; data curation, P.E.P. and A.G.; writing—original draft preparation, P.E.P., A.G. and G.M.d.A.; writing—review and editing, P.E.P. and A.G.; visualization, P.E.P., A.G., G.M.d.A., M.A.S.R. and A.M.B.; supervision, A.G., M.A.S.R. and A.M.B.; project administration, A.M.B.; funding acquisition, M.A.S.R. and A.M.B. All authors have read and agreed to the published version of the manuscript.

Funding: The authors are grateful to the Brazilian funding agencies CAPES, CNPq, FINEP, and FAPERGS and to the Programa Iberoamericano de Ciencia y Tecnología para el Desarollo (CYTED; grant 323RT0143) for the financial support.

Institutional Review Board Statement: Not applicable.

Informed Consent Statement: Not applicable.

Data Availability Statement: No new data were created or analyzed in this study. Data sharing is not applicable to this article.

Acknowledgments: The authors thank the four anonymous reviewers and two academic editors for their constructive comments that significantly improved the manuscript. A.G. thanks FAPERGS-RITEs (process 22/2551-0000393-1) for the fellowship he received.

Conflicts of Interest: The authors declare no conflicts of interest.

References

1. Birley, A.W.; Heath, R.J.; Scott, M.J. *Plastic Materials*; Springer: Berlin/Heidelberg, Germany, 1991.
2. OCDE. *Global Plastics Outlook: Economic Drivers, Environmental Impacts and Policy Options*; OECD: Paris, France, 2022; ISBN 9789264654945.
3. Ritchie, H.; Roser, M. *Plastic Pollution—Our World in Data*; Our World in Data: Oxford, UK, 2022.
4. Singh, S.; Kalyanasundaram, M.; Diwan, V. Removal of Microplastics from Wastewater: Available Techniques and Way Forward. *Water Sci. Technol.* **2021**, *84*, 3689–3704. [CrossRef] [PubMed]

5. Smyth, K.; Tan, S.; Van Seters, T.; Henderson, V.; Passeport, E.; Drake, J. Pavement Wear Generates Microplastics in Stormwater Runoff. *J. Hazard. Mater.* **2025**, *481*, 136495. [CrossRef] [PubMed]
6. Olivatto, G.P.; Carreira, R.; Tornisielo, V.L.; Montagner, C.C. Microplastics: Contaminants of Global Concern in the Anthropocene. *Rev. Virtual Quim.* **2018**, *10*, 1968–1989. [CrossRef]
7. Ho, K.T.; Bjorkland, R.; Burgess, R.M. Comparing the Definitions of Microplastics Based on Size Range: Scientific and Policy Implications. *Mar. Pollut. Bull.* **2024**, *207*, 116907. [CrossRef]
8. Barbier, J.S.; Dris, R.; Lecarpentier, C.; Raymond, V.; Delabre, K.; Thibert, S.; Tassin, B.; Gasperi, J. Microplastic Occurrence after Conventional and Nanofiltration Processes at Drinking Water Treatment Plants: Preliminary Results. *Front. Water* **2022**, *4*, 886703. [CrossRef]
9. Mohana, A.A.; Rahman, M.; Sarker, S.K.; Haque, N.; Gao, L.; Pramanik, B.K. Nano/Microplastics: Fragmentation, Interaction with Co-Existing Pollutants and Their Removal from Wastewater Using Membrane Processes. *Chemosphere* **2022**, *309*, 136682. [CrossRef]
10. Hachemi, C.; Enfrin, M.; Rashed, A.O.; Jegatheesan, V.; Hodgson, P.D.; Callahan, D.L.; Lee, J.; Dumée, L.F. The Impact of PET Microplastic Fibres on PVDF Ultrafiltration Performance—A Short-Term Assessment of MP Fouling in Simple and Complex Matrices. *Chemosphere* **2023**, *310*, 136891. [CrossRef]
11. Lee, M.; Choi, W.; Lim, G. Electrokinetic-Assisted Filtration for Fast and Highly Efficient Removal of Microplastics from Water. *Chem. Eng. J.* **2023**, *452*, 139152. [CrossRef]
12. Enfrin, M.; Lee, J.; Le-Clech, P.; Dumée, L.F. Kinetic and Mechanistic Aspects of Ultrafiltration Membrane Fouling by Nano- and Microplastics. *J. Memb. Sci.* **2020**, *601*, 117890. [CrossRef]
13. Pramanik, B.K.; Pramanik, S.K.; Monira, S. Understanding the Fragmentation of Microplastics into Nano-Plastics and Removal of Nano/Microplastics from Wastewater Using Membrane, Air Flotation and Nano-Ferrofluid Processes. *Chemosphere* **2021**, *282*, 131053. [CrossRef]
14. Duncan, T.V.; Khan, S.A.; Patri, A.K.; Wiggins, S. Regulatory Science Perspective on the Analysis of Microplastics and Nanoplastics in Human Food. *Anal. Chem.* **2024**, *96*, 4343–4358. [CrossRef]
15. Lee, J.; Choi, Y.J.; Jeong, J.; Chae, K.J. Eye-Glass Polishing Wastewater as Significant Microplastic Source: Microplastic Identification and Quantification. *J. Hazard. Mater.* **2021**, *403*, 123991. [CrossRef] [PubMed]
16. Ma, B.; Xue, W.; Hu, C.; Liu, H.; Qu, J.; Li, L. Characteristics of Microplastic Removal via Coagulation and Ultrafiltration during Drinking Water Treatment. *Chem. Eng. J.* **2019**, *359*, 159–167. [CrossRef]
17. WWF. *No Plastic in Nature: Assessing Plastic Ingestion from Nature to People*; WWF: Gland, Switzerland, 2019.
18. Jenner, L.C.; Rotchell, J.M.; Bennett, R.T.; Cowen, M.; Tentzeris, V.; Sadofsky, L.R. Detection of Microplastics in Human Lung Tissue Using MFTIR Spectroscopy. *Sci. Total Environ.* **2022**, *831*, 154907. [CrossRef] [PubMed]
19. Ali, N.; Katsouli, J.; Marczylo, E.L.; Gant, T.W.; Wright, S.; Bernardino de la Serna, J. The Potential Impacts of Micro-and-Nano Plastics on Various Organ Systems in Humans. *EBioMedicine* **2024**, *99*, 104901. [CrossRef]
20. Amato-Lourenço, L.F.; Dantas, K.C.; Júnior, G.R.; Paes, V.R.; Ando, R.A.; de Oliveira Freitas, R.; da Costa, O.M.M.M.; Rabelo, R.S.; Soares Bispo, K.C.; Carvalho-Oliveira, R.; et al. Microplastics in the Olfactory Bulb of the Human Brain. *JAMA Netw. Open* **2024**, *7*, e2440018. [CrossRef]
21. Senathirajah, K.; Attwood, S.; Bhagwat, G.; Carbery, M.; Wilson, S.; Palanisami, T. Estimation of the Mass of Microplastics Ingested—A Pivotal First Step towards Human Health Risk Assessment. *J. Hazard. Mater.* **2021**, *404*, 124004. [CrossRef]
22. Diaz-Basantes, M.F.; Conesa, J.A.; Fullana, A. Microplastics in Honey, Beer, Milk and Refreshments in Ecuador as Emerging Contaminants. *Sustainability* **2020**, *12*, 5514. [CrossRef]
23. Cherian, A.G.; Liu, Z.; McKie, M.J.; Almuhtaram, H.; Andrews, R.C. Microplastic Removal from Drinking Water Using Point-of-Use Devices. *Polymers* **2023**, *15*, 1331. [CrossRef]
24. Lambert, S.; Wagner, M. Microplastics Are Contaminants of Emerging Concern in Freshwater Environments: An Overview. In *Handbook of Environmental Chemistry*; Wagner, M., Lambert, S., Eds.; Springer Nature: Cham, Switzerland, 2018; Volume 58, pp. 1–23; ISBN 978-3-319-61615-5.
25. Malankowska, M.; Echaide-Gorriz, C.; Coronas, J. Microplastics in Marine Environment: A Review on Sources, Classification, and Potential Remediation by Membrane Technology. *Environ. Sci.* **2021**, *7*, 243–258. [CrossRef]
26. Rasheed, A.; Sharma, N.; Surampalli, R.Y.; Das, S. Evaluating Treatment Solutions: Critical Review on Technologies Employed for Microplastic Removal from Water Matrices. *Curr. Opin. Environ. Sci. Health* **2023**, *36*, 100516. [CrossRef]
27. Talvitie, J.; Mikola, A.; Koistinen, A.; Setälä, O. Solutions to Microplastic Pollution—Removal of Microplastics from Wastewater Effluent with Advanced Wastewater Treatment Technologies. *Water Res.* **2017**, *123*, 401–407. [CrossRef] [PubMed]
28. Das, A.A.; Mishra, S.; Pradhan, P.; Nayak, L.; Raulo, S.; Das, K.; Sahoo, M.; Sahoo, P.K.; Naik, B.; Rout, P.R.; et al. Recent Advancements in Microplastics Treatments: Characteristics, Occurrence, and Removal Technologies. *Mater. Today Proc.* **2022**, *67*, 1211–1217. [CrossRef]

29. Hidayaturrahman, H.; Lee, T.G. A Study on Characteristics of Microplastic in Wastewater of South Korea: Identification, Quantification, and Fate of Microplastics during Treatment Process. *Mar. Pollut. Bull.* **2019**, *146*, 696–702. [CrossRef] [PubMed]
30. Von Sperling, M. Comparison among the Most Frequently Used Systems for Wastewater Treatment in Developing Countries. *Water Sci. Technol.* **1996**, *33*, 59–72. [CrossRef]
31. Thompson, R.C.; Olson, Y.; Mitchell, R.P.; Davis, A.; Rowland, S.J.; John, A.W.G.; McGonigle, D.; Russell, A.E. Lost at Sea: Where Is All the Plastic? *Science* **2004**, *304*, 838. [CrossRef]
32. Golgoli, M.; Khiadani, M.; Shafieian, A.; Sen, T.K.; Hartanto, Y.; Johns, M.L.; Zargar, M. Microplastics Fouling and Interaction with Polymeric Membranes: A Review. *Chemosphere* **2021**, *283*, 131185. [CrossRef]
33. Sanguanpak, S.; Chiemchaisri, W.; Chiemchaisri, C. Membrane Fouling and Micro-Pollutant Removal of Membrane Bioreactor Treating Landfill Leachate. *Rev. Environ. Sci. Biotechnol.* **2019**, *18*, 715–740. [CrossRef]
34. Zhang, Z.; Su, Y.; Zhu, J.; Shi, J.; Huang, H.; Xie, B. Distribution and Removal Characteristics of Microplastics in Different Processes of the Leachate Treatment System. *Waste Manag.* **2021**, *120*, 240–247. [CrossRef]
35. Siddiqui, M.U.; Arif, A.F.M.; Bashmal, S. Permeability-Selectivity Analysis of Microfiltration and Ultrafiltration Membranes: Effect of Pore Size and Shape Distribution and Membrane Stretching. *Membranes* **2016**, *6*, 40. [CrossRef]
36. Fijoł, N.; Aguilar-Sánchez, A.; Ruiz-Caldas, M.X.; Redlinger-Pohn, J.; Mautner, A.; Mathew, A.P. 3D Printed Polylactic Acid (PLA) Filters Reinforced with Polysaccharide Nanofibers for Metal Ions Capture and Microplastics Separation from Water. *Chem. Eng. J.* **2023**, *457*, 141153. [CrossRef]
37. Giacobbo, A.; Bernardes, A.M. Membrane Separation Process in Wastewater and Water Purification. *Membranes* **2022**, *12*, 259. [CrossRef] [PubMed]
38. Giacobbo, A.; Pasqualotto, I.F.; Filho, R.C.d.C.M.; Minhalma, M.; Bernardes, A.M.; de Pinho, M.N. Ultrafiltration and Nanofiltration for the Removal of Pharmaceutically Active Compounds from Water: The Effect of Operating Pressure on Electrostatic Solute—Membrane Interactions. *Membranes* **2023**, *13*, 743. [CrossRef] [PubMed]
39. Giacobbo, A.; do Prado, J.M.; Meneguzzi, A.; Bernardes, A.M.; de Pinho, M.N. Microfiltration for the Recovery of Polyphenols from Winery Effluents. *Sep. Purif. Technol.* **2015**, *143*, 12–18. [CrossRef]
40. Bai, W.; Wang, F.; Yan, L.; Sun, H.; Zhu, Z.; Chen, L.; Li, J.; Liang, W.; Li, A. Porous Organic Polymers (POPs) Membrane via Thiol-Yne Click Chemistry for Efficient Particulate Matter Capture and Microplastics Separation. *Microporous Mesoporous Mater.* **2022**, *329*, 111509. [CrossRef]
41. Yang, L.; Cao, X.; Cui, J.; Wang, Y.; Zhu, Z.; Sun, H.; Liang, W.; Li, J.; Li, A. Holey Ti_3C_2 Nanosheets Based Membranes for Efficient Separation and Removal of Microplastics from Water. *J. Colloid Interface Sci.* **2022**, *617*, 673–682. [CrossRef]
42. Yogarathinam, L.T.; Usman, J.; Othman, M.H.D.; Ismail, A.F.; Goh, P.S.; Gangasalam, A.; Adam, M.R. Low-Cost Silica Based Ceramic Supported Thin Film Composite Hollow Fiber Membrane from Guinea Corn Husk Ash for Efficient Removal of Microplastic from Aqueous Solution. *J. Hazard. Mater.* **2022**, *424*, 127298. [CrossRef]
43. Acarer, S. A Review of Microplastic Removal from Water and Wastewater by Membrane Technologies. *Water Sci. Technol.* **2023**, *88*, 199–219. [CrossRef]
44. Acarer, S. Abundance and Characteristics of Microplastics in Drinking Water Treatment Plants, Distribution Systems, Water from Refill Kiosks, Tap Waters and Bottled Waters. *Sci. Total Environ.* **2023**, *884*, 163866. [CrossRef]
45. Pizzichetti, A.R.P.; Pablos, C.; Álvarez-Fernández, C.; Reynolds, K.; Stanley, S.; Marugán, J. Evaluation of Membranes Performance for Microplastic Removal in a Simple and Low-Cost Filtration System. *Case Stud. Chem. Environ. Eng.* **2021**, *3*, 100075. [CrossRef]
46. Hyeon, Y.; Kim, S.; Ok, E.; Park, C. A Fluid Imaging Flow Cytometry for Rapid Characterization and Realistic Evaluation of Microplastic Fiber Transport in Ceramic Membranes for Laundry Wastewater Treatment. *Chem. Eng. J.* **2023**, *454*, 140028. [CrossRef]
47. Kook, H.; Cha, M.; Park, C. Transport of Emerging Organic Ultraviolet (UV) Filters in Ceramic Membranes: Role of Polyethylene (PE) Microplastics. *Chemosphere* **2022**, *309*, 136570. [CrossRef] [PubMed]
48. Kook, H.; Park, C. Engineered Approaches to Facile Identification of Tiny Microplastics in Polymeric and Ceramic Membrane Filtrations for Wastewater Treatment. *Membranes* **2022**, *12*, 565. [CrossRef] [PubMed]
49. Sriani, T.; Mahardika, M.; Miki, N.; Wulandari, C.P.; Prihandana, G.S. Impact of Polyimide on the Recycling of Waste Expanded Polystyrene into Flat-Sheet Filtration Membrane. *J. Mater. Cycles Waste Manag.* **2024**, *26*, 3745–3756. [CrossRef]
50. Yaranal, N.A.; Subbiah, S.; Mohanty, K. Identification, Extraction of Microplastics from Edible Salts and Its Removal from Contaminated Seawater. *Environ. Technol. Innov.* **2021**, *21*, 101253. [CrossRef]
51. Bitter, H.; Krause, L.; Kirchen, F.; Fundneider, T.; Lackner, S. Semi-Crystalline Microplastics in Wastewater Plant Effluents and Removal Efficiencies of Post-Treatment Filtration Systems. *Water Res. X* **2022**, *17*, 100156. [CrossRef]
52. Hyeon, Y.; Kim, S.; Park, C. Exploring the Transformation of Polyethylene and Polyamide Microplastics during Membrane Filtration through FlowCam Analysis. *Sep. Purif. Technol.* **2024**, *334*, 126036. [CrossRef]
53. Akarsu, C.; Kumbur, H.; Kideys, A.E. Removal of Microplastics from Wastewater through Electrocoagulation-Electroflotation and Membrane Filtration Processes. *Water Sci. Technol.* **2021**, *84*, 1648–1662. [CrossRef]

54. Pizzichetti, A.R.P.; Pablos, C.; Álvarez-Fernández, C.; Reynolds, K.; Stanley, S.; Marugán, J. Kinetic and Mechanistic Analysis of Membrane Fouling in Microplastics Removal from Water by Dead-End Microfiltration. *J. Environ. Chem. Eng.* **2023**, *11*, 109338. [CrossRef]
55. Lee, J.; Wang, J.; Oh, Y.; Jeong, S. Highly Efficient Microplastics Removal from Water Using In-Situ Ferrate Coagulation: Performance Evaluation by Micro-Fourier-Transformed Infrared Spectroscopy and Coagulation Mechanism. *Chem. Eng. J.* **2023**, *451*, 138556. [CrossRef]
56. Kim, H.J.; Lee, U.; Lee, I.; Hong, Y.; Lee, J. Enhancement of Polystyrene Microplastic Removal by near Dissolved Organic Matter Microfiltration (NDOM MF) Coupled with Cold Plasma Treatment. *J. Water Process Eng.* **2023**, *54*, 103901. [CrossRef]
57. Takeuchi, H.; Tanaka, S.; Koyuncu, C.Z.; Nakada, N. Removal of Microplastics in Wastewater by Ceramic Microfiltration. *J. Water Process Eng.* **2023**, *54*, 104010. [CrossRef]
58. Minh Tran, H.; Kwak, D.; Lee, U.; Chang, S.; Thanh Tran, D.; Lee, J. A Hydrophilic near Dissolved Organic Matter Microfiltration (NDOM MF) Membrane Prepared Using Multifunctional Porogen Synthesized via Metal-Free Atom Transfer Radical Polymerization for Highly Efficient Microplastic Removal. *Chem. Eng. J.* **2024**, *480*, 147564. [CrossRef]
59. Fundneider, T.; Alejo, L.; Lackner, S. Tertiary Phosphorus Removal to Extremely Low Levels by Coagulation-Flocculation and Cloth-Filtration. *Water Sci. Technol.* **2020**, *82*, 131–143. [CrossRef]
60. Song, Y.; Pan, J.; Chen, M.; Wang, Y.; Li, Z.; Ge, Y. Chitosan-Modified Geopolymer Sub-Microparticles Reinforced Multifunctional Membrane for Enhanced Removal of Multiple Contaminants in Water. *J. Memb. Sci.* **2022**, *658*, 120704. [CrossRef]
61. Li, Z.; Xie, W.; Zhang, Z.; Wei, S.; Chen, J.; Li, Z. Multifunctional Sodium Alginate/Chitosan-Modified Graphene Oxide Reinforced Membrane for Simultaneous Removal of Nanoplastics, Emulsified Oil, and Dyes in Water. *Int. J. Biol. Macromol.* **2023**, *245*, 125524. [CrossRef]
62. Yang, J.; Monnot, M.; Sun, Y.; Asia, L.; Wong-Wah-Chung, P.; Doumenq, P.; Moulin, P. Microplastics in Different Water Samples (Seawater, Freshwater, and Wastewater): Removal Efficiency of Membrane Treatment Processes. *Water Res.* **2023**, *232*, 119673. [CrossRef]
63. Cuartucci, M. Ultrafiltration, a Cost-Effective Solution for Treating Surface Water to Potable Standard. *Water Pr. Technol.* **2020**, *15*, 426–436. [CrossRef]
64. Baresel, C.; Harding, M.; Fang, J. Ultrafiltration/Granulated Active Carbon-Biofilter: Efficient Removal of a Broad Range of Micropollutants. *Appl. Sci.* **2019**, *9*, 710. [CrossRef]
65. Le Minh, T.; Van Ha, N.T. Surveying the Possibility of Recovering Microplastic in Waste Water from Plastic Processing and Recycling Facilities by Electrocoagulation and Ultrafiltration at Lab-Scale. *Vietnam J. Sci. Technol.* **2022**, *60*, 50–62. [CrossRef]
66. Wang, W.; Yang, M.; Ma, H.; Liu, Z.; Gai, L.; Zheng, Z.; Ma, H. Removal Behaviors and Mechanism of Polystyrene Microplastics by Coagulation/Ultrafiltration Process: Co-Effects of Humic Acid. *Sci. Total Environ.* **2023**, *881*, 163408. [CrossRef] [PubMed]
67. Ghasemi, S.; Yan, B.; Zargar, M.; Ling, N.N.A.; Fridjonsson, E.O.; Johns, M.L. Impact of Microplastics on Organic Fouling of Hollow Fiber Membranes. *Chem. Eng. J.* **2023**, *467*, 143320. [CrossRef]
68. Yuan, C.; Almuhtaram, H.; McKie, M.J.; Andrews, R.C. Assessment of Microplastic Sampling and Extraction Methods for Drinking Waters. *Chemosphere* **2022**, *286*, 131881. [CrossRef] [PubMed]
69. González-Camejo, J.; Morales, A.; Peña-Lamas, J.; Lafita, C.; Enguídanos, S.; Seco, A.; Martí, N. Feasibility of Rapid Gravity Filtration and Membrane Ultrafiltration for the Removal of Microplastics and Microlitter in Sewage and Wastewater from Plastic Industry. *J. Water Process Eng.* **2023**, *51*, 103452. [CrossRef]
70. Xiong, X.; Siddique, M.S.; Graham, N.J.D.; Yu, W. Towards Microplastics Contribution for Membrane Biofouling and Disinfection By-Products Precursors: The Effect on Microbes. *J. Hazard. Mater.* **2022**, *426*, 127797. [CrossRef]
71. Tadsuwan, K.; Babel, S. Microplastic Abundance and Removal via an Ultrafiltration System Coupled to a Conventional Municipal Wastewater Treatment Plant in Thailand. *J. Environ. Chem. Eng.* **2022**, *10*, 107142. [CrossRef]
72. Tadsuwan, K.; Babel, S. Unraveling Microplastics Removal in Wastewater Treatment Plant: A Comparative Study of Two Wastewater Treatment Plants in Thailand. *Chemosphere* **2022**, *307*, 135733. [CrossRef]
73. Enfrin, M.; Lee, J.; Fane, A.G.; Dumée, L.F. Mitigation of Membrane Particulate Fouling by Nano/Microplastics via Physical Cleaning Strategies. *Sci. Total Environ.* **2021**, *788*, 147689. [CrossRef]
74. Luogo, B.D.P.; Salim, T.; Zhang, W.; Hartmann, N.B.; Malpei, F.; Candelario, V.M. Reuse of Water in Laundry Applications with Micro-and Ultrafiltration Ceramic Membrane. *Membranes* **2022**, *12*, 223. [CrossRef]
75. Fryczkowska, B.; Przywara, L. Removal of Microplastics from Industrial Wastewater Utilizing an Ultrafiltration Composite Membrane Rgo/Pan Application. *Desalination Water Treat.* **2021**, *214*, 252–262. [CrossRef]
76. Ma, B.; Xue, W.; Ding, Y.; Hu, C.; Liu, H.; Qu, J. Removal Characteristics of Microplastics by Fe-Based Coagulants during Drinking Water Treatment. *J. Environ. Sci.* **2019**, *78*, 267–275. [CrossRef] [PubMed]
77. Akarsu, C.; Isik, Z.; M'barek, I.; Bouchareb, R.; Dizge, N. Treatment of Personal Care Product Wastewater for Reuse by Integrated Electrocoagulation and Membrane Filtration Processes. *J. Water Process Eng.* **2022**, *48*, 102879. [CrossRef]

78. Synder Filtration BY (PVDF 100,000Da) Sanitary UF Membrane Data Sheet. Available online: https://synderfiltration.com/2014/wp-content/uploads/2018/05/BY-PVDF-100kD-Sanitary-Specsheet.pdf (accessed on 26 January 2025).
79. Microdyn-Nadir NADIR UP150 P Flat Sheet Membrane Data Sheet. Available online: https://water-membrane-solutions.mann-hummel.com/content/dam/water-membrane-solutions/download/technical-data-sheets/nadir/nadir-up150-p-flat-sheet-membrane-data-sheet.pdf/_jcr_content/renditions/original./nadir-up150-p-flat-sheet-membrane-data-sheet.pdf (accessed on 26 January 2025).
80. Lin, D.; Lai, C.; Wang, X.; Wang, Z.; Kuang, K.; Wang, Z.; Du, X.; Liu, L. Enhanced Membrane Fouling by Microplastics during Nanofiltration of Secondary Effluent Considering Secretion, Interaction and Deposition of Extracellular Polymeric Substances. *Sci. Total Environ.* **2024**, *906*, 167110. [CrossRef] [PubMed]
81. Chen, J.; Huo, L.; Yuan, Y.; Jiang, Y.; Wang, H.; Hui, K.; Li, Y.; Huang, Z.; Xi, B. Interactions between Microplastics and Heavy Metals in Leachate: Implications for Landfill Stabilization Process. *J. Hazard. Mater.* **2024**, *480*, 135830. [CrossRef]
82. Kara, N.; Sari Erkan, H.; Onkal Engin, G. Characterization and Removal of Microplastics in Landfill Leachate Treatment Plants in Istanbul, Turkey. *Anal. Lett.* **2023**, *56*, 1535–1548. [CrossRef]
83. Fortin, S.; Song, B.; Burbage, C. Quantifying and Identifying Microplastics in the Effluent of Advanced Wastewater Treatment Systems Using Raman Microspectroscopy. *Mar. Pollut. Bull.* **2019**, *149*, 110579. [CrossRef]
84. Lin, D.; Lai, C.; Shen, X.; Wang, Z.; Xu, D.; Nie, J.; Song, W.; Du, X.; Liu, L. Electro-Coagulation Pretreatment for Improving Nanofiltration Membrane Performance during Reclamation of Microplastic-Contaminated Secondary Effluent: Unexpectedly Enhanced Membrane Fouling and Mechanism Analysis by MD-DFT Simulation. *Chem. Eng. J.* **2024**, *498*, 155779. [CrossRef]
85. Dalmau-Soler, J.; Ballesteros-Cano, R.; Boleda, M.R.; Paraira, M.; Ferrer, N.; Lacorte, S. Microplastics from Headwaters to Tap Water: Occurrence and Removal in a Drinking Water Treatment Plant in Barcelona Metropolitan Area (Catalonia, NE Spain). *Environ. Sci. Pollut. Res.* **2021**, *28*, 59462–59472. [CrossRef]
86. de Souza, D.I.; Dottein, E.M.; Giacobbo, A.; Rodrigues, M.A.S.; de Pinho, M.N.; Bernardes, A.M. Nanofiltration for the Removal of Norfloxacin from Pharmaceutical Effluent. *J. Environ. Chem. Eng.* **2018**, *6*, 6147–6153. [CrossRef]
87. Zakmout, A.; Sadi, F.; Portugal, C.A.M.; Crespo, J.G.; Velizarov, S. Tannery Effluent Treatment by Nanofiltration, Reverse Osmosis and Chitosan Modified Membranes. *Membranes* **2020**, *10*, 378. [CrossRef]
88. Venzke, C.D.; Rodrigues, M.A.S.; Giacobbo, A.; Bacher, L.E.; Lemmertz, I.S.; Viegas, C.; Striving, J.; Pozzebon, S. Application of Reverse Osmosis to Petrochemical Industry Wastewater Treatment Aimed at Water Reuse. *Manag. Environ. Qual. Int. J.* **2017**, *28*, 70–77. [CrossRef]
89. Ziajahromi, S.; Neale, P.A.; Rintoul, L.; Leusch, F.D.L. Wastewater Treatment Plants as a Pathway for Microplastics: Development of a New Approach to Sample Wastewater-Based Microplastics. *Water Res.* **2017**, *112*, 93–99. [CrossRef] [PubMed]
90. Xu, Z.; Bai, X.; Ye, Z. Removal and Generation of Microplastics in Wastewater Treatment Plants: A Review. *J. Clean. Prod.* **2021**, *291*, 125982. [CrossRef]
91. Giacobbo, A.; Soares, E.V.; Bernardes, A.M.; Rosa, M.J.; de Pinho, M.N. Atenolol Removal by Nanofiltration: A Case-Specific Mass Transfer Correlation. *Water Sci. Technol.* **2020**, *81*, 210–216. [CrossRef] [PubMed]
92. Farahbakhsh, J.; Najafi, M.; Golgoli, M.; Haeri, S.Z.; Khiadani, M.; Razmjou, A.; Zargar, M. Dual Modification of Reverse Osmosis Membranes with NH2-MIL-125 and Functionalised Multiwalled Carbon Nanotubes for Enhanced Nanoplastic Removal. *Chemosphere* **2024**, *361*, 142401. [CrossRef]
93. Lin, W.; Zhang, Y.; Li, D.; Wang, X.; Huang, X. Roles and Performance Enhancement of Feed Spacer in Spiral Wound Membrane Modules for Water Treatment: A 20-Year Review on Research Evolvement. *Water Res.* **2021**, *198*, 117146. [CrossRef]
94. Karabelas, A.J.; Koutsou, C.P.; Sioutopoulos, D.C. Comprehensive Performance Assessment of Spacers in Spiral-Wound Membrane Modules Accounting for Compressibility Effects. *J. Memb. Sci.* **2018**, *549*, 602–615. [CrossRef]

membranes

MDPI

Article

Electrospun Membranes Loaded with Melanin Derived from Pecan Nutshell (*Carya illinoinensis*) Residues for Skin-Care Applications

Michell García-García [1,†], Jesús Salvador Jaime-Ferrer [1,*], Fernanda Nayeli Medrano-Lango [2,†], Elizabeth Quintana-Rodríguez [1], Tonatiu Campos-García [3], Erika Rodríguez-Sevilla [4] and Domancar Orona-Tamayo [1,*]

1 CIATEC A.C., Centro de Innovación Aplicada en Tecnologías Competitivas, Omega 201, Industrial Delta, León C.P. 37545, Guanajuato, Mexico; mgarcia.picyt@ciatec.mx (M.G.-G.); equintana@ciatec.mx (E.Q.-R.)
2 Unidad Profesional Interdisciplinaria de Ingeniería Campus Guanajuato, Instituto Politécnico Nacional, Av. Mineral de Valenciana 200, Col. Fraccionamiento Industrial Puerto Interior, Silao de la Victoria C.P. 36275, Guanajuato, Mexico; fernay_medrano@yahoo.com
3 CONACYT Research Fellow, CIATEJ. A.C., Centro de Investigación y Asistencia en Tecnología y Diseño del Estado de Jalisco, Av. de los Normalistas No. 800, Colinas de La Normal, Guadalajara C.P. 44270, Jalisco, Mexico; tocampos_pos@ciatej.edu.mx
4 CIO. A.C., Centro de Investigaciones en Óptica, Loma del Bosque 105, Lomas del Campestre, León C.P. 37150, Guanajuato, Mexico; erika.rodriguez@cio.mx
* Correspondence: jjaime@ciatec.mx (J.S.J.-F.); dorona@ciatec.mx (D.O.-T.)
† These authors contributed equally to this work.

Abstract: This study investigates the incorporation of melanin extracted from pecan nutshell residues into a polyacrylonitrile (PAN) matrix during the electrospinning of microfiber membranes. Melanin concentrations of 0.5, 2.0, and 5.0% *w*/*w* were incorporated to enhance the physicochemical and biological properties of the fibers. The melanin-loaded PAN fibers exhibited significant antioxidant activity against DPPH and ABTS radicals, with scavenging rates ranging from 46.58% to 62.77% and 41.02% to 82.36%, respectively, while unmodified PAN fibers showed no activity. Furthermore, the melanin-loaded membranes demonstrated antimicrobial effects. The membranes also exhibited an important enzyme inhibition activity against collagenase (37%), hyaluronidase (22%), tyrosinase (36%), and elastase (33%). Molecular docking studies reveal different potential amino acids of the active sites of aging enzymes that interact strongly with melanin pigment, particularly collagenase, followed by hyaluronidase, tyrosinase, and elastase. These results suggest that the novel melanin-loaded PAN membranes possess promising bioactive properties with potential applications in different skin-care applications.

Keywords: spinning fibers; molecular docking; bioactive compound; antioxidant activity; antimicrobial activity; antiaging effect

1. Introduction

The skin is one of the human body's largest and most versatile organs. In that regard, the skin is exposed continuously to different environmental factors (temperature variations, tobacco smoke, chemical pollutants, and radiation UV), and also, this organ has the ability to reduce negative environmental effects, decreasing mechanical damage, regulating critical physiological and biochemical processes, is sensitive to stimuli, regulates vitamin synthesis, promotes immune activity, and is sensitive to humidity [1,2]. Many of these stressful events trigger degenerative processes such as skin cancer, cellular aging, spots, and burns, as

https://doi.org/10.3390/membranes15020044

well as microbial infections [3]. Faced with these stressful events on the skin, people have sought solutions in cosmetic products made with ingredients of natural origin to feel better and to prevent damage, as well as aging, skin infections, and other aspects that can harm the skin. In that regard, in recent years, cosmetic industries, particularly in skin care, have been looking for new products with natural ingredients that help people who seek to have young and healthy skin to feel well.

The skin-care industry has shifted towards sustainability and the more efficient use of natural resources [4]. In this context, bioactive compounds obtained from agricultural biomasses have emerged as an innovative and environmentally friendly alternative for the formulation of cosmetic products. These compounds, derived from fruit and vegetable peels, herbs, coffee grounds, leaves, prunes, shells, and flower petals, benefit the skin and the environment [4–6]. Using natural compounds not only helps reduce the environmental impact by giving a second life to waste but also incorporates antioxidant, antiaging, antimicrobial, moisturizing, and regenerative properties that benefit skin health [6]. In addition, this practice supports the circular economy, aligning with the growing demands of conscious consumers looking for responsible and practical solutions in their personal care routines. Natural plant molecules remain particularly interesting for new development research to produce skin-care products with fewer side effects and rich sources of beneficial compounds for the skin.

Plant bioactive compounds can be used in new technologies that combine polymers and molecule bioactivities. One of these technologies is electrospinning, which can be used versatilely due to the use of biodegradable polymeric materials and uses electricity to produce functional biomaterials that can fuse bioactive molecules into their membranes with different target uses [7]. For example, in recent years, electrospinning technology has emerged as a promising avenue to overcome the limitations inherent in conventional skin-care approaches. The electrospinning technique has garnered significant attention due to its ability to fabricate nanofibrous or microfibrous membranes with exceptional properties, including high porosity, controlled drug release capabilities, and enhanced biocompatibility [8]. By combining the versatility of electrospinning with the potential of active compounds, researchers aim to develop materials that not only provide a physical barrier but also deliver active agents to the application site [9]. Active compounds such as antimicrobial and antioxidant agents, growth factors, antiaging compounds, and anti-inflammatory drugs can be incorporated into the electrospun fibers [10,11]. Several reviews have highlighted the potential advantages of electrospinning in skin-care applications. These reviews mainly summarize aspects of fiber technologies, including the synthesis methodologies [12–15], dressing composition [16,17], and nature of the materials employed, encompassing synthetic and natural polymers as well as self-responsive matter [18–22]. Furthermore, these reviews delved into the morphology, structure, and surface functionalization of the fibers [23,24], the bioactive constituents [25–28], the healing process of wounds and the therapeutic approach to reduce healing time, and antiaging effects [29,30].

Nowadays, the integration of active compounds derived from plant extracts [31,32], fruit peels [33,34], or agricultural by-product [35–37] with fibers has garnered significant attention in the field of skin-care applications. These approaches address environmental concerns associated with agricultural waste and the need for advanced materials with antimicrobial, antioxidant, and antiaging bioactivities. In this scenario, certain processing industries generate a significant amount of shell waste annually, most of which is disposed, leading to considerable environmental challenges [38,39]. In that sense, there is a comprehensive overview of various alternatives that have been investigated using electrospinning techniques, with the aim of imparting added value to shell wastes. These efforts represent a promising avenue for transforming environmental liabilities into valuable resources for skin-care applications.

The incorporation of pecan nuts (*Carya illinoinensis*) into the bakery and snack industry poses significant waste management challenges. This is primarily due to the substantial proportion of the original nut that is discarded during processing, which amounts to approximately 40 to 50% of its total mass. Most of this waste consists of the outer protective layer, or shell, that encases the edible kernel [40]. Pecan shells are also known for containing phenolic compounds (ranging from 32 to 117 mg GAE g^{-1}), condensed tannins (with values between 130 and 357 mg CE g^{-1}), and total anthocyanins (measuring 1 to 3 mg 3-glucoside cyanidin g^{-1}) [41]. In addition to the previously listed compounds, melanin was recovered from the pecan shell with antioxidant and antifungal capacities [42], enhancing the value of an otherwise underutilized resource. Extracted melanin exhibits robust free-radical-scavenging properties, effectively neutralizing reactive oxygen species (ROS) and protecting cells from oxidative damage. This antioxidant activity plays a crucial role in mitigating the harmful effects of oxidative stress, which has been associated with various diseases and aging processes [43,44]. Ultraviolet (UV) radiation is a primary contributor to skin aging. Prolonged exposure to high levels of UV radiation induces the generation of ROS, leading to oxidative stress. This oxidative stress subsequently damages collagen and elastin fibers in the extracellular matrix (ECM), contributing significantly to the visible signs of skin aging. For example, the hyaluronidase enzyme regulates the hyaluronic acid degradation essential in skin hydration, while the tyrosinase enzyme is crucial for skin pigmentation; however, under oxidative stress, increased tyrosinase activity can produce skin hyperpigmentation [45]. Additionally, the ROS can increase the activity of the different aging enzymes, such as elastase, hyaluronidase, collagenase, and tyrosinase enzymes, that promote ECM degradation and the formation of photoaging signs. Therefore, bioactive compounds loaded into fibers could protect the skin from environmental pollutants to combat oxidative stress. Particularly, melanin extracts have demonstrated the ability to increase antioxidant enzyme activities, decrease lipid peroxidation, and potentially inhibit enzymes involved in cellular aging processes in a mice model [46].

In view of the bioactive properties of melanin outlined earlier, this new study aims to investigate the potential melanin-incorporated polyacrylonitrile (PAN) fibers as a novel skin-care material. Melanin extracted from Pecan nut (*C. illinoinensis*) shells was integrated into PAN fibers at varying concentrations (0.5%, 2.0%, and 5.0% *w*/*w*). This research encompasses a comprehensive evaluation of electrospinning parameters and extensive physico-chemical characterization of the resulting membranes. Furthermore, this study examines the degradation kinetics of the membranes and the release profile of melanin. This investigation extends to assessing the antioxidant properties of the fibers, their bacterial inhibition against *Escherichia coli*, *Staphylococcus aureus*, and *Pseudomonas aeruginosa* strains, and the capacity of melanin to inhibit aging-related enzymes including elastase, collagenase, tyrosinase, and hyaluronidase. Additionally, a molecular docking study was performed to determine the binding interactions between catalytic center amino acids and melanin pigment. This multifaceted approach aims to elucidate the potential applications of melanin-infused PAN membranes in advanced skin-care formulations.

2. Materials and Methods

2.1. Reagents

Polyacrylonitrile (PAN) (Mw 150,000), DPPH (1,1-diphenyl-2-picrylhydrazyl), ABTS (2,2′-azino-bis (3-ethylbenzothiazoline-6-sulfonic acid), tyrosinase enzyme (from mushroom, 1.14.18.1), 3,4-dihydroxy-L-phenylalanine (≥98%, CAS 59-92-7), collagenase enzyme (from *Clostridium histolyticum*, 3.4.24.3), Z-Gly-Pro-Gly-Gly-Pro-Ala-OH (collagenase substrate, CAS 13075-38-2), hyaluronidase enzyme (from Bovine testes, Type I-S, 3.2.1.35), Poly(β-glucuronic acid-[1→3]-β-N-acetylglucosamine-[1→4]) (alternating hyaluronic acid sodium salt from rooster comb, CAS 9067-32-7), elastase enzyme (from porcine pancreas, type IV, 3.4.21.36), and N-succinyl-L-alanyl-L-alanyl-L-alanine 4-nitroanilide (CAS 52299-14-6) were purchased from Sigma-Aldrich Chemical Reagent Co., Ltd. (Saint Louis, MO, USA). N,N-dimethylformamide (N,N-DMF, 99.8%), chloroform ($CHCl_3$, >99.5%), ammonium hydroxide (NH_4OH, 28.0–30.0%), and hydrochloric acid (HCl, 36.5–38.0%) were purchased from Karal^{TM} (Guanajuato, Mexico). *Escherichia coli*, *Staphylococcus aureus*, and *Pseudomonas aeruginosa* were obtained from the American Type Culture Collection (ATCC, Manassas, VA, USA).

2.2. Melanin Extraction

The method for obtaining melanin from the pecan shell was performed following the method described by García et al. [42]. Briefly, the shells were washed with running water by immersing them in water at 25 °C (1:10) and adding $NH_3 \bullet H_2O$ to adjust the pH to 10.5 (10 g/100 g) to a final concentration of 2 g/100 g. They were then incubated at room temperature (24 h). Following incubation, the mixture was filtered, and the filtrate was adjusted to a pH of 2.5 with 2 mol L^{-1} HCl. The acidified solution was then filtered again to obtain raw melanin.

2.3. Melanin Purification

The raw melanin was hydrolyzed with 7 mol L^{-1} HCl for 6 h at 100 °C; the residues that were not hydrolyzed were centrifuged at 10,000 rpm for 10 min and resuspended at 10% $NH_3 \bullet H_2O$. Sequentially, three washes were made with chloroform and ethyl acetate. The aqueous phase was acidified with 1 mol L^{-1} HCl, and the filtered residue was washed with water. The residue (solid matter) was again dissolved in 1 mol L^{-1} $NH_3 \bullet H_2O$ and filtered. The supernatant was acidified with 1 mol L^{-1} HCl, and the filtered residue was washed with H_2O to obtain the purified melanin.

2.4. Fiber Preparation

Solutions for electrospinning were prepared by solubilizing polyacrylonitrile (PAN) in N,N-dimethylformamide (N,N-DMF) in a ratio of 12:88% (*w*/*w*) [47]. The mixtures were agitated for 24 h. The solutions were prepared by adding the powdered and sieved melanin powder (SIEVE TEST No. 230; 63 μm), in concentrations of 0.5% (PM0.5), 2.0% (PM2.0), and 5.0% (PM5.0) (*w*/*w*). Membranes were fabricated using the NE100 Single Nozzle Electrospinning/Spraying Machine (INOVENSO Ltd., Istanbul, Turkey). A total of 6 mL of each polymeric solution was injected at a flow rate of 0.1 mL h^{-1}. The solution was then projected onto a rotating drum covered with an aluminum film, spinning at 300 rpm. The distance between the nozzle and the collector was 17 cm, and a voltage of 15 kV was applied.

2.5. Fiber Characterization

2.5.1. Morphology Analysis

Fiber morphologies were analyzed by a field-emitting scanning electron microscope (FESEM) (Hitachi S-4800, Tokyo, Japan) at an acceleration voltage of 2 kV and a current flow of 10 μA. For characterization, a small part of each sample was coated with 9 nm gold/palladium to reduce the loading effect before imaging (Polaron equipment, SEM E5100 coating unit, Kontron AG, Switzerland). The software ImageJ 1.54 (National Institutes of Health, Bethesda, MD, USA) was used to measure the diameter of the fiber.

2.5.2. Chemical Composition Analysis

To identify the functional groups and chemical bonds present in PAN–melanin fibers, Fourier transform infrared spectroscopy (FTIR) analysis was conducted using a Nicolet iS10 instrument from Thermo Scientific (Waltham, MA, USA) equipped with an attenuated total reflectance (ATR) diamond accessory. The analysis followed the guidelines outlined in ASTM E573-01(2013) [48] and covered a range of 4000 cm^{-1} to 400 cm^{-1}.

2.5.3. Wettability Analysis

To evaluate the water absorption capability of each sample, the 1 cm^2 membranes were immersed in phosphate buffer (PBS) (37 °C) or water (room temperature) for 24 h, and both the dry (W_d) and wet (W_s) weight of the membranes were recorded.

Subsequently, the water uptake was determined using Equation (1) [49]:

$$Water\ uptake = \frac{W_s - W_d}{W_d} \times 100 \quad (1)$$

The contact angle was determined following the methodology proposed by Ruiz-Rocha et al. [50], using the OCA 15EC equipment (Dataphysic, Filderstadt, Germany). Ten reference points were selected to examine the interaction between water and both ground melanin and electrospun fibers.

2.5.4. Mechanical Characterization

In compliance with the ASTM-882 standard governing thin plastic films, the Young's modulus, tensile strength, and elongation percentage of the membranes will be assessed using 100 mm × 10 mm specimens [51], at a rate of 10 mm min^{-1}, as determined by Equation (2) and executed using the INSTRON 5565 (Canton, MA, USA). Polyacrylonitrile (PAN) is a highly versatile polymer due to its high carbon content. PAN exists in semi-crystalline and amorphous polymer forms and has a degree of crystallinity of 35.05%. The physical, chemical, electrical, mechanical, and thermal properties strongly depend on the polymer's crystalline structures [52,53]. Crystalline solids are inherently anisotropic, whereas amorphous solids are isotropic. Therefore, being an anisotropic material [54], ten specimens, five normal and five parallel with the principal axis of anisotropy, were tested from each sample. The rate of grip separation will be determined for the purpose of these test methods based on the initial strain rate as follows:

$$A = B \times C \quad (2)$$

where A denotes the separation speed in mm min^{-1}, B represents the initial distance between the jaws in mm, and C signifies the initial deformation rate obtained from the standard and is utilized as specified by the standard for speed calculation.

2.5.5. Liberation of Melanin

The methodology of Kalantary et al. [55] was used, with some modifications. A total of 1 mg of the membrane was weighed, subsequently immersed in 3 mL of PBS pH 7.4, and finally incubated with shaking at 180 rpm and 37 °C. Samples of 1 mL were taken every 2 h, for a total period of 10 h, and the same volume of solution was applied to the incubation solution. The samples were read on the Evolution 300 from Thermo Scientific (Madison, WI, USA) equipment at 220 nm, determining the concentration using a previously created calibration curve.

2.6. Fiber Bioactivities

2.6.1. Antioxidant Activity Against DPPH

The DPPH radical scavenging assay was conducted following a modified version of the methodologies described by Orona-Tamayo et al. [56] and Kalantary et al. [55]. A DPPH radical solution of 0.000180 mol L^{-1} was prepared in 80/100 mL of methanol. For sample preparation, 1 mg of each fiber was dissolved in 1 mL of N,N-DMF solution, maintaining a 1:1 ratio of the sample to the antioxidant in a 96-well plate. The plate was then placed in darkness and incubated at 37 °C for 30 min. After incubation, the absorbance was measured at 517 nm using the Multiskan SkyHigh microplate spectrophotometer from Thermo Fisher Scientific (Waltham, MA, USA). The results were computed using Equation (3).

$$RSA(\%) = \left(1 - \frac{A_i - A_j}{A_c}\right) \times 100 \quad (3)$$

where A_c is the absorbance of the radical without the sample, A_i is the absorbance of the tested sample, and A_j is the absorbance of the blank group ($n = 3$).

2.6.2. Antioxidant Activity Against ABTS

The experiment followed the methodology proposed by Orona-Tamayo et al. [56], albeit with some modifications. Initially, an ABTS solution (0.007 mol L^{-1}, 3 mL) and APS (0.00245 mol L^{-1}, 15 mL) were dissolved in distilled water and mixed in darkness at 25 °C for 16 h. The resulting blue/green ABTS radical solution was then freshly prepared and adjusted with 100% methanol to achieve an absorbance of 0.700 ± 0.02, measured at 734 nm. Sample preparation involved dissolving 1 mg of each membrane in 1 mL of N,N-DMF solution, maintaining a 1:1 ratio of sample to antioxidant. Subsequently, the plate was incubated at room temperature for 6 min, followed by an absorbance measurement at 734 nm using the Multiskan SkyHigh microplate spectrophotometer from Thermo Fisher Scientific (Waltham, MA, USA). A decrease in absorbance indicates higher antioxidant activity. The results were calculated using Equation (3).

2.6.3. Antimicrobial Activity

The antimicrobial efficacy of the electrospun fibers was evaluated using the disk diffusion method, performed in triplicate. Circular samples with a diameter of 5 mm were prepared from each fiber. *E. coli* was used (gram-negative, enteropathogenic) along with two opportunistic bacteria, *S. aureus* (gram-positive skin bacterium) and *P. aeruginosa* (a gram-negative respiratory tract pathogen). The strains were activated by inoculation on nutrient agar and then reseeded in nutrient broth. As a positive control, gentamicin and ampicillin (0.1 mg mL^{-1}) embedded in filter paper (diameter of 5.0 mm; thickness of 1.0 mm) were used. Therefore, 100 µL of each strain was inoculated and measured at 600 nm (OD_{600}); samples were read on the Evolution 300 from Thermo Scientific (Madison, WI, USA) to estimate cell density or bacterial growth in liquid culture. The plates were then incubated at 37 °C for 24 h. Antimicrobial capacity was reported by the inhibition halo (mm) generated for each sample on the plates.

2.7. Antiaging Effect

The enzymatic activities were performed based on the methodology of Aguilar-Toalá and Liceaga [57], with some modifications, using tyrosinase, collagenase, elastase, and hyaluronidase enzymes. Samples consisted of melanin dissolved in saline phosphate buffer (0.5 mM) at pH 7.2 in the different concentrations of solution (0.5%, 2.0%, and 5.0% w/w) in which the membranes were made.

2.7.1. Elastase Inhibition Assay

The assay was performed in a 96-well microplate format, and 5 µL of the melanin solution was combined with 5 µL of 10 mM substrate *N*-succinyl-Ala-Ala-Pro-Val-*p*-nitroanilide (S4760, Sigma-Aldrich) dissolved in DMSO. This mixture was incubated for 15 min at 37 °C. Afterward, 50 µL of type IV porcine pancreatic elastase (10 mU in 100 mM Tris-HCl buffer; pH 8.0) was added and pre-incubated 5 min at 37 °C, after this time, 140 µL of 100 mM Tris-HCl buffer (pH 8.0) was added, and the mixture was incubated at 37 °C. A change in absorbance was recorded every 10 min for 60 min with measurements at 405 nm using a microplate spectrophotometer (Multiskan SkyHigh, Thermo Fisher Scientific). The percentage of inhibition was calculated using Equation (4), where $OD_{control}$ and OD_{sample} represent the optical density of the control and samples, respectively [57].

$$Enzymatic\ inhibition\ (\%) = \frac{OD_{control} - OD_{sample}}{OD_{control}} \times 100 \tag{4}$$

2.7.2. Tyrosinase Inhibition Assay

The tyrosinase inhibition assay was conducted in a 96-well microplate format. The procedure involved combining 5 µL of melanin samples with 5 µL of 10 mM 3,4-dihydroxy-L-phenylalanine (L-DOPA; D9628, Sigma-Aldrich) substrate dissolved in 0.5 mM phosphate-buffered saline (PBS; pH 7.2). This mixture was incubated for 15 min at 30 °C. Subsequently, 50 µL of pre-incubated mushroom tyrosinase (75 mU in 0.5 mM PBS; pH 7.2) was added and incubated for an additional 5 min at 37 °C. Following this incubation, 140 µL of 0.5 mM PBS (pH 7.2) was added to each well. Absorbance measurements were recorded at 405 nm every 10 min for a total duration of 60 min using a Multiskan SkyHigh microplate spectrophotometer (Thermo Fisher Scientific). The percentage of tyrosinase inhibition was calculated using Equation (4), as described in the literature [57].

2.7.3. Collagenase Inhibition Assay

The inhibition enzyme assay was performed in a 96-well microplate format, and 5 µL of the melanin samples were combined with 5 µL of 1 mM substrate Z-Gly-Pro-Gly-Gly-Pro-Ala-OH (27673, Sigma-Aldrich) dissolved in 50 mM Tris-HCl buffer; pH 7.5. This mixture was incubated for 15 min at 30 °C. Afterward, 50 µL of pre-incubated collagenase from *Clostridium histolyticum* (5 min, 30 °C, and 50 mU; 50 mM Tris-HCl buffer; pH 7.5) was added. After this time, 140 µL of 50 mM Tris-HCl, pH 7.5, was added. A change in absorbance was recorded every 10 min for 60 min with measurements at 340 nm using a microplate spectrophotometer (Multiskan SkyHigh, Thermo Fisher Scientific). The percentage of inhibition was calculated using Equation (4) [57].

2.7.4. Hyaluronidase Inhibition Assay

The assay was performed using a 96-well microplate format, for which 5 µL of the melanin samples were combined with 5 µL of 1.4 mg mL^{-1} substrate hyaluronic acid sodium salt (H5388, Sigma-Aldrich) dissolved in acetate buffer (50 mM; pH 4.5). This mixture was incubated for 15 min at 37 °C. Afterward, 50 µL of pre-incubated hyaluronidase (5 min; 37 °C) in acetate buffer (50 mM; pH 4.5) was added. After this time, 140 µL of 50 mM acetate buffer, pH 4.5, was added. A change in absorbance was recorded every 10 min for 60 min with measurements at 550 nm using a microplate spectrophotometer (Multiskan SkyHigh, Thermo Fisher Scientific). The percentage of inhibition was calculated using Equation (4) [57].

2.8. Docking In Silico of Each Tested Enzymes with Melanin Results

For this analysis, the core structure of melanin was fetched from the PubChem database (https://pubchem.ncbi.nlm.nih.gov/; accessed on 29 October 2024). The 3D crystal structures of the peptidase domain of collagenase G from *Clostridium histolyticum* (PDB ID 7Z5U) [58], human hyaluronidase 1 (PDB ID 2PE4) [59], human neutrophil elastase (PDB ID 3Q77) [60], and tyrosinase from *Bacillus megaterium* (PDB ID 6EI4) [61] were downloaded from the Protein Data Bank (PDB) at the Research Collaboratory for Structural Bioinformatics (RCSB) (https://www.rcsb.org/; accessed on 29 October 2024) [62]. Molecular docking simulations were conducted using the "Ligand Docker" module within the CHARMM-GUI web interface (https://www.charmm-gui.org/; accessed on 29 October 2024) [63]. The AutoDock Vina algorithm, which employs the Lamarckian Genetic Algorithm and Empirical Free Energy Scoring Function [64], was used for docking. Protein–ligand interactions within a 4 Å radius were analyzed and visualized using the web-based tool PLIP (Protein–Ligand Interaction Profiler) [65]. Images for the figures were generated using the molecular visualization software PyMOL (Version PyMOL(TM) 2.5.8) [66]. Simultaneously, experimentally crystallized inhibitors were re-docked under identical conditions to facilitate a comparative analysis of binding poses and energies.

2.9. Statistical Analysis

ANOVA (Analysis of Variance) was conducted following a multivariate test using Tukey's test for post hoc analysis. All graphs and analyses were performed using RStudio 2024.12.0+467 software for Windows.

3. Results and Discussions

3.1. Membrane Elaboration

Four distinct membranes were successfully fabricated using varying melanin concentrations to obtain the physical fibers (Figure 1). The electrospinning parameters were maintained constant for each membrane, while non-controllable parameters, such as temperature and humidity, were closely monitored. The resulting membranes exhibited a visible color transition turning from white to brown, with the shade intensifying proportionally to the concentration (Figure 1). This observation aligns with previous studies on polycaprolactone (PCL) and polyurethane (PUR) fibers loaded with different melanin concentrations, which similarly demonstrated an increase in color intensity as melanin content increased, resulting in progressively darker fibers [67,68]. Our fabricated materials displayed comparable color changes, indicating successful incorporation of melanin into the fiber structure.

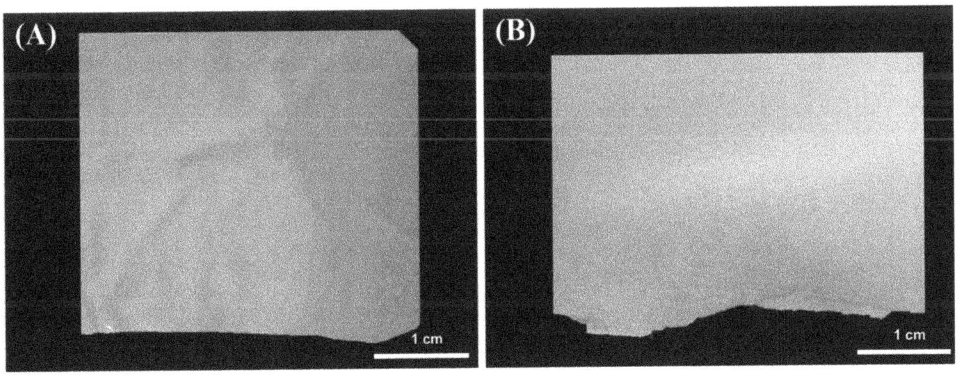

Figure 1. *Cont.*

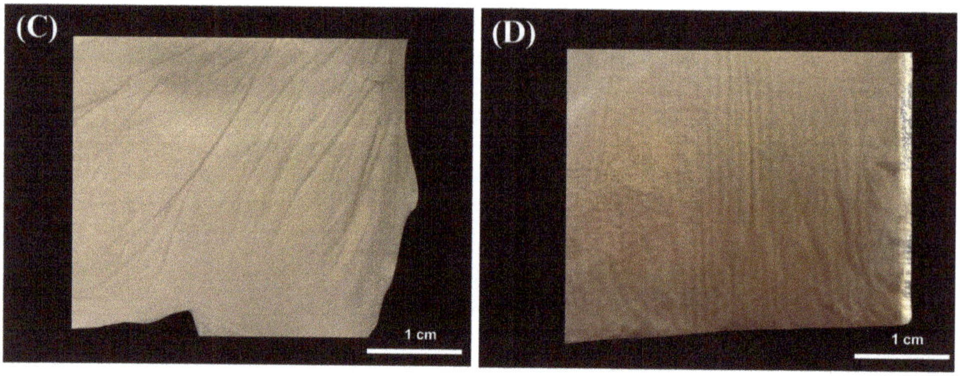

Figure 1. Fibers composed of polyacrylonitrile (PAN) and melanin pigment. (**A**) Control (PAN), (**B**) PAN + 0.5% of melanin (PM0.5), (**C**) PAN + 2.0% of melanin (PM2.0), and (**D**) PAN + 5.0% of melanin (PM5.0).

3.2. *Light and Scanning Electron Microscopy*

The light microscopy analysis of fiber morphology revealed distinct characteristics across different melanin concentrations (Figure 2A). Control PAN fibers without melanin exhibited uniform and clean extension, free from dark particles or inclusions, indicating a pure and homogeneous structure. PAN fibers with 0.5% melanin (PM0.5, Figure 2B) appeared like the control, with a slight difference in density and distribution. These fibers remained mostly clean with minimal dispersed particles, suggesting initial melanin incorporation into the fiber matrix.

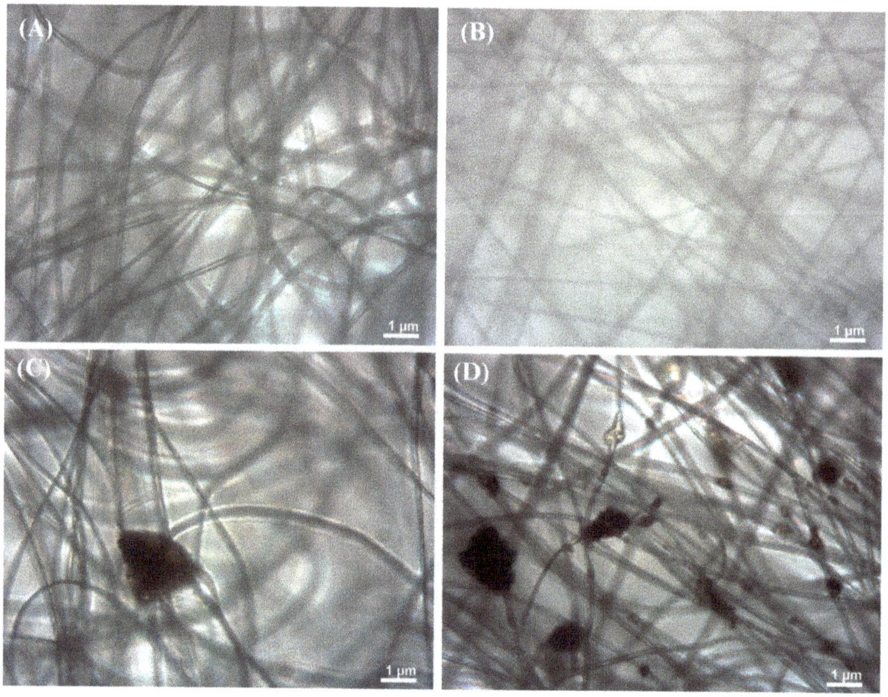

Figure 2. Light microscopy images of fiber membranes composed by PAN and melanin pigment. (**A**) Control (PAN), (**B**) PAN + 0.5% of melanin (PM0.5), (**C**) PAN + 2.0% of melanin (PM2.0), and (**D**) PAN + 5.0% of melanin (PM5.0).

PAN fibers with 2% melanin concentration (PM2.0, Figure 2C) showed the emergence of dark particles distributed along the fibrous structure. These inclusions indicated a higher presence of melanin within the fibers, resulting in a more complex and heterogeneous texture compared to lower concentrations. The highest melanin concentration of 5% (PM5.0, Figure 2D) revealed a considerable number of dark particles and aggregates dispersed among the fibers. This micrograph displayed a densely packed network of fibers with visible melanin inclusions, potentially influencing the fibers' mechanical and functional properties. This progressive series of images clearly demonstrates how increasing melanin concentrations alter the morphology and composition of polyacrylonitrile fibers, providing valuable insights into the integration of bioactive compounds in fibrous materials.

The electrospun fibers in the fabricated membranes were characterized using scanning electron microscopy (SEM) micrographs. The micrographs of the PAN (Figure 3) and PAN–melanin membranes at different concentrations (Figure 3B–D) show the alignment of fibers due to the equipment configuration and the type of collector used.

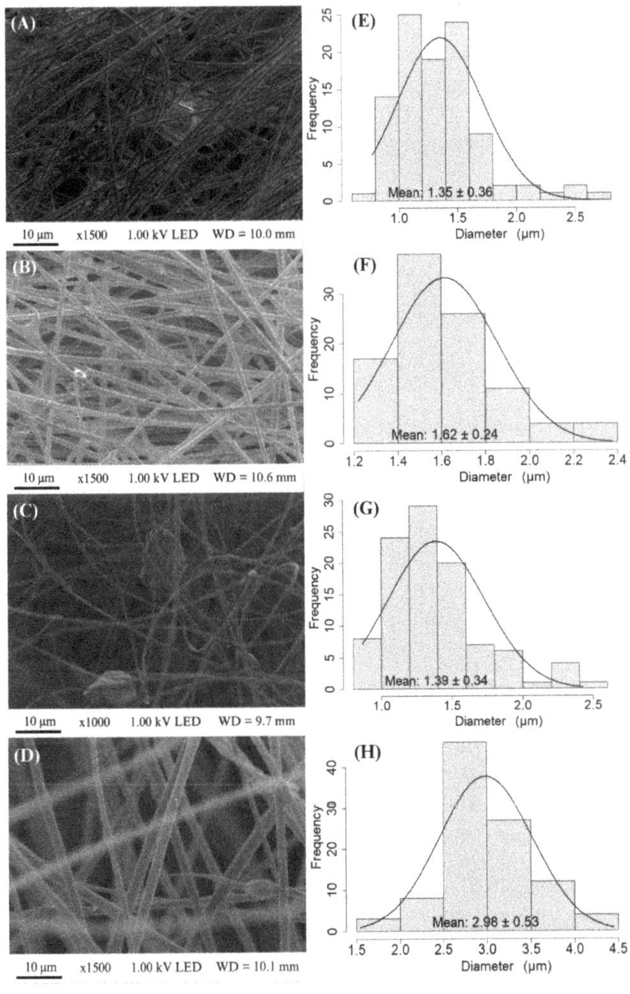

Figure 3. SEM micrographs of fiber composed by PAN and melanin pigment. (**A**) Control (PAN), (**B**) PAN + 0.5% of melanin (PM0.5), (**C**) PAN + 2.0% of melanin (PM2.0), and (**D**) PAN + 5.0% of melanin (PM5.0) and diameter size distribution histogram of samples labeled as (**E–H**), respectively, $n = 100$.

The fibers exhibit surface roughness in their morphological structure. Additionally, the presence of particles is notable in the membranes of the PAN + 0.5% melanin (PM0.5), PAN + 2.0% melanin (PM2.0), and PAN + 5.0% melanin (PM5.0) treatments. The number of particles increases as the melanin concentration rises, along with an increase in the fiber diameter. The electrospun PAN fibers have a diameter around 1.35 μm ± 0.36 μm, while fibers containing melanin range in diameter from 1.39 μm ± 0.34 μm to 2.98 μm ± 0.53 μm. As the concentration of melanin increases in the PAN fibers, the diameter of the fibers also increases.

This phenomenon can be attributed to several interconnected factors associated with the electrospinning process and the intrinsic properties of the material. As the melanin concentration increases, the viscosity of the polymer solution correspondingly rises. A more viscous solution is less prone to stretching during the electrospinning process, leading to the formation of thicker fibers [69,70]. Additionally, highly concentrated polymer solutions experience accelerated solvent vaporization, which restricts polymer extension and accelerates compaction, resulting in the formation of thicker fibers [71]. Melanin can alter the electrical conductivity of the solution. These changes in conductivity can affect the stability of the electrospinning jet and the stretching forces applied to the fibers, often resulting in thicker fibers. Therefore, a decrease in conductivity, likely caused by the melanin, results in fibers with lower conductivity and greater thickness [69]. Finally, the mechanical properties of the polymer solution, such as elasticity and plasticity, are affected by the concentration of melanin. These changes impact how the solution behaves under the electric field during electrospinning, leading to variations in fiber thickness.

3.3. Chemical Properties of Fibers

Figure 4 shows the infrared spectra (FTIR) of membranes obtained with the PAN formulations PAN + 0.5% melanin (PM0.5), PAN + 2.0% melanin (PM2.0), and PAN + 5.0% melanin (PM5.0). The FTIR spectra of the PAN fibers shows characteristic peaks at 2931 cm^{-1} (γ CH_2), 2242 cm^{-1} (γ C≡N), and 1664 cm^{-1} (γ C=O) of the residual N,N-DMF and 1453 cm^{-1} (δ CH_2), where γ represents a stretching vibration and δ denotes a bending vibration. Additionally, it is possible to observe the peak located at 1249 cm^{-1} related to the methine vibration, and the peak at 1071 cm^{-1} corresponds to the C-H interaction [72,73]. The absorption peaks of these characteristics are still present in the fibers added with melanin, indicating that melanin does not destroy or weaken the original molecular structure of PAN fibers and that some characteristics of the original PAN fibers are preserved, highly comparable peaks to those obtained by Naragund and Panda [74] and Yao et al. [75]. The wavenumber region between 1000 cm^{-1} and 500 cm^{-1} exhibits a significant amount of noise, which prevents the identification of characteristic peaks in this specific region.

Furthermore, Figure 4 shows the comparison of the infrared spectra obtained from melanin derived from walnut shells [42] and electrospun PAN fibers, along with their infusions at different melanin concentrations (PM2.0 and PM5.0). The melanin extracted from pecan shells [42] is an allomelanin that exhibits characteristic peaks at 3352 cm^{-1} associated with the hydroxyl group and at 1106 cm^{-1} and 1056 cm^{-1} attributed to C-O bonds, and in the spectral range between 700 cm^{-1} and 600 cm^{-1}, C-S bonds are observed. However, when melanin is infused into the polymeric fibers of polyacrylonitrile, these characteristic peaks are not present. The absence of the characteristic melanin peaks when infused into polyacrylonitrile polymer fibers can occur for several reasons: (i) through a chemical interaction, melanin may chemically interact with PAN, which could alter the structure of melanin or change its spectroscopic properties; (ii) in the matrix environment, melanin could be dispersed differently or adopt a different structure when incorporated into the PAN matrix, which could affect the observation of its characteristic peaks; (iii) with regard to the dispersion state, the way melanin is dispersed in PAN fibers can influence the accessibility of functional groups to interact with infrared radiation; (iv) with regard to structural modifications, during the infusion process into PAN fibers, melanin could undergo structural or conformational changes that affect the position or intensity of its spectral bands [76–78].

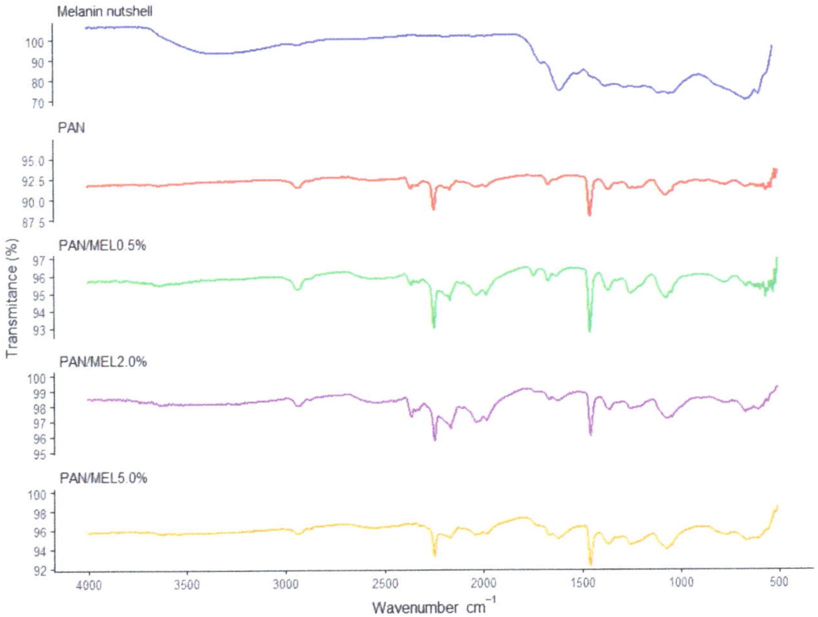

Figure 4. Infrared spectra (FTIR) within a wavenumber range of 4000 cm^{-1} to 500 cm^{-1} of melanin from nutshell and membranes obtained with melanin pigment, PAN control, PAN + 0.5% of melanin (PM0.5), PAN + 2.0% of melanin (PM2.0), and PAN + 5.0% of melanin (PM5.0) fibers.

3.4. Wettability Results

Figure 5 shows that the incorporation of melanin into the PAN membranes significantly reduces their water uptake capacity. The recorded absorption percentages were 23.25 ± 1.16% for control PAN fibers, 5.69 ± 0.57% for PM0.5 fibers, 1.58 ± 0.68% for PM2.0 fibers, and 1.63 ± 0.70% for PM5.0 fibers. PAN is inherently hydrophobic. However, the lone-pair orbital located on nitrogen and oriented 180° relative to the C≡N bond is associated with hydrogen bonding that could interact with water [79]. The overall structure of PAN and the arrangement of its molecules do not favor an affinity for water. This characteristic makes it resistant to water absorption and other polar solvents. The higher the melanin concentration, the lower the degree of swelling, with PM2.0 and PM5.0 fibers showing the least water uptake. This suggests that melanin effectively reduces the hydrophilicity of the PAN fibers, making them less prone to swelling in the presence of water.

The incorporation of melanin into PAN fibers reduces their hydrophilicity and swelling in the presence of water due to the hydrophobic nature of melanin, the creation of a denser and less porous fiber structure, increased cross-linking density, and the formation of a hydrophobic phase that acts as a barrier to water absorption. Surfaces with a water contact angle below 60° are considered hydrophilic, while those with contact angles above 90° are considered hydrophobic. When the contact angle exceeds 120°, the surface is classified as super-hydrophobic [80].

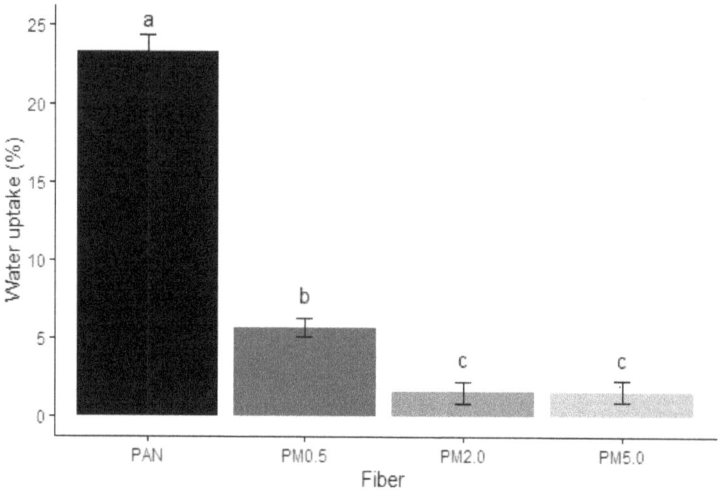

Figure 5. Degree of swelling of electrospun membranes composed of PAN and melanin pigment. Control (PAN), PAN + 0.5% of melanin (PM0.5), PAN + 2.0% of melanin (PM2.0), and PAN + 5.0% of melanin (PM5.0). Bars represent means ± SE ($n = 3$) of the water uptake in %. Different letters indicate significant differences between all factor combinations (univariate ANOVA after post hoc Tukey test, $p < 0.05$).

PAN fibers exhibit hydrophobic behavior, with a contact angle of 110.59° ± 1.95°, which constitute very similar results to those reported by Sanchaniya and Kanukuntla [80] for a 12% (w/w) PAN concentration. It is possible to observe an increase in the hydrophobicity of the fibers as the concentration of melanin that is present increases. Melanin is a hydrophobic compound (right contact angle 136.16° ± 0.77°; left contact angle 136.00° ± 0.96°) [42], which, when combined with PAN in high concentrations, considerably increases its hydrophobicity (Figure 6).

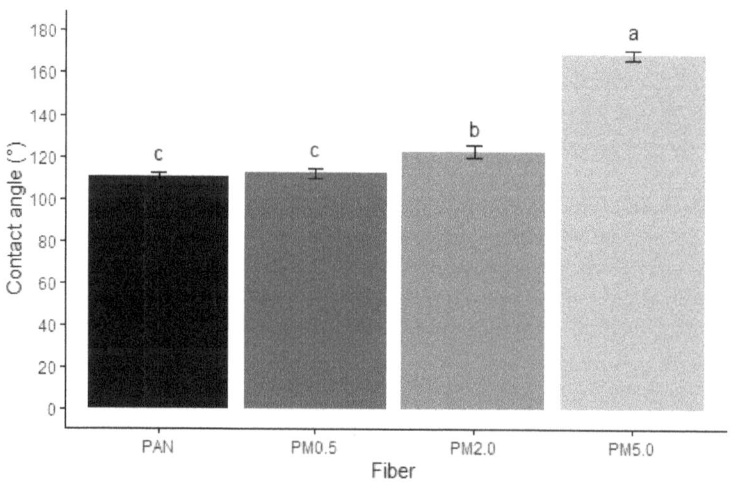

Figure 6. Contact angle of electrospun membranes composed of PAN and melanin pigment. Control (PAN), PAN + 0.5% of melanin (PM0.5), PAN + 2.0% of melanin (PM2.0), and PAN + 5.0% of melanin (PM5.0). Bars represent means ± SE ($n = 10$) of the contact angle in degrees. Different letters indicate significant differences between all factor combinations (univariate ANOVA after post hoc Tukey test, $p < 0.05$).

3.5. Fibers Degradation

The degradation rates of the control PAN and PAN membranes loaded with melanin were monitored over 7 and 14 days (Figure 7). The data show distinct differences in degradation percentages across various melanin concentrations and time periods. The control PAN membranes exhibited a degradation percentage of 26.6 ± 6.6% after 7 days, increasing to 32.02 ± 4.8% after 14 days. For the PAN + 0.5% melanin (PM0.5) membrane, the degradation percentage was notably lower at 7.7 ± 2.9% after 7 days and rising to 20.6 ± 2.2% after 14 days. The PAN + 2.0% melanin (PM2.0) membrane showed a degradation of 30.5 ± 4.8% after 7 days and 24.0 ± 1.6% after 14 days. The PAN + 5.0% melanin (PM5.0) membrane had the lowest degradation rates, around 9.4 ± 0.5% after 7 days and 19.3 ± 2.3% after 14 days. Melanin's inherent stability and resistance to environmental degradation likely contribute to the lower degradation rates observed in PAN–melanin membranes. As the concentration of melanin increases, the protective effect becomes more pronounced, resulting in a lower degree of fiber degradation. Also, the incorporation of melanin can enhance the structural stability of the membranes. By increasing cross-linking density and forming a more robust polymer matrix, melanin reduces the susceptibility of the fibers to degradation processes. Finally, melanin imparts hydrophobic properties to the PAN fibers, reducing water absorption and thus limiting hydrolytic degradation. The dominant characteristic of the PAN molecule is the presence of strong polar nitrile groups. The CN groups have a wide range of possibilities for interacting with their surroundings. The high dipole moment can cause strong attraction or repulsion (depending on orientation) of other molecules or substituents in molecules that also have a high dipole moment [79]. PAN differs in many ways from common commercial polymers. Its typical properties include hardness, rigidity, and resistance to most solvents and chemicals. PAN only dissolves in (i) aprotic polar organic solvents such as dimethylformamide, dimethylacetamide, dimethyl sulfoxide, sulfolane, ethylene carbonate, and N-methylpyrrolidone; (ii) concentrated sulfuric acid and nitric acid; and (iii) concentrated aqueous solutions of certain inorganic salts, such as lithium bromide, sodium thiocyanate, and zinc chloride. Acrylonitrile copolymers are often soluble in less polar organic solvents, such as dioxane, tetrahydrofuran, chlorobenzene, cyclohexanone, and acetone [81], which makes them highly resistant. Then, less water uptake means slower breakdown of the polymer chains, leading to lower degradation rates.

Figure 8 shows the melanin release kinetics that were expressed as a percentage over the time for control PAN and loaded fibers with different melanin concentrations. The release of the melanin from the PAN fibers occurred gradually during the 10 h kinetics assessment, with the PAN membrane with 5.0% of melanin (PM5.0) exhibiting a higher percentage of release (66.26 ± 4.29%) compared to the control PAN membranes. However, during the first two hours, the PAN–melanin membranes at 2.0 (PM2.0) and 5.0% (PM5.0) have the highest melanin release (23.66 ± 5.04% and 24.42 ± 1.91%, respectively), while the PAN–melanin membrane at 0.5% (PM0.5) had the maximum release from 2 to 4 h. All PAN–melanin membranes showed a constant release of melanin from the point of maximum release. Gradually, a more significant release is produced by PAN membranes with higher melanin contents. This results from the increased presence of melanin within the fiber matrix, which can disperse with greater ease. Higher melanin content may disrupt the fiber matrix, facilitating easier diffusion of melanin particles.

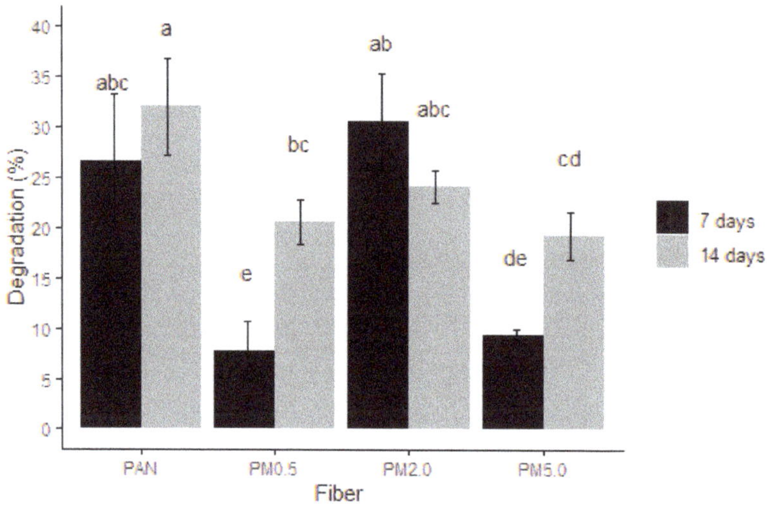

Figure 7. PAN and PAN–melanin fiber degradations for two weeks. Control (PAN), PAN + 0.5% of melanin (PM0.5), PAN + 2.0% of melanin (PM2.0), and PAN + 5.0% of melanin (PM5.0). Bars represent means ± SE ($n = 3$) of the fiber degradation in %. Different letters indicate significant differences between all factor combinations (univariate ANOVA after post hoc Tukey test, $p < 0.05$).

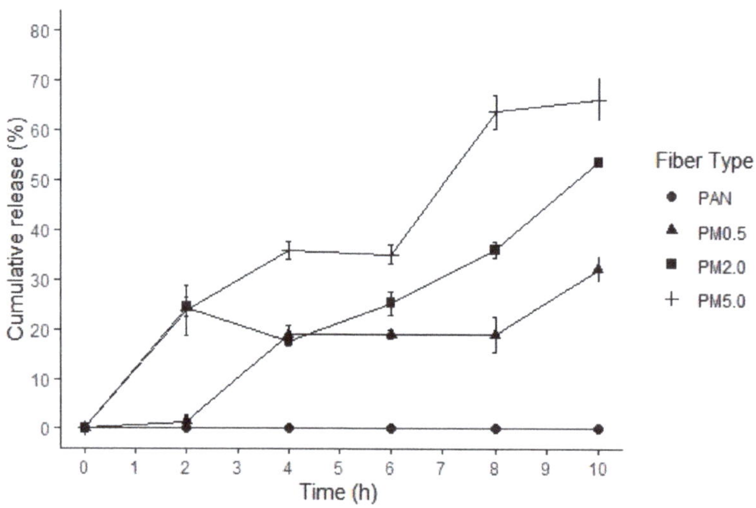

Figure 8. The in vitro release profiles of melanin from different PAN and PAN–melanin fibers over time. Control (PAN), PAN + 0.5% of melanin (PM0.5), PAN + 2.0% of melanin (PM2.0), and PAN + 5.0% of melanin (PM5.0). Points represent means ± SE ($n = 3$) of the melanin liberation from fibers.

3.6. Mechanical Properties of the Membrane

The physical mechanical properties of PAN and PAN–melanin membranes (Figure 9) are significantly influenced by the addition of melanin and the orientation of the fibers. Melanin concentrations impact the differential tensile and rupture strengths and the Young's modulus differently. For example, the higher melanin concentrations (2.0% and 5.0%) added to the fibers enhance the tensile strength in the horizontal orientation (Figure 9A). This improvement can be attributed to the reinforcing effect of melanin particles within the polymer matrix. Melanin acts as a filler, providing additional support and resistance against tensile forces applied in the horizontal direction. However, the vertical orientation shows variable effects because the alignment and distribution of fibers differ, and melanin may not provide the same level of reinforcement due to differences in stress distribution. The addition of melanin generally decreases the rupture strength (Figure 9B), with the most significant reduction observed at the 0.5% and 5.0% melanin concentrations. This reduction can be explained by the introduction of heterogeneities and potential stress concentrators within the fiber matrix. Melanin particles can create points of weakness where cracks can initiate and propagate with ease, leading to a lower rupture strength. At lower concentrations (0.5%), the melanin may not be well-distributed, leading to uneven stress distribution. At higher concentrations (5.0%), the excessive amount of melanin can disrupt the polymer network, further compromising the material's integrity. Higher melanin concentrations (2.0%) significantly increase the stiffness of the fibers, as indicated by the increased Young's modulus (Figure 9C). This increase in stiffness can be attributed to the rigid nature of melanin particles, which enhance the overall rigidity of the composite material. The melanin particles restrict the mobility of the polymer chains, making the material more resistant to deformation under applied stress. However, at very high concentrations (5.0%), the stiffness may not increase further or could even decrease slightly due to potential aggregation of melanin particles, which might lead to an uneven distribution and potential flaws within the matrix. The orientation of the fibers plays a crucial role in determining the mechanical properties. In the horizontal orientation, the alignment of fibers may allow for better load distribution and more effective reinforcement by melanin particles. In contrast, the vertical orientation may not benefit as much from melanin addition due to differences in how the fibers bear the applied loads and how the melanin is distributed throughout the matrix.

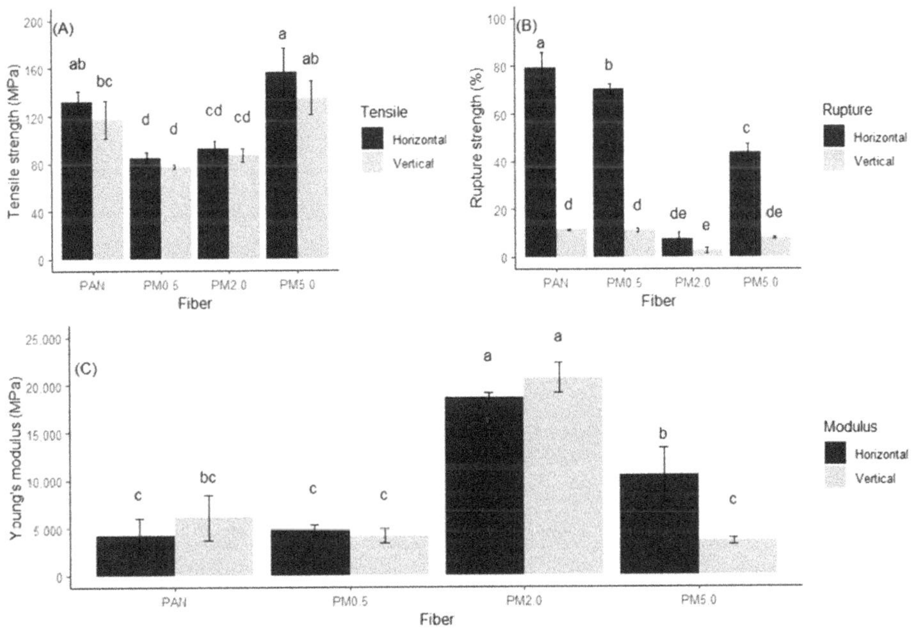

Figure 9. Analysis of mechanical properties of the different produced membranes, (**A**) tensile strength (MPa), (**B**) rupture strength (%), and (**C**) Young's modulus (MPa) in PAN electrospun membranes in vertical and horizontal direction. Control (PAN), PAN + 0.5% of melanin (PM0.5), PAN + 2.0% of melanin (PM2.0), and PAN + 5.0% of melanin (PM5.0). Bars represent means ± SE (n = 5) for all physical properties. Different letters indicate significant differences between all factor combinations (univariate ANOVA after post hoc Tukey test, $p < 0.05$).

3.7. Antioxidant Activities

Melanins are efficient free radical scavengers and protect different organisms against oxidative damage by neutralizing free radicals. The antioxidant capacity of fibers containing varying concentrations of melanin was evaluated against two chemical oxidants such as DPPH and ABTS Figure 10, Supplementary Figure S1).

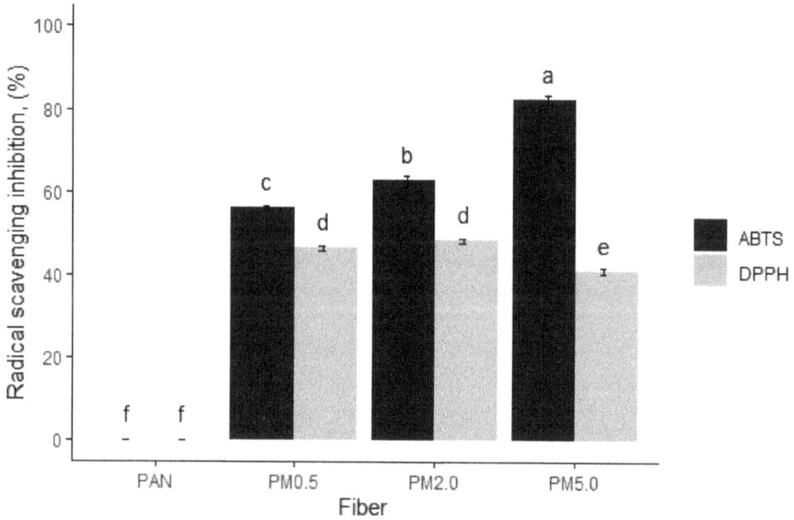

Figure 10. Free radical scavenging activity of melanin membranes against DPPH and ABTS. Control (PAN), PAN + 0.5% of melanin (PM0.5), PAN + 2.0% of melanin (PM2.0), and PAN + 5.0% of melanin (PM5.0). Bars represent means ± SE (n = 3) for radical scavenging in %. Different letters indicate significant differences between all factor combinations (univariate ANOVA after post hoc Tukey test, $p < 0.05$).

It is possible to observe that control PAN fibers do not possess antiradical scavenging activity against the DPPH and ABTS radicals. However, fibers loaded with melanin PM0.5, PM2.0, and PM5.0 show activity with values of 56.2% ± 0.3%, 62.7% ± 1.2%, and 82.3% ± 1.0% against ABTS, and 46.5% ± 0.5%, 48.4% ± 0.5%, and 41.0% ± 0.7% against DPPH, respectively (Figure 10). Melanin's capacity to effectively scavenge free radicals is the mechanism behind its antioxidant activity. Free radicals are extremely reactive chemicals that have the potential to destroy cells. Melanin can neutralize them. Melanin reduces oxidative stress and protects cells and tissues by retaining free radicals. Additionally, melanin can act as an extracellular redox buffer, helping to neutralize oxidants generated by environmental stress. This antioxidant mechanism contributes to melanin's ability to protect against cellular damage and promote health in living organisms [82–85]. In that sense, the PAN with melanin demonstrated a considerable antioxidant capacity against ABTS or DPPH assays. The antioxidant capacity of the fibers loaded with melanin likely attributed to their molecular structure, in particular the different functional groups in the heterocyclic rings, o-quinones, and other reducing groups with oxidative capacity such as o-hydroquinones; therefore, melanin has a high potential to donate or capture electrons from different electronegative elements such as hydroxyl radicals, superoxide anions, singlet oxygen, or peroxyanions [86]. Melanins from different microorganisms contain a high antiradical activity against different oxidants such as DPPH and ABTS [83,87,88].

3.8. Antibacterial Activity

The control PAN fiber does not possess antimicrobial activity; however, the addition of any melanin concentration to PAN fibers inhibits microbial growth comparably to the positive controls such as the antibiotics gentamicin and ampicillin. Furthermore, a higher amount of melanin incorporated into the fibers produces a greater inhibitory effect [42]. Specifically, the inhibition halos against bacteria were proportional to the melanin concentration; for example, when the nanomembrane was loaded with 5% of melanin, we found halos of 7.68 mm against *S. aureus*, 7.61 mm for *E. coli*, and 8.0 mm for *P. aeruginosa* (Table 1, Supplementary Figure S2).

Table 1. Antimicrobial capacity of PAN membranes with melanin pigment against bacterial strains.

Fiber	Antibacterial Activity Mean Diameter of Inhibition Zone (mm)		
	E. coli	*P. aeruginosa*	*S. aureus*
PAN	0.0 ± 0.0 [a]	0.0 ± 0.0 [a]	0.0 ± 0.0 [a]
PM0.5	7.1 ± 0.4 [b]	$7.3 \pm 0.1.5$ [b]	6.7 ± 0.5 [b]
PM2.0	7.4 ± 0.5 [b]	7.8 ± 0.7 [bc]	6.8 ± 0.5 [b]
PM5.0	7.6 ± 0.7 [bc]	8.0 ± 0.8 [bcd]	7.6 ± 0.3 [bc]
Gentamicine 10 mg	9.8 ± 0.6 [de]	9.3 ± 0.3 [cde]	9.9 ± 1.4 [de]
Ampicillin 100 mg	9.8 ± 0.1 [de]	11.3 ± 2.0 [e]	9.5 ± 1.8 [cde]

Values represent means ± SE ($n = 3$) for diameter of inhibition zone (mm). Control fiber (PAN), PAN + 0.5% of melanin (PM0.5), PAN + 2.0% of melanin (PM2.0), and PAN + 5.0% of melanin (PM5.0). Different letters indicate significant differences (univariate ANOVA after post hoc Tukey test, $p < 0.05$).

The inhibition results are aligned with previously reported findings, indicating that melanin exhibits antimicrobial activity against gram-positive and gram-negative pathogens such as *Bacillus pumilus, Staphylococcus warneri, Pseudomonas aeruginosa, Bacillus altitudinis, Brevibacterium casei, Micrococcus luteus, Bacillus* sp., *Geobacillus stearothermophilus* and *Streptococcus mutans, Escherichia coli, Staphylococcus aureus, Bacillus megaterium*, and *Salmonella* [89,90]. The presence of the reactive oxygen species (ROS) in cells has garnered interest due to their ability to decompose organic compounds and inhibit bacterial proliferation. ROS includes oxygen free radicals and any other oxygen-containing molecule in which the oxygen atom has greater reactivity than molecular oxygen (O_2). The elevated reactivity enables ROS to extract electrons from organic compounds, leading to the degradation of these compounds into water and carbon dioxide. This similar reactivity damages the cell membrane of bacteria, which results in their death. Allomelanin derived from walnut shells contains catecholic compounds that, when oxidized, generate reactive oxygen species. The oxidation of these catecholic compounds involves electron transfer that converts O_2 into superoxide (O_2^-) and hydrogen peroxide (H_2O_2). The release of H_2O_2 interacts with the cell membrane or cell wall of microorganisms, resulting in cellular destruction [4,42,91]. Finally, compounds synthesized by plants as a defense mechanism can act by interacting with the cell membrane or cell wall of microorganisms, causing changes in membrane permeability, which results in cell destruction. They are also capable of penetrating bacterial cells and promoting the coagulation of their contents [4].

3.9. Antiaging Activity

The evaluation of enzymatic inhibition against skin-related aging enzymes, including tyrosinase, hyaluronidase, elastase, and collagenase, was conducted to assess the efficacy of melanin-loaded nanomembranes (Table 2). This analysis was crucial in determining the potential antiaging properties of the fabricated materials. Notably, polyacrylonitrile (PAN) fibers without melanin incorporation demonstrated no significant enzymatic inhibition potential. However, the addition of melanin to the polymer mixture demonstrated effective inhibition. Additionally, we observed that all concentrations of melanin have inhibitory effects against the four enzymes (Table 2). Enzymatic inhibition is dependent on the concentration of melanin, and we observed that the membranes with 5.0% of melanin (PM5.0), inhibited the activity of tyrosinase (36%), hyaluronidase (22%), collagenase (25%), and elastase (28%). The fibers loaded with 2.0% (PM2.0) of melanin, dropped the enzymatic activities in tyrosinase (35%), hyaluronidase (15%), collagenase (25%), and elastase (28%), respectively, and the fiber with the lowest melanin concentration (PM0.5) lightly diminished the enzymatic activity for tyrosinase (0.8%), hyaluronidase (9.8%), and collagenase (10.8%) but not for elastase (27%) (Table 2). There are no reports on inhibiting antiaging enzymes using melanin-loaded fibers or the pigment alone. However, several studies have used polyphenols and peptides with high inhibition of enzymatic activities for tyrosinase, hyaluronidase, collagenase, and elastase [92–96]. For example, peptides derived from chia seed hydrolysates showed 38% enzymatic inhibition against collagenase, similar to our results, as well as enzymatic inhibition for elastase (43%) and tyrosinase (61%) [57]. Similarly, polyphenolic extracts from grape pomace revealed enzymatic inactivation for elastase (17–47% inhibition) and collagenase (24–43%) [45]. Our results, in conjunction with these findings, suggest promising potential for the development of cosmetic formulations. Specifically, these bioactive compound-loaded nanomembranes could serve as a foundation for innovative facial masks with enhanced antiaging properties.

Table 2. Enzymatic inhibition of PAN fibers loaded with melanin against aging enzymes.

	Enzyme Inhibition (%)			
Fiber	**Tyrosinase**	**Hyaluronidase**	**Collagenase**	**Elastase**
PAN	0.0 ± 0.0 [a]	0.0 ± 0.0 [a]	0.0 ± 0.0 [a]	0.0 ± 0.0 [a]
PM0.5	0.8 ± 13.3 [b]	9.8 ± 0.7 [b]	10.9 ± 3.8 [b]	27.3 ± 9.5 [b]
PM2.0	35.2 ± 8.7 [c]	15.3 ± 3.4 [c]	25.3 ± 4.3 [c]	28.2 ± 7.4 [b]
PM5.0	36.0 ± 5.9 [c]	21.9 ± 1.4 [d]	37.2 ± 3.0 [d]	33.0 ± 9.6 [b]

Values represent means ± SE ($n = 3$) of enzymatic inhibition in %. Control fiber (PAN), PAN + 0.5% of melanin (PM0.5), PAN + 2.0% of melanin (PM2.0), and PAN + 5.0% of melanin (PM5.0). Different letters indicate significant differences (univariate ANOVA after post hoc Tukey test, $p < 0.05$).

3.10. Docking In Silico of Each Tested Enzyme with Melanin

Molecular docking results indicate that melanin exhibits a stronger binding affinity for collagenase compared to the re-docked inhibitor, as evidenced by the lower binding energy of -10.4 kcal mol^{-1} for melanin versus -9.3 kcal mol^{-1} for the inhibitor (Table 3). This suggests that melanin could potentially be a more competitive inhibitor of collagenase. Further, results show that melanin exhibits similar binding affinities for elastase and tyrosinase compared to the re-docked inhibitors. The binding energies of melanin and the re-docked inhibitors for elastase are -7.1 and -7.6 kcal mol^{-1}, respectively, and for tyrosinase, they are -6.3 kcal mol^{-1} for both. These comparable binding energies suggest that melanin might have a similar inhibitory potential for these enzymes as the reference inhibitors. Distinct collagenase inhibitors, often featuring a zinc-binding group (ZBG), have been identified. These inhibitors target the catalytic zinc ion, rendering the enzyme inactive. However, their lack of selectivity over human metalloproteinases (MMPs) hinders their clinical development [58].

Table 3. Chemical structures and binding energies of docked melanin and re-docked experimental inhibitors against their respective target proteins, as determined by molecular docking simulations.

Ligand	PubChem CID	Binding Energy (Kcal/mol)	Target Protein	Ligand Chemical Structure
2HY	11539563	−7.6	Elastase	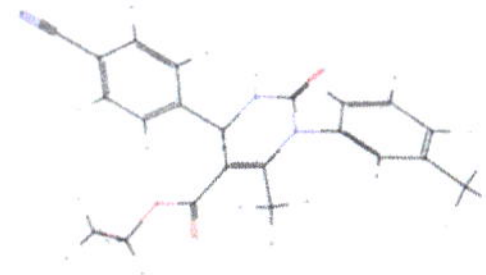
B5N	17249811	−6.3	Tyrosinase	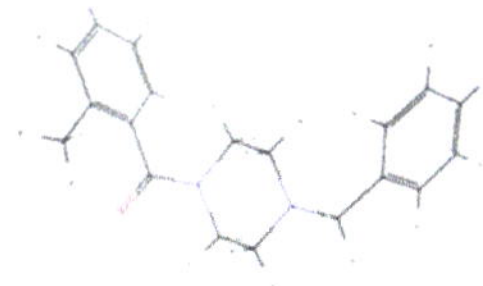
IFW	165416375	−9.3	Collagenase	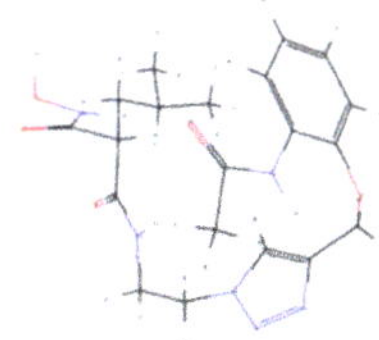
Melanin	-	−10.4 −8.6 −7.1 6.3	Collagenase Hyaluronidase Elastase Tyrosinase	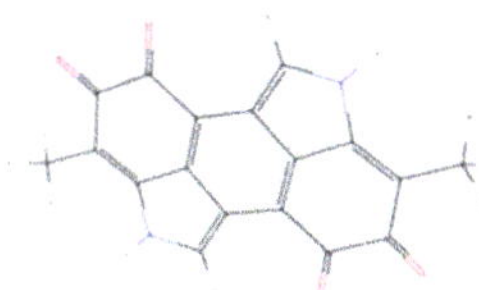

The experimental crystal structure of collagenase reveals that the ortho-acetamide group of the inhibitor IFW forms hydrogen bonds with Glu498 and engages in π-π stacking interactions with Trp539. The triazole ring interacts with Glu555 and Tyr599, while the hydroxamate group coordinates with the zinc ion. Our molecular docking results suggest that melanin interacts with key residues within the active site of the target protein, similarly to the known inhibitor IFW (Figure 11). Specifically, melanin appears to form hydrophobic interactions with Glu498 and Trp539 and hydrogen bonds with Tyr599. The crystal structure of hyaluronidase (2PE4) reveals a molecule composed of two distinct domains: a catalytic domain and a novel EGF-like domain, which is often involved in protein–protein interactions and regulatory processes [59].

(A) **(E)**

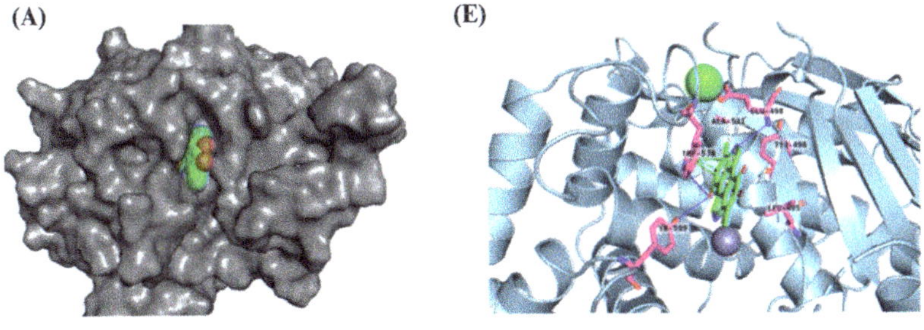

Figure 11. *Cont.*

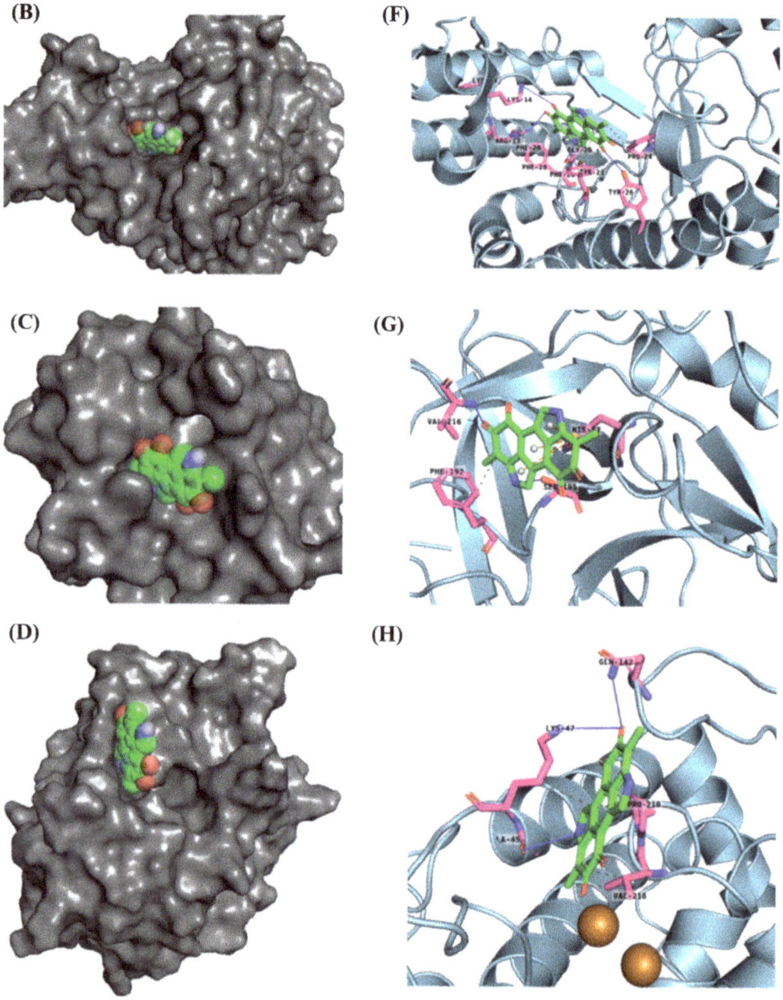

Figure 11. Tridimensional structure of the molecular docking between melanin pigment and the catalytic sites of various aging-related enzymes. Surface (**A–D**) and cartoon (**E–H**) representations of the predicted models illustrate interactions between melanin and the enzymes: collagenase (**A**,**E**), hyaluronidase (**B**,**F**), elastase (**C**,**G**), and tyrosinase (**D**,**H**). The surface models depict the most accurate predictions, while the cartoon representations highlight the amino acids (pink) involved in the protein–ligand interactions within the enzyme binding pocket and melanin (green). In (**E**), green and gray dots represent calcium (Ca) and zinc (Zn) atoms, respectively. Brown dots in (**H**) depict copper (Cu) atoms.

While no experimentally determined crystal structures of hyaluronidase–inhibitor complexes are currently available, our computational results suggest that melanin can form a stable complex within the enzyme's active site, with a predicted binding energy of -8.6 kcal mol^{-1}. This interaction is stabilized by a combination of hydrophobic interactions, hydrogen bonds, π-π stacking, and π-cation interactions. Previous studies have demonstrated that garcinol binds to human hyaluronidase (2PE4) with a binding affinity of -4.41 kcal mol^{-1}, forming interactions with key residues Leu381, Asn383, Leu379, Lys377, and His305, as well as two hydrogen bonds with Asp408 [97]. Our results indicate that melanin interacts within a binding pocket, forming specific bonds with Phe204, Pro249, and

Tyr210. Additionally, conventional hydrogen bonds are observed with Gly203, Arg134, and Lys144. Computational modeling indicates that melanin can form a stable complex within the reported enzyme's active site, stabilized by various interactions. Our inhibitory assay, along with the stronger interaction results compared to garcinol, supports melanin's inhibitory effect. The human neutrophil elastase (HNE) used in this analysis was crystallized with a non-covalently bounded dihydropyrimidone inhibitor (DHPI) [60]. Previous studies have reported that the primary binding interactions between DHPI and elastase involve van der Waals forces with hydrophobic residues Val190, Ala213, and Val216. Additionally, π-π stacking occurs between the phenyl ring of Phe192 and the trifluoromethylphenyl group of DHPI, while Asp226 participates in hydrogen bonding networks, and Val216 appears to be a key residue for binding peptidic inhibitors [60,98]. Our results show that the molecular docking of melanin with HNE shares similar binding energies of 7.6 kcal mol^{-1} for melanin and 7.1 kcal mol^{-1} for DHPI. Furthermore, the predicted binding pose suggests interactions with key residues, like those described above, such as Phe192, Ser195, and Val216 (Table 4).

Table 4. Residue numbers and interaction types for the best binding pose of melanin with its respective target protein, as determined by molecular docking simulations.

Interaction Type	Enzyme Inhibition (%)			
	Collagenase	**Hyaluronidase**	**Elastase**	**Tyrosinase**
Hydrophobic	Leu495 Tyr496 Glu498 Ala531	Phe204 Pro249	Phe192	Lys47 Val218 Pro219
	Trp539			
Hydrogen Bonds	Glu498	Arg134	Ser195	Gly46
	Trp539 Tyr599	Lys144 Gly203	Val216	Lys47 Gln142
π-π stacking π-cation	Trp539	Tyr210		Trp539

The experimental data on the binding pose of B5N in the tyrosinase active site show that the 4-fluorobenzyl moiety is oriented towards CuA and stabilized by stacking interactions with His208. Additionally, a polar interaction with Arg209 is observed; however, the flexibility of Arg209 appears to contribute to accommodating different ligands [61]. Interestingly, our results show that melanin binds to a different cavity in the protein's binding site compared to B5N; however, the predicted binding energy is similar, at −6.3 kcal mol^{-1} for both ligands. Our melanin–tyrosinase docking analysis revealed two key interaction types: hydrophobic interactions involving Lys47, Val218, and Pro219 and hydrogen bonds formed with Gly46, Lys47, and Gln142. Overall, the computational analysis revealed similar binding modes and affinities between melanin and known inhibitors for each target protein, corroborating the experimental observations that melanin-loaded PAN nanofibers possess significant bioactive potential.

4. Conclusions

This study demonstrates that PAN fibers infused with melanin exhibit a dense and homogeneous structure, with the fiber diameter increasing proportionally to the melanin concentration. Based on the above results, membranes' characteristics have a direct relationship with a higher concentration of melanin (5.0%) since membranes with a high concentration have excellent antioxidant and antimicrobial activity and a strong inhibitory activity against aging-related enzymes. The incorporation of melanin preserves the original molecular structure of the PAN fibers while enhancing their mechanical properties, particularly the tensile strength at higher melanin concentrations. Release kinetics reveal that higher melanin concentrations in PAN membranes result in a more substantial and sustained release over time, attributed to the greater availability of melanin within the fibrous matrix. The PAN–melanin fibers display significant antioxidant activity, with radical scavenging capacity increasing in proportion to melanin concentration. Furthermore, these fibers exhibit antimicrobial properties. Notably, higher melanin concentrations correlate with stronger inhibitory effects against aging-related enzymes, including tyrosinase, hyaluronidase, collagenase, and elastase. The molecular docking of proteins with the melanin ligand presented good binding energies with all aging-related enzymes. These findings highlight the potential of melanin-infused PAN fibers for applications requiring enhanced mechanical properties, controlled release, antioxidant activity, antimicrobial effects, and the inhibition of aging-related enzymatic processes.

Supplementary Materials: The following supporting information can be downloaded at https://www.mdpi.com/article/10.3390/membranes15020044/s1: Figure S1: Free radical scavenging activity of melanin membranes against DPPH and ABTS experiments; Figure S2: Antibacterial activity of membranes loaded with melanin pigment.

Author Contributions: Methodology, M.G.-G., F.N.M.-L., E.Q.-R., T.C.-G. and E.R.-S.; Conceptualization, J.S.J.-F. and D.O.-T.; Funding Acquisition, D.O.-T.; Writing—original draft preparation: M.G.-G.; Writing—Review and Editing, M.G.-G., J.S.J.-F. and D.O.-T. All authors have read and agreed to the published version of the manuscript.

Funding: This research was funded by Empuje Científico y Tecnológico (SICES) in SICES/CONV/121/2020 from Instituto de Innovación, Ciencia y Emprendimiento para la Competitividad del Estado de Guanajuato (IDEA GTO) to Domancar Orona-Tamayo.

Data Availability Statement: The authors confirm that the data supporting the findings of this study are available within this article.

Conflicts of Interest: No potential conflicts of interest are reported by the author(s).

References

1. Hadgraft, J. Skin, the Final Frontier. *Int. J. Pharm.* **2001**, *224*, 1–18. [CrossRef] [PubMed]
2. Dini, I.; Falanga, D.; Di Lorenzo, R.; Tito, A.; Carotenuto, G.; Zappelli, C.; Grumetto, L.; Sacchi, A.; Laneri, S.; Apone, F. An Extract from Ficus Carica Cell Cultures Works as an Anti-Stress Ingredient for the Skin. *Antioxidants* **2021**, *10*, 515. [CrossRef] [PubMed]
3. Eidt, L.M. Cutaneous Aging and Dermatosis in Geriatric Patients. In *Dermatology in Public Health Environments*; Springer International Publishing: Cham, Switzerland, 2023; pp. 967–1001. [CrossRef]
4. Ribeiro, A.; Estanqueiro, M.; Oliveira, M.; Sousa Lobo, J. Main Benefits and Applicability of Plant Extracts in Skin Care Products. *Cosmetics* **2015**, *2*, 48–65. [CrossRef]
5. Kumar, V. Perspective of Natural Products in Skincare. *Pharm. Pharmacol. Int. J.* **2016**, *4*, 339–341. [CrossRef]
6. Krzyżostan, M.; Wawrzyńczak, A.; Nowak, I. Use of Waste from the Food Industry and Applications of the Fermentation Process to Create Sustainable Cosmetic Products: A Review. *Sustainability* **2024**, *16*, 2757. [CrossRef]
7. Rostamabadi, H.; Assadpour, E.; Tabarestani, H.S.; Falsafi, S.R.; Jafari, S.M. Electrospinning Approach for Nanoencapsulation of Bioactive Compounds; Recent Advances and Innovations. *Trends Food Sci. Technol.* **2020**, *100*, 190–209. [CrossRef]
8. Luraghi, A.; Peri, F.; Moroni, L. Electrospinning for Drug Delivery Applications: A Review. *J. Control. Release* **2021**, *334*, 463–484. [CrossRef] [PubMed]
9. Shahriar, S.M.S.; Mondal, J.; Hasan, M.N.; Revuri, V.; Lee, D.Y.; Lee, Y.-K. Electrospinning Nanofibers for Therapeutics Delivery. *Nanomaterials* **2019**, *9*, 532. [CrossRef] [PubMed]
10. Sethuram, L.; Thomas, J. Therapeutic Applications of Electrospun Nanofibers Impregnated with Various Biological Macromolecules for Effective Wound Healing Strategy—A Review. *Biomed. Pharmacother.* **2023**, *157*, 113996. [CrossRef]

11. Serra, D.; Garroni, G.; Cruciani, S.; Coradduzza, D.; Pashchenko, A.; Amler, E.; Pintore, G.; Satta, R.; Montesu, M.A.; Kohl, Y.; et al. Electrospun Nanofibers Encapsulated with Natural Products: A Novel Strategy to Counteract Skin Aging. *Int. J. Mol. Sci.* **2024**, *25*, 1908. [CrossRef]
12. Gul, A.; Gallus, I.; Tegginamath, A.; Maryska, J.; Yalcinkaya, F. Electrospun Antibacterial Nanomaterials for Wound Dressings Applications. *Membranes* **2021**, *11*, 908. [CrossRef]
13. Liu, X.; Xu, H.; Zhang, M.; Yu, D.-G. Electrospun Medicated Nanofibers for Wound Healing: Review. *Membranes* **2021**, *11*, 770. [CrossRef] [PubMed]
14. Gao, C.; Zhang, L.; Wang, J.; Jin, M.; Tang, Q.; Chen, Z.; Cheng, Y.; Yang, R.; Zhao, G. Electrospun Nanofibers Promote Wound Healing: Theories, Techniques, and Perspectives. *J. Mater. Chem. B* **2021**, *9*, 3106–3130. [CrossRef] [PubMed]
15. Chen, K.; Hu, H.; Zeng, Y.; Pan, H.; Wang, S.; Zhang, Y.; Shi, L.; Tan, G.; Pan, W.; Liu, H. Recent Advances in Electrospun Nanofibers for Wound Dressing. *Eur. Polym. J.* **2022**, *178*, 111490. [CrossRef]
16. Mulholland, E.J. Electrospun Biomaterials in the Treatment and Prevention of Scars in Skin Wound Healing. *Front. Bioeng. Biotechnol.* **2020**, *8*, 481. [CrossRef]
17. Gruppuso, M.; Turco, G.; Marsich, E.; Porrelli, D. Polymeric Wound Dressings, an Insight into Polysaccharide-Based Electrospun Membranes. *Appl. Mater. Today* **2021**, *24*, 101148. [CrossRef]
18. Naomi, R.; Bahari, H.; Ridzuan, P.M.; Othman, F. Natural-Based Biomaterial for Skin Wound Healing (Gelatin vs. Collagen): Expert Review. *Polymers* **2021**, *13*, 2319. [CrossRef]
19. Zhang, X.; Wang, Y.; Gao, Z.; Mao, X.; Cheng, J.; Huang, L.; Tang, J. Advances in Wound Dressing Based on Electrospinning Nanofibers. *J. Appl. Polym. Sci.* **2024**, *141*, e54746. [CrossRef]
20. Zahedi, P.; Rezaeian, I.; Ranaei-Siadat, S.; Jafari, S.; Supaphol, P. A Review on Wound Dressings with an Emphasis on Electrospun Nanofibrous Polymeric Bandages. *Polym. Adv. Technol.* **2010**, *21*, 77–95. [CrossRef]
21. Ji, Y.; Song, W.; Xu, L.; Yu, D.G.; Bligh, S.W.A. A Review on Electrospun Poly(Amino Acid) Nanofibers and Their Applications of Hemostasis and Wound Healing. *Biomolecules* **2022**, *12*(6), 794. [CrossRef] [PubMed]
22. Adamu, B.F.; Gao, J.; Gebeyehu, E.K.; Beyene, K.A.; Tadesse, M.G.; Liyew, E.Z. Self-Responsive Electrospun Nanofibers Wound Dressings: The Future of Wound Care. *Adv. Mater. Sci. Eng.* **2022**, *2022*, 2025170. [CrossRef]
23. Abdelhakeem, E.; Monir, S.; Teaima, M.H.M.; Rashwan, K.O.; El-Nabarawi, M. State-of-the-Art Review of Advanced Electrospun Nanofiber Composites for Enhanced Wound Healing. *AAPS PharmSciTech* **2023**, *24*, 246. [CrossRef]
24. Miguel, S.P.; Figueira, D.R.; Simões, D.; Ribeiro, M.P.; Coutinho, P.; Ferreira, P.; Correia, I.J. Electrospun Polymeric Nanofibres as Wound Dressings: A Review. *Colloids Surf. B Biointerfaces* **2018**, *169*, 60–71. [CrossRef] [PubMed]
25. Ajith, G.; Tamilarasi, G.P.; Sabarees, G.; Gouthaman, S.; Manikandan, K.; Velmurugan, V.; Alagarsamy, V.; Solomon, V.R. Recent Developments in Electrospun Nanofibers as Delivery of Phytoconstituents for Wound Healing. *Drugs Drug Candidates* **2023**, *2*, 148–171. [CrossRef]
26. Sharma, A.; Dheer, D.; Singh, I.; Puri, V.; Kumar, P. Phytoconstituent-Loaded Nanofibrous Meshes as Wound Dressings: A Concise Review. *Pharmaceutics* **2023**, *15*, 1058. [CrossRef] [PubMed]
27. Guo, S.; Wang, P.; Song, P.; Li, N. Electrospinning of Botanicals for Skin Wound Healing. *Front. Bioeng. Biotechnol.* **2022**, *10*, 1006129. [CrossRef] [PubMed]
28. Gaspar-Pintiliescu, A.; Stanciuc, A.-M.; Craciunescu, O. Natural Composite Dressings Based on Collagen, Gelatin and Plant Bioactive Compounds for Wound Healing: A Review. *Int. J. Biol. Macromol.* **2019**, *138*, 854–865. [CrossRef] [PubMed]
29. Xu, L.; Liu, Y.; Zhou, W.; Yu, D. Electrospun Medical Sutures for Wound Healing: A Review. *Polymers* **2022**, *14*, 1637. [CrossRef]
30. Ambekar, R.S.; Kandasubramanian, B. Advancements in Nanofibers for Wound Dressing: A Review. *Eur. Polym. J.* **2019**, *117*, 304–336. [CrossRef]
31. Adamu, B.F.; Gao, J.; Jhatial, A.K.; Kumelachew, D.M. A Review of Medicinal Plant-Based Bioactive Electrospun Nano Fibrous Wound Dressings. *Mater. Des.* **2021**, *209*, 109942. [CrossRef]
32. Zhang, W.; Ronca, S.; Mele, E. Electrospun Nanofibres Containing Antimicrobial Plant Extracts. *Nanomaterials* **2017**, *7*, 42. [CrossRef]
33. Wang, Y.; Wang, X.; Zhou, D.; Xia, X.; Zhou, H.; Wang, Y.; Ke, H. Preparation and Characterization of Polycaprolactone (PCL) Antimicrobial Wound Dressing Loaded with Pomegranate Peel Extract. *ACS Omega* **2023**, *8*, 20323–20331. [CrossRef]
34. Al-Naymi, H.A.S.; Mahmoudi, E.; Kamil, M.M.; Almajidi, Y.Q.; Al-Musawi, M.H.; Mohammadzadeh, V.; Ghorbani, M.; Mortazavi Moghadam, F. A Novel Designed Nanofibrous Mat Based on Hydroxypropyl Methyl Cellulose Incorporating Mango Peel Extract for Potential Use in Wound Care System. *Int. J. Biol. Macromol.* **2024**, *259*, 129159. [CrossRef]
35. Zdraveva, E.; Gaurina Srček, V.; Kraljić, K.; Škevin, D.; Slivac, I.; Obranović, M. Agro-Industrial Plant Proteins in Electrospun Materials for Biomedical Application. *Polymers* **2023**, *15*, 2684. [CrossRef] [PubMed]
36. Ortega, F.; Versino, F.; López, O.V.; García, M.A. Biobased Composites from Agro-Industrial Wastes and by-Products. *Emergent Mater.* **2022**, *5*, 873–921. [CrossRef]

37. Abd Elkodous, M.; El-Husseiny, H.M.; El-Sayyad, G.S.; Hashem, A.H.; Doghish, A.S.; Elfadil, D.; Radwan, Y.; El-Zeiny, H.M.; Bedair, H.; Ikhdair, O.A.; et al. Recent Advances in Waste-Recycled Nanomaterials for Biomedical Applications: Waste-to-Wealth. *Nanotechnol. Rev.* **2021**, *10*, 1662–1739. [CrossRef]
38. Topić Popović, N.; Lorencin, V.; Strunjak-Perović, I.; Čož-Rakovac, R. Shell Waste Management and Utilization: Mitigating Organic Pollution and Enhancing Sustainability. *Appl. Sci.* **2023**, *13*, 623. [CrossRef]
39. Ngasotter, S.; Xavier, K.A.M.; Meitei, M.M.; Waikhom, D.; Madhulika; Pathak, J.; Singh, S.K. Crustacean Shell Waste Derived Chitin and Chitin Nanomaterials for Application in Agriculture, Food, and Health—A Review. *Carbohydr. Polym. Technol. Appl.* **2023**, *6*, 100349. [CrossRef]
40. Villasante, J.; Pérez-Carrillo, E.; Heredia-Olea, E.; Metón, I.; Almajano, M.P. In Vitro Antioxidant Activity Optimization of Nut Shell (*Carya illinoinensis*) by Extrusion Using Response Surface Methods. *Biomolecules* **2019**, *9*, 883. [CrossRef]
41. Ferrari, V.; Gil, G.; Heinzen, H.; Zoppolo, R.; Ibáñez, F. Influence of Cultivar on Nutritional Composition and Nutraceutical Potential of Pecan Growing in Uruguay. *Front. Nutr.* **2022**, *9*, 868054. [CrossRef] [PubMed]
42. García-García, M.; Orona-Tamayo, D.; Estrada-Monje, A.; Quintana-Rodríguez, E.; Navarro-Mendoza, R.; Hernández-Perales, L.; Lozoya-Perez, N.E.; Jaime-Ferrer, J.S. Melanin from Pecan Nut Shell Waste as an Antioxidant and Antifungal Additive in Membranes for Food Packaging. *J. Polym. Environ.* **2025**. [CrossRef]
43. Fu, X.; Xie, M.; Lu, M.; Shi, L.; Shi, T.; Yu, M. Characterization of the Physicochemical Properties, Antioxidant Activity, and Antiproliferative Activity of Natural Melanin from S. Reiliana. *Sci. Rep.* **2022**, *12*, 2110. [CrossRef]
44. Mavridi-Printezi, A.; Menichetti, A.; Mordini, D.; Amorati, R.; Montalti, M. Recent Applications of Melanin-like Nanoparticles as Antioxidant Agents. *Antioxidants* **2023**, *12*, 863. [CrossRef] [PubMed]
45. Khojah, H.; Ahmed, S.R.; Alharbi, S.Y.; AlSabeelah, K.K.; Alrayyes, H.Y.; Almusayyab, K.B.; Alrawiliy, S.R.; Alshammari, R.M.; Qasim, S. Skin Anti-Aging Potential of Launaea Procumbens Extract: Antioxidant and Enzyme Inhibition Activities Supported by ADMET and Molecular Docking Studies. *Saudi Pharm. J.* **2024**, *32*, 102107. [CrossRef] [PubMed]
46. Lu, Y.; Ye, M.; Song, S.; Li, L.; Shaikh, F.; Li, J. Isolation, Purification, and Anti-Aging Activity of Melanin from Lachnum Singerianum. *Appl. Biochem. Biotechnol.* **2014**, *174*, 762–771. [CrossRef] [PubMed]
47. Reyes-Aguilera, J.A.; Villafaña-López, L.; Rentería-Martínez, E.C.; Anderson, S.M.; Jaime-Ferrer, J.S. Electrospinning of Polyepychlorhydrin and Polyacrylonitrile Anionic Exchange Membranes for Reverse Electrodialysis. *Membranes* **2021**, *11*, 717. [CrossRef] [PubMed]
48. *ASTM E573-01(2013)*; Standard Practices for Internal Reflection Spectroscopy. ASTM International: West Conshohocken, PA, USA, 2013. [CrossRef]
49. Liu, S.-J.; Kau, Y.-C.; Chou, C.-Y.; Chen, J.-K.; Wu, R.-C.; Yeh, W.-L. Electrospun PLGA/Collagen Nanofibrous Membrane as Early-Stage Wound Dressing. *J. Memb. Sci.* **2010**, *355*, 53–59. [CrossRef]
50. Ruiz Rocha, J.E.; Moreno Tovar, K.R.; Navarro Mendoza, R.; Gutiérrez Granados, S.; Cavaliere, S.; Giaume, D.; Barboux, P.; Jaime Ferrer, J.S. Critical Electrospinning Parameters for Synthesis Control of Stabilized Polyacrylonitrile Nanofibers. *Nanomaterials* **2023**, *13*, 2648. [CrossRef]
51. *ASTM D882-18*; Standard Test Method for Tensile Properties of Thin Plastic Sheeting. ASTM International: West Conshohocken, PA, USA, 2018. [CrossRef]
52. Adegbola, T.A.; Agboola, O.; Fayomi, O.S.I. Review of Polyacrylonitrile Blends and Application in Manufacturing Technology: Recycling and Environmental Impact. *Results Eng.* **2020**, *7*, 100144. [CrossRef]
53. Zhang, H.; Quan, L.; Gao, A.; Tong, Y.; Shi, F.; Xu, L. Thermal Analysis and Crystal Structure of Poly(Acrylonitrile-Co-Itaconic Acid) Copolymers Synthesized in Water. *Polymers* **2020**, *12*, 221. [CrossRef] [PubMed]
54. Isaac, B.; Taylor, R.M.; Reifsnider, K. Anisotropic Characterizations of Electrospun PAN Nanofiber Mats Using Design of Experiments. *Nanomaterials* **2020**, *10*, 2273. [CrossRef]
55. Kalantary, S.; Golbabaei, F.; Latifi, M.; Shokrgozar, M.A.; Yaseri, M. Feasibility of Using Vitamin E-Loaded Poly(ε-Caprolactone)/Gelatin Nanofibrous Mat to Prevent Oxidative Stress in Skin. *J. Nanosci. Nanotechnol.* **2020**, *20*, 3554–3562. [CrossRef] [PubMed]
56. Orona-Tamayo, D.; Valverde, M.E.; Nieto-Rendón, B.; Paredes-López, O. Inhibitory Activity of Chia (*Salvia hispanica* L.) Protein Fractions against Angiotensin I-Converting Enzyme and Antioxidant Capacity. *LWT-Food Sci. Technol.* **2015**, *64*, 236–242. [CrossRef]
57. Aguilar-Toalá, J.E.; Liceaga, A.M. Identification of Chia Seed (*Salvia hispanica* L.) Peptides with Enzyme Inhibition Activity towards Skin-Aging Enzymes. *Amino Acids* **2020**, *52*, 1149–1159. [CrossRef] [PubMed]
58. Alhayek, A.; Abdelsamie, A.S.; Schönauer, E.; Camberlein, V.; Hutterer, E.; Posselt, G.; Serwanja, J.; Blöchl, C.; Huber, C.G.; Haupenthal, J.; et al. Discovery and Characterization of Synthesized and FDA-Approved Inhibitors of Clostridial and Bacillary Collagenases. *J. Med. Chem.* **2022**, *65*, 12933–12955. [CrossRef]
59. Chao, K.L.; Muthukumar, L.; Herzberg, O. Structure of Human Hyaluronidase-1, a Hyaluronan Hydrolyzing Enzyme Involved in Tumor Growth and Angiogenesis. *Biochemistry* **2007**, *46*, 6911–6920. [CrossRef]

60. Hansen, G.; Gielen-Haertwig, H.; Reinemer, P.; Schomburg, D.; Harrenga, A.; Niefind, K. Unexpected Active-Site Flexibility in the Structure of Human Neutrophil Elastase in Complex with a New Dihydropyrimidone Inhibitor. *J. Mol. Biol.* **2011**, *409*, 681–691. [CrossRef]
61. Ferro, S.; Deri, B.; Germanò, M.P.; Gitto, R.; Ielo, L.; Buemi, M.R.; Certo, G.; Vittorio, S.; Rapisarda, A.; Pazy, Y.; et al. Targeting Tyrosinase: Development and Structural Insights of Novel Inhibitors Bearing Arylpiperidine and Arylpiperazine Fragments. *J. Med. Chem.* **2018**, *61*, 3908–3917. [CrossRef]
62. Berman, H.M. The Protein Data Bank. *Nucleic Acids Res.* **2000**, *28*, 235–242. [CrossRef]
63. Jo, S.; Kim, T.; Iyer, V.G.; Im, W. CHARMM-GUI: A Web-based Graphical User Interface for CHARMM. *J. Comput. Chem.* **2008**, *29*, 1859–1865. [CrossRef] [PubMed]
64. Trott, O.; Olson, A.J. AutoDock Vina: Improving the Speed and Accuracy of Docking with a New Scoring Function, Efficient Optimization, and Multithreading. *J. Comput. Chem.* **2010**, *31*, 455–461. [CrossRef] [PubMed]
65. Adasme, M.F.; Linnemann, K.L.; Bolz, S.N.; Kaiser, F.; Salentin, S.; Haupt, V.J.; Schroeder, M. PLIP 2021: Expanding the Scope of the Protein–Ligand Interaction Profiler to DNA and RNA. *Nucleic Acids Res.* **2021**, *49*, W530–W534. [CrossRef]
66. PyMOL | pymol.org. Available online: https://www.pymol.org/ (accessed on 6 November 2024).
67. Tran-Ly, A.N.; Ribera, J.; Schwarze, F.W.M.R.; Brunelli, M.; Fortunato, G. Fungal Melanin-Based Electrospun Membranes for Heavy Metal Detoxification of Water. *Sustain. Mater. Technol.* **2020**, *23*, e00146. [CrossRef]
68. Micheal, H.S.R.; Thyagarajan, D.; Govindaraj, M.; Saravanakumar, V.K.; Mohammed, N.B.; Murugasamy Maheswari, K. Biosorption of Halophilic Fungal Melanized Membrane—PUR/Melanin Polymer for Heavy Metal Detoxification with Electrospinning Technology. *Environ. Technol.* **2024**, *45*, 5865–5877. [CrossRef] [PubMed]
69. El Fawal, G. Polymer Nanofibers Electrospinning: A Review. *Egypt. J. Chem.* **2019**, *63*(4), 1279–1303. [CrossRef]
70. Guo, Y.; Wang, X.; Shen, Y.; Dong, K.; Shen, L.; Alzalab, A.A.A. Research Progress, Models and Simulation of Electrospinning Technology: A Review. *J. Mater. Sci.* **2022**, *57*, 58–104. [CrossRef]
71. Rist, M.; Greiner, A. Bio-based electrospun polyamide membrane—Sustainable multipurpose filter membranes for microplastic filtration. *RSC Appl. Polym.* **2024**, *2*, 642–655. [CrossRef]
72. Street, R.M.; Minagawa, M.; Vengrenyuk, A.; Schauer, C.L. Piezoelectric Electrospun Polyacrylonitrile with Various Tacticities. *J. Appl. Polym. Sci.* **2019**, *136*, 47530. [CrossRef]
73. Alarifi, I.; Alharbi, A.; Khan, W.; Swindle, A.; Asmatulu, R. Thermal, Electrical and Surface Hydrophobic Properties of Electrospun Polyacrylonitrile Nanofibers for Structural Health Monitoring. *Materials* **2015**, *8*, 7017–7031. [CrossRef]
74. Shekharagouda, V.; Kumar Panda, P. Electrospinning of Polyacrylonitrile Nanofiber Membrane for Bacteria Removal. *J. Mater. Sci. Appl.* **2018**, *4*, 68–74. [CrossRef]
75. Yao, Y.; Liang, Y.; Navik, R.; Dong, X.; Cai, Y.; Zhang, P. Modification of Polyacrylonitrile Fibers by Coupling to Thiosemicarbazones. *Materials* **2019**, *12*, 3980. [CrossRef]
76. Wang, Y.; Wang, X.; Li, T.; Ma, P.; Zhang, S.; Du, M.; Dong, W.; Xie, Y.; Chen, M. Effects of Melanin on Optical Behavior of Polymer: From Natural Pigment to Materials Applications. *ACS Appl. Mater. Interfaces* **2018**, *10*, 13100–13106. [CrossRef]
77. Wei, H.; Ren, J.; Han, B.; Xu, L.; Han, L.; Jia, L. Stability of Polydopamine and Poly(DOPA) Melanin-like Films on the Surface of Polymer Membranes under Strongly Acidic and Alkaline Conditions. *Colloids Surf. B Biointerfaces* **2013**, *110*, 22–28. [CrossRef] [PubMed]
78. Wakamatsu, K.; Ito, S. Recent Advances in Characterization of Melanin Pigments in Biological Samples. *Int. J. Mol. Sci.* **2023**, *24*, 8305. [CrossRef]
79. Henrici-Olivé, G.; Olivé, S. Molecular Interactions and Macroscopic Properties of Polyacrylonitrile and Model Substances. In *Chemistry. Advances in Polymer Science*; Springer: Berlin/Heidelberg, Germany, 1979; pp. 123–152. [CrossRef]
80. Sanchaniya, J.V.; Kanukuntla, S. Morphology and mechanical properties of PAN nanofiber mat. *J. Phys. Conf. Ser.* **2023**, *2423*, 012018. [CrossRef]
81. Sebesta, F.; John, J.; Motl, A.; Stamberg, K. *Evaluation of Polyacrylonitrile (PAN) as a Binding Polymer for Absorbers Used to Treat Liquid Radioactive Wastes*; Sandia National Labs.: Albuquerque, NM, USA, 1995. [CrossRef]
82. Taborda, C.P.; da Silva, M.B.; Nosanchuk, J.D.; Travassos, L.R. Melanin as a Virulence Factor of Paracoccidioides Brasiliensis and Other Dimorphic Pathogenic Fungi: A Minireview. *Mycopathologia* **2008**, *165*, 331–339. [CrossRef] [PubMed]
83. Goncalves, R.d.C.R.; Pombeiro-Sponchiado, S.R. Antioxidant Activity of the Melanin Pigment Extracted from Aspergillus Nidulans. *Biol. Pharm. Bull.* **2005**, *28*, 1129–1131. [CrossRef] [PubMed]
84. Wang, Y.; Casadevall, A. Susceptibility of Melanized and Nonmelanized Cryptococcus Neoformans to Nitrogen- and Oxygen-Derived Oxidants. *Infect. Immun.* **1994**, *62*, 3004–3007. [CrossRef] [PubMed]
85. Manini, P.; Lino, V.; Franchi, P.; Gentile, G.; Sibillano, T.; Giannini, C.; Picardi, E.; Napolitano, A.; Valgimigli, L.; Chiappe, C.; et al. A Robust Fungal Allomelanin Mimic: An Antioxidant and Potent Π-Electron Donor with Free-Radical Properties That Can Be Tuned by Ionic Liquids. *Chempluschem* **2019**, *84*, 1331–1337. [CrossRef]

86. McGraw, K.J. The Antioxidant Function of Many Animal Pigments: Are There Consistent Health Benefits of Sexually Selected Colourants? *Anim. Behav.* **2005**, *69*, 757–764. [CrossRef]
87. Peng, Z.; Luo, S.; Zhao, D.; Zhang, J. Biosynthetic Melanin with Excellent Performance Can Be Used for Heavy Metal Adsorption. *J. Clean. Prod.* **2023**, *385*, 135655. [CrossRef]
88. Sunil Kumar, B.T.; Sathyendra Rao, B.V.; Swathi, S.; Vanajakshi, V.; Hebbar, H.U.; Singh, S.A. Characterization, Antioxidant, and Antimicrobial Activity of Melanin Extracted from Nigerseed Hulls. *Food Biosci.* **2024**, *61*, 104929. [CrossRef]
89. Bhat, S.G.; Laxmi, M.; Kurian, N.K.; Smitha, S. Melanin and Bacteriocin from Marine Bacteria Inhibit Biofilms of Foodborne Pathogens. *Indian J. Biotechnol.* **2016**, *15*, 392–399.
90. Yeroslavsky, G.; Richman, M.; Dawidowicz, L.; Rahimipour, S. Sonochemically Produced Polydopamine Nanocapsules with Selective Antimicrobial Activity. *Chem. Commun.* **2013**, *49*, 5721. [CrossRef] [PubMed]
91. Eliato, T.R. Melanin Pigments as Antibacterial Agents. Doctoral Dissertations, University of New Hampshire, Durham, NH, USA, 2022.
92. Mustafa, Y.F. Synthesis of 7,8-Dihydroxy-4-Phenylbenzo[g]Coumarins as Potential Multitarget Anti-Skin-Aging Candidates. *J. Mol. Struct.* **2025**, *1321*, 139806. [CrossRef]
93. Widowati, W.; Ginting, C.N.; Lister, I.N.E.; Girsang, E.; Amalia, A.; Wibowo, S.H.B.; Kusuma, H.; Rizal, R. Anti-Aging Effects of Mangosteen Peel Extract and Its Phytochemical Compounds: Antioxidant Activity, Enzyme Inhibition and Molecular Docking Simulation. *Trop. Life Sci. Res.* **2020**, *31*, 127–144. [CrossRef] [PubMed]
94. Aguilar-Toalá, J.E.; Hernández-Mendoza, A.; González-Córdova, A.F.; Vallejo-Cordoba, B.; Liceaga, A.M. Potential Role of Natural Bioactive Peptides for Development of Cosmeceutical Skin Products. *Peptides* **2019**, *122*, 170170. [CrossRef] [PubMed]
95. Maleki, M.; Khelghati, N.; Alemi, F.; Bazdar, M.; Asemi, Z.; Majidinia, M.; Sadeghpoor, A.; Mahmoodpoor, A.; Jadidi-Niaragh, F.; Targhazeh, N.; et al. Stabilization of Telomere by the Antioxidant Property of Polyphenols: Anti-Aging Potential. *Life Sci.* **2020**, *259*, 118341. [CrossRef] [PubMed]
96. de Lima Cherubim, D.J.; Buzanello Martins, C.V.; Oliveira Fariña, L.; da Silva de Lucca, R.A. Polyphenols as Natural Antioxidants in Cosmetics Applications. *J. Cosmet. Dermatol.* **2020**, *19*, 33–37. [CrossRef] [PubMed]
97. Thoyajakshi, R.S.; Megha, G.T.; Ravi Kumar, H.; Mathad, S.N.; Khan, A.; Nagaraju, S.; Mahmoud, M.H.; Ansari, A.Z. Garcinol: A Novel and Potent Inhibitor of Hyaluronidase Enzyme. *Int. J. Biol. Macromol.* **2024**, *266 Pt 2*, 131145. [CrossRef] [PubMed]
98. Navia, M.A.; McKeever, B.M.; Springer, J.P.; Lin, T.Y.; Williams, H.R.; Fluder, E.M.; Dorn, C.P.; Hoogsteen, K. Structure of Human Neutrophil Elastase in Complex with a Peptide Chloromethyl Ketone Inhibitor at 1.84-A Resolution. *Proc. Natl. Acad. Sci. USA* **1989**, *86*, 7–11. [CrossRef] [PubMed]

membranes

MDPI

Article

Asymmetric Membranes Obtained from Sulfonated HIPS Waste with Potential Application in Wastewater Treatment

Marcial Alfredo Yam-Cervantes [1], Rita Sulub-Sulub [1], Mauricio Hunh-Ibarra [1], Santiago Duarte [1], Erik Uc-Fernandez [1], Daniel Pérez-Canales [2], Manuel Aguilar-Vega [1] and Maria Ortencia González-Díaz [3,*]

[1] Centro de Investigación Científica de Yucatán, A.C., Calle 43 No. 130, Chuburná de Hidalgo, Mérida 97200, Yucatán, Mexico; mjav@cicy.mx (M.A.-V.)

[2] Independent Researcher, Calle Florida 9 Int. 17, Col. Noche Buena, Del. Benito Juárez, Ciudad de México 03720, Distrito Federal, Mexico

[3] CONAHCYT—Centro de Investigación Científica de Yucatán, A.C., Calle 43 No. 130, Chuburná de Hidalgo, Mérida 97200, Yucatán, Mexico

* Correspondence: maria.gonzalez@cicy.mx

Abstract: The recovery and reuse of high-impact polystyrene (HIPS) into high-value products is crucial for reducing environmental thermoplastics waste and promoting sustainable materials for various applications. In this study, asymmetric membranes obtained from sulfonated HIPS waste were used for salt and dye removals. The incorporation of sulfonic acid (-SO_3H) groups into HIPS waste by direct chemical sulfonation with chlorosulfonic acid (CSA), at two different concentrations, was investigated to impart antifouling properties in membranes for water treatment. Asymmetric membranes from recycled HIPS, R-HIPS, R-HIPS-3, and R-HIPS-5 with 3 and 5% sulfonation degrees, respectively. Sulfonated HIPS shows a decrease in water contact angle (WCA) from 83.8° for recycled R-HIPS to 66.1° for R-HIPS-5, respectively. A WCA decrease leads to an increase in antifouling properties for R-HIPS-5, compared to non-sulfonated R-HIPS, which leads to a higher flux recovery ratio (FRR) and enhanced separation properties for sulfonated membranes. The HIPS-5 membrane exhibited the highest rejection rates for Reactive Black 5 dye (94%) and divalent salts (72% for $MgSO_4$ and 67% for Na_2SO_4). The performance of the recycled HIPS asymmetric membranes is well correlated with porosity, water uptake, and the higher negative charge from the sulfonic acid groups present, which enhance the electrostatic repulsions of salts and dyes.

Keywords: asymmetric membranes; high-impact polystyrene waste; dye removal; water treatment; sulfonated high-impact polystyrene

1. Introduction

In recent years, there has been a substantial rise in global plastic waste, leading to serious environmental and public health challenges [1]. In 2018, annual plastic production reached 359 million tons, and it is expected to double by 2050. In Mexico, the per capita plastic waste generation is 59 kg annually, with less than 10% being recycled [2]. In 2023, the Mexican Secretariat of Environment and Natural Resources (SEMARNAT) identified polypropylene (PP), low-density polyethylene (LDPE), and polystyrene (PS) as the primary polymers contributing to environmental pollution, with PS having the lowest recycling potential based on a cost–benefit ratio [2]. In response, the Mexican government, following international initiatives, proposed a policy for sustainable development and a circular economy to enhance waste management practices. A key initiative within this framework is the redesign and reuse of plastic products. Specifically, recycling and transforming polystyrene—available in several forms, such as expanded polystyrene (EPS), general-purpose polystyrene (GPPS), and high-impact polystyrene (HIPS)—is crucial for reducing PS waste in the environment and promoting sustainable materials for various applications.

Recently, single-use polystyrene (PS) waste has been identified as a valuable material for the circular economy [3–5]. For example, nanofiber membranes from expanded polystyrene (EPS) waste with different poly(vinylpyrrolidone) (PVP) concentrations were elaborated by electrospinning for their use in water filtration [6]. PVP introduction reduced EPS's high hydrophobicity on the membrane surface, changing their water contact angle from 140.46° to 84.70° and increasing pure water flux from 526.86 L $m^{-2}h^{-1}$ to 1901.35 L $m^{-2}h^{-1}$. The introduction of carbon nanotubes is also a strategy to obtain a more hydrophilic membrane surface with lower fouling properties [7]. Aquino Lima et al. developed an ultra-fast method to produce recycled EPS membranes using DL-limonene as a natural solvent, with adequate characteristics for its applications in air filtration [8]. Moreover, the production of ion exchange resins from polystyrene waste, sulfonated with sulfuric acid, has been reported for heavy metals (Cu^{2+}, Zn^{2+}, and Cd^{2+}) removal in wastewater [9–11], as well as for the adsorption of fluoroquinolone, a second-generation synthetic antibiotic [12]. Membranes from sulfonated PSE blended with 1% of polybutylene succinate (PBS) exhibited a higher methylene blue uptake and good regeneration efficiency, compared with membranes containing a higher PBS amount, following a pseudo-second order kinetic model in the adsorption process [13].

The utilization of HIPS waste has been less extensively studied, despite its notable advantages over brittle and rigid EPS and GPPS [14,15]. HIPS offers superior impact resistance, flexibility, and stability, making it a promising candidate for various applications. Sharma et al. investigated the potential use of recycled HIPS as a partial replacement for fine aggregates in lightweight concrete and evaluated the impact on key properties, such as compressive strength, flexural strength, and bond strength. The authors concluded that recycled HIPS at 10% concentration is an optimal level for achieving lightweight concrete with desirable mechanical properties, offering a sustainable solution for recycling plastic waste in construction materials [16]. Moreover, cation exchange resins for the removal of heavy metal ions (Zn^{2+}, Pb^{2+}) were prepared from sulfonated polystyrene waste (WPS) containing HIPS and EPS. These cation exchange resins were capable of reducing Zn^{2+} metal concentration from 80 mg/L to 38.3 mg/L and Pb^{2+} from 100 mg/L to 33 mg/L by an absorption process [17]. Air filter membranes were also fabricated from a HIPS waste solution in a D-limonene:DMF mixture using needleless electrospinning. A fine fiber morphology was achieved by increasing DMF concentration in the solvent mixture. The HIPS membrane demonstrated excellent performance as an air filtration medium, with a filtration efficiency up to 98.5%, as shown in filtration tests with $PM_{2.5}$ particles [18]. Khaled et al. elaborated micro-porous membranes from recycled HIPS, which presented contact angle values between 81° and 91°, with surface pore sizes ranging from 150 nm to 550 nm and with macrovoids in their cross-sections [19]. These membranes presented Rhodamine B (RhB) rejection values between 95 and 97% and from 85 to 87% for humic acid (HA) [20]. They exhibited a slight increase in tensile strength values as HIPS concentration increased from 20 to 35% [15]. Waste HIPS's feasibility as membranes for gas separation was also investigated [21]. The authors reported that HIPS membranes exhibited better gas transport properties than EPS and GPPS, due to their butadiene rubber content in the matrix, with a gas separation factor that was well correlated with the difference in the gas penetrant kinetic diameter. Recently, a method for the total revalorization of HIPS was investigated by Giakoumakis et al. These authors employed HIPS waste as potential feedstock for styrene monomer recovery using a green fractionation process using ethyl acetate (EtOAc). This strategy consists of EtOAc addition as a pre-fractionation solvent to isolate the free PS matrix from the rubber phase, followed by thermal degradation of the isolated free PS matrix, which leads to an enhanced styrene yield, compared to the non-fractionated HIPS sample, and finally, an ethylene metathesis for the valorization of the rubber phase [22].

Here, we present the development of asymmetric sulfonated HIPS membranes from recycled HIPS (R-HIPS) for water treatment. The incorporation of sulfonic pendant groups in R-HIPS by a simple and efficient electrophilic substitution reaction was directed to enhance the hydrophilicity, morphology, permeability, and antifouling properties of HIPS membranes [23,24]. R-HIPS sulfonated asymmetric membranes' general properties and performance in removing salts and dyes under low-pressure conditions and the effect of R-HIPS sulfonation degree on membrane fouling are discussed. Overall, this study has the potential to make a significant impact on both reducing plastic pollution and improving wastewater treatment systems through the use of recycled HIPS. With further research focused on the scalability as spiral-wound filtration modules and addressing potential challenges, sulfonated high-impact polystyrene membranes could emerge as a cost-effective and sustainable alternative for polystyrene recycling and wastewater treatment.

2. Materials and Methods

2.1. Chemicals

Recycled high-impact polystyrene (R-HIPS) was obtained from a Cole-Parmer tray. Ethyl acetate (99.5%, M TEDIA, Fairfield, CA, USA), chlorosulfonic acid (CSA, 99.0%), 1-2-dichloroethane (DCE, 99.8%), bovine serum albumin (BSA, 67 kDa), potassium phosphate di-basic (K_2HPO_4, 98%), magnesium sulfate ($MgSO_4$, ≥97%), sodium sulfate (Na_2SO_4, ≥99%), Reactive Black 5, 1methyl-2-pyrrolidinone anhydrous (NMP, 99.5%), toluene (HPLC, 99.9%), methanol (MeOH, 99.8%), and phenolphthalein (ACS reagent) were all supplied by Sigma Aldrich S.A. de C.V. (Toluca, México) and used as received. Potassium phosphate monobasic (KH_2PO_4, 99.5%, Fluka chemical, GmbH, Buchs, Switzerland), 2-propanol (99.8%, J.T. Baker, Edo. de México, México), acetic acid 5% solution in water (commercial vinegar, @La Costeña, Yucatán, México), ethanol (EtOH, 96%, CTR Scientific, Nuevo León, México), sodium bicarbonate ($NaHCO_3$, Arm & Harmmer, Church & Dwight, Ewing, NJ, USA), and sodium hydroxide (NaOH, J.T. Baker, 99.5%, Edo. de México, México) were also used as received.

2.2. R-HIPS Sulfonation

R-HIPS sulfonation was carried out by direct sulfonation in solution with chlorosulfonic acid (CSA) at two different concentrations (3 and 5%). In a typical sulfonation reaction, in a 500 mL three-neck flask with a mechanical stirrer, 10 g of R-HIPS was dissolved in 100 mL of DCE. Then, the homogeneous solution was cooled in an ice bath, and the required amount of chlorosulfonic acid was added dropwise, using an addition funnel to achieve the desired sulfonation degree (see Table 1). After this point, the sulfonation reaction was stirred at room temperature for 20 h. The resulting product was then gradually poured into ethanol while being stirred vigorously. The sulfonated R-HIPS was filtered, washed with EtOH, and dried in a vacuum oven at 60 °C for 24 h [25].

Table 1. Amounts of each reactant used based on 10 g of sulfonated R-HIPS.

Membranes	Theoretical SD (%)	1,2, Dichloroethne (mL)	Chlorosulfonic Acid (mL)	Experimental SD (%)
R-HIPS-3	3	100	0.19	2.7
R-HIPS-5	5	100	0.33	5.9

Sulfonation degree (SD) was determined by a titration method as follows: 100 mg of R-HIPS (in triplicate) was solubilized in 10 mL of a toluene/methanol mixture (90/10 v/v) and subsequently titrated with a 0.01 M NaOH/methanol solution, using phenolphthalein as an indicator. The SD value (%) was calculated by Equation (1) [26]:

$$SD = \frac{104 \times N_{NaOH} \times V_{NaOH}}{(w \times 1000) - (81 \times N_{NaOH} \times V_{NaOH})} \quad (1)$$

where N_{NaOH} and V_{NaOH} are the concentration (mol/L) and volume of methanolic NaOH solution (mL), and W is the R-HIPS weight (g). The numbers 104 and 81 are the molar mass of the styrene unit and sulfonic, $-SO_3H$, groups, respectively.

2.3. Preparation of Asymmetric Membranes

The membranes were prepared by the phase inversion method using a 20% (*w*/*v*) doping solution in NMP. The polymer solution was cast on a glass support using a knife blade previously graduated at 33 μm, and it was immersed in the coagulation bath, which consisted of a mixture of commercial vinegar (acetic acid 5% solution in water)/isopropanol (15/85% *v*/*v*) with 0.12% $NaHCO_3$ at room temperature, at 50% relative humidity. Finally, the membranes were stored in distilled water for further characterization. Based on the sulfonation degree (SD), the membranes were identified as R-HIPS (non-sulfonated), R-HIPS-3 (3% SD), and R-HIPS-5 (5% SD).

2.4. Characterization

A scanning electron microscope (SEM) (JEOL JSM-6360LV SEM, Tokyo, Japan) was used to obtain cross-sectional and surface micrographs of membrane samples previously coated with gold in a DENTON VACUUM DESK II equipment (Philadelphia, NJ, USA).

FTIR-ATR spectra of R-HIPS and sulfonated R-HIPS were recorded on a Nicolet 8700 FTIR (Waltham, MA, USA) in the 4000–650 cm^{-1} range. Contact angle measurements on R-HIPS membrane surface were determined on a Ramé-Hart 590 goniometer equipped (Suççasunna, NJ, USA) with a high-resolution drop image software.

Water uptake was measured by immersing dry samples (1.5 cm × 1.5 cm) in water for 7 days at room temperature. The dry weight was determined after drying the samples at 60 °C for 24 h. The percentage of water content (WC %) was calculated with the following Equation (2).

$$WC(\%) = \frac{W_0 - W_1}{W_0} \times 100 \tag{2}$$

where W_0 = is the membrane wet weight, and W_1 = is the membrane dry weight.

2.5. Membrane Performance Assessment

Membrane porosity (ε) was calculated from the difference between the bulk membrane volume (V_m) and the polymer volume (V_p). The porosity was calculated using Equation (3) [27].

$$\varepsilon(\%) = \frac{V_m - V_p}{V_m} \times 100 \tag{3}$$

where V_m is the bulk membrane volume, and V_p represents the polymer volume. V_m was obtained by multiplying the membrane area by its thickness. The polymer volume V_p was obtained by the ratio $\frac{W_m}{\rho_p}$, where W_m is the membrane weight, and ρ_p is the polymer's density.

The mean pore radius was calculated using the Guerout–Elfort–Ferry equation (Equation (4)).

$$r_m = \sqrt{\frac{(2.9 - 1.75\varepsilon)8\eta\delta_0 Q}{\varepsilon A \Delta P}} \tag{4}$$

where η is the pure water viscosity (8.9 × 10^{-4} Pa s), Q is the pure water volume permeated per unit of time (mm^3s^{-1}), A is the effective area of the membrane (mm^2), δ_0 is the thickness of the membrane (mm), and ΔP is the transmembrane pressure (Pa).

2.6. Membrane Filtration Experiments

Membrane filtration experiments were conducted using a dead-end filtration module (Sterlitech HP4750 Stirred Cell) with magnetic stirring at 500 rpm, using an effective membrane area of 14.6 cm^2. Initially, the membrane was compacted at 30 psi for 1 h; after this time, 100 mL of deionized pure water was permeated through the membrane to determine an initial flux J_{W1}, followed by three 100 mL filtration cycles of a 1.0 mg/mL BSA solution to obtain J_{BSA}. Finally, the membrane was cleaned with distilled water for 30 min, and the first permeate after cleaning was taken as water flux J_{w2}. This procedure allowed for the assessment of the flux recuperation ratio (FRR) and BSA flux recovery ratio (Rr), which were calculated using Equations (5) and (6) [27].

$$\text{FRR}(\%) = \frac{J_{w2}}{J_{w1}} \times 100 \quad (5)$$

$$\text{Rr}(\%) = \left(\frac{J_{w2} - J_{BSA}}{J_{w1}}\right) \times 100 \quad (6)$$

Moreover, total flux loss caused by the total fouling (Rt) and percentage of irreversible flux loss caused by irreversible fouling (Rir) were calculated using Equations (7) and (8), respectively.

$$\text{Rt}(\%) = \left(1 - \frac{J_{BSA}}{J_{w1}}\right) \times 100 \quad (7)$$

$$\text{Rir}(\%) = \left(\frac{J_{w1} - J_{w2}}{J_{w1}}\right) \times 100 \quad (8)$$

2.7. Salt and Dye Rejection Experiments

The salt and dye rejection methodologies were conducted in a similar way to the one described for the BSA filtration experiment in the Sterlitech HP4750 Stirred Cell (Auburn, WA, USA). For salt rejection, 100 mL of 0.5 mg/mL $MgSO_4$ solution or Na_2SO_4 solution (500 ppm) was permeated through R-HIPS membranes at 30 psi, whereas for dye rejection, a 100 mL solution containing 0.05 mg/mL of Reactive Black 5 (50 ppm) was used. The permeation and retention fluxes of salt solution were calculated using Equations (9) and (10):

$$J = \frac{V}{A \cdot t} \quad (9)$$

$$R(\%) = \left(1 - \frac{C_p}{C_f}\right) \times 100 \quad (10)$$

where C_p is the concentration of salt or dye in the permeate, and C_f is the salt or dye concentration in the feed. Salt rejection was measured using a conductometer. In the dye rejection experiments, concentrations were determined using a UV-Vis spectrometer at 595 nm.

3. Results and Discussion

The introduction of sulfonic groups (-SO_3H) into the polymer chain can significantly enhance important properties, such as hydrophilicity, permeability, morphology, and antifouling capacity [28,29]. In this study, the primary goal in modifying R-HIPS membranes was to make them less susceptible to fouling by changing them from a hydrophobic to hydrophilic nature through varying concentrations of -SO_3H. The experimental sulfonation degrees (SDs) of R-HIPS membranes ranged from 0% to 5.9%, closely aligning with the theoretical SD, as can be seen in Table 1. Initial experimental tests for membrane fabrication were conducted using a higher SD (10, 15, and 20%) of recycled HIPS. However, at these sulfonation levels, HIPS becomes partially soluble in water and alcohols, which complicates both the membrane formation process and their application in water treatment.

The reaction mechanism for HIPS sulfonation with chlorosulfonic acid is a typical electrophilic aromatic substitution, which introduces a sulfonic acid group on the benzene rings of the polystyrene fragment, as depicted in Scheme 1 [25].

FTIR spectroscopy was employed to verify the chemical structures of R-HIPS and sulfonated R-HIPS-3 and R-HIPS-5. Figure 1 shows that the aromatic C-H stretching vibrations appeared at 3082, 3058, and 3024 cm^{-1}. Additionally, the peaks centered at 2921 cm^{-1} and 2850 cm^{-1} corresponded to the symmetric and asymmetric stretching of C-H bonds of vinyl groups, respectively [14]. The spectra also displayed vibrations at 1601 cm^{-1} and 1492 cm^{-1}, which were attributed to aromatic C=C stretching. Furthermore, all spectra exhibited characteristic polybutadiene signals between 994 and 720 cm^{-1}, attributed to the deformation bands of vinyl-1,2, *trans*-1,4, and cis-1,4 configurations [30]. The peaks around 1184 and 1042 cm^{-1}, assigned to the antisymmetric and symmetric vibrations of the SO_3^- group [31], may overlap with absorption peaks from additives in R-HIPS, complicating the identification of those signals. Moreover, a broad peak between 3100 and 3650 cm^{-1} was observed, corresponding to O–H stretching vibrations from hydrogen bonds. This peak was intensified with an increasing sulfonation degree, indicating that higher sulfonation enhances polymer hydrophilicity.

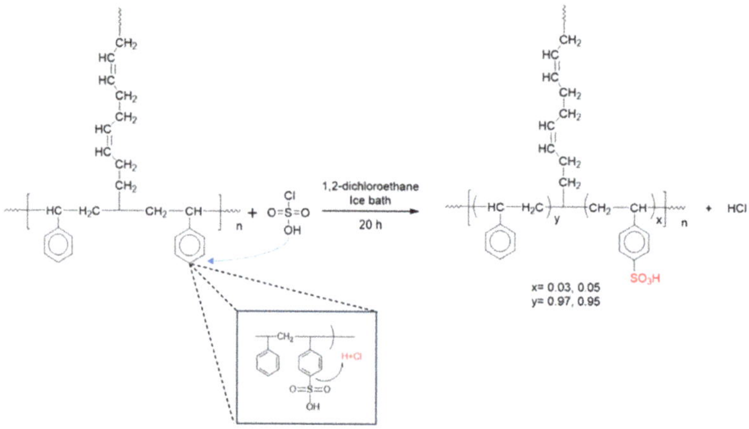

Scheme 1. Proposed reaction mechanism for HIPS sulfonation with CSA.

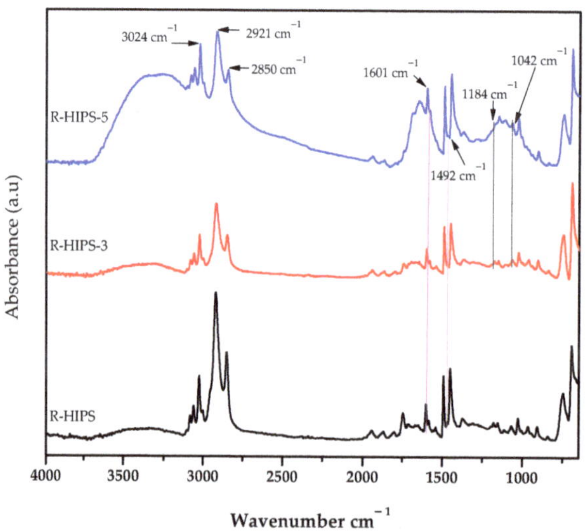

Figure 1. ATR-FTIR spectra of R-HIPS and sulfonated R-HIPS.

Changes in the hydrophobic character of R-HIPS due to sulfonation could modify the thermodynamic equilibrium for the R-HIPS solution behavior between the polymer/solvent/non-solvent system during the membrane formation process [25]. As reported in the literature, R-HIPS membranes with macrovoids in their structure and larger cavities in the sublayer are obtained using deionized water as non-solvent in asymmetric membrane preparation, due to an instantaneous demixing [14,15,19,20]. As a result, the formation of macrovoids and larger cavities can indeed compromise the mechanical stability of R-HIPS membranes.

Thus, it is important to consider alternative non-solvents or look into adjusting the casting conditions to obtain delayed demixing processes to avoid the formation of macrovoids and large cavities. In this case, after several tests, commercial vinegar/isopropanol (15/85% *v*/*v*) containing 0.12% $NaHCO_3$ was selected as the coagulation bath to reduce macropore formation in the membrane's cross-section and promote the formation of the asymmetric membrane with a sponge-like porous structure. This formulation closely follows those reported by Yam-Cervantes et al. for asymmetric membranes preparation from sulfonated polymers [24]. Additionally, this coagulation bath was chosen because it was not favorable to use water as a non-solvent for sulfonated membranes formation.

3.1. Membrane Morphology

Figure 2 displays the SEM images of R-HIPS, R-HIPS-3, and R-HIPS-5. The sublayer surfaces of R-HIPS (Figure 2a) exhibited micro-sized pores, while the sulfonated membranes R-HIPS-3 (Figure 2c) and R-HIPS-5 (Figure 2e) revealed completely dense surfaces. The cross-section morphologies of R-HIPS and sulfonated R-HIPS-3 and R-HIPS-5 revealed a sponge-like matrix with rounded and elongated macropores (Figure 2b,d,f), with this trend becoming more pronounced as the sulfonation degree increased. This phenomenon can be attributed to a delayed demixing process in the coagulation bath. This effect is due to hydrogen bonding interactions between sulfonic acid groups and water molecules present in the vinegar. According to Feng et al. [32], the presence of hydrogen bonding increases dope viscosity, which slows down the convection and diffusion processes between the solvent and nonsolvent during membrane formation. The increased viscosity can lead to a reduced droplet growth rate, resulting in a smaller pore size distribution within the membrane [33]. The morphology of sulfonated R-HIPS allows for better stability in the as-prepared membranes.

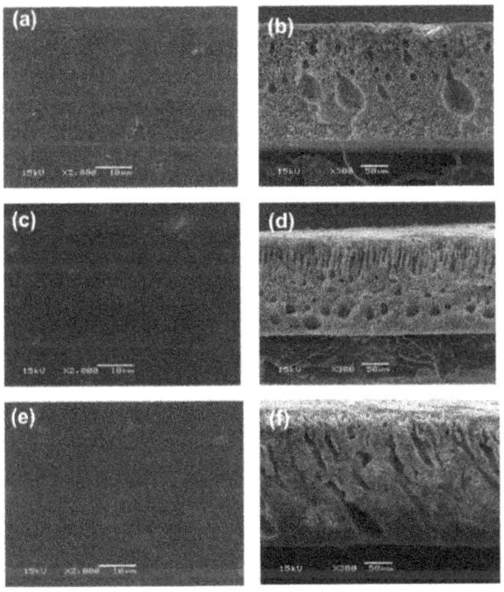

Figure 2. SEM images of surface and cross section at 2000× and 300× of (**a**,**b**) R-HIPS, (**c**,**d**) R-HIPS-3, and (**e**,**f**) R-HIPS-5, respectively.

Table 2 presents the % porosity (ε) and pore radius (r_m) values for R-HIPS and sulfonated R-HIPS membranes prepared in this work and their comparison with some recycled HIPS membranes reported in literature [15,21]. This study confirmed that the pore radius decreased slightly from 4.0 nm to 3.3 nm with an increasing sulfonation degree (Table 2), which also led to a reduction in membrane porosity (ε). However, no significant decrease in the finger-like structures within the cross-section of R-HIPS sulfonated membranes was observed, which may be attributed to $NaHCO_3$ addition that weakened the hydrogen bonding interactions between -SO_3H groups and water molecules [34]. The membrane porosity (ε), calculated using the dry–wet weight method, decreased from 75 for R-HIPS to 63.9% in R-HIPS-3. Increasing sulfonation for R-HIPS-5 showed a slight porosity increase (67.0%) in comparison with R-HIPS-3, which was attributed to the presence of elongated macropores. As shown in Table 2, for membranes from recycled HIPS reported in literature, termed HIPS-R [15], 30% and 35% [21] exhibited higher pore radii that led to membranes for ultrafiltration and microfiltration, resulting in increased flux and lower solutes retention. Thus, this larger pore size compromised the membrane selectivity and overall performance. The increase in pore size is attributed to the instantaneous demixing formation that occurs during the fabrication process when water is used as a non-solvent.

Table 2. Water contact angle, water uptake, weight loss, % porosity, radius pore size, and flux of R-HIPS membranes.

Membrane	Contac Angle (°)	Water Uptake (%)	Weight Loss (%)	Porosity (ε) (%)	r_m (nm)	Flux (L/m^{2}h)
R-HIPS	83.8 ± 1.034	2.2 ± 0.5	1.2± 0.33	75.0	4.0 ± 0.32	6.0 ± 0.84
R-HIPS-3	74.8 ± 0.26	3.6 ± 2.9	3.2 ± 0.61	63.9	3.8 ± 0.08	4.3 ± 0.13
R-HIPS-5	66.1 ± 0.87	10.6 ± 0.4	6.0 ± 0.95	67.0	3.3 ± 0.67	3.1 ± 1.02
HIPS-R [15]	81.7 ± 3.42	51.7 ± 5.5	-	52.3	32.4 ± 1.12	180.6 ± 0.71
30 wt% [21]	83.0	-	-	62.5	50–500	8.5
35 wt% [21]	91.0	-	-	61.6	50–250	8.3

3.2. Hydrophilicity

Water contact angle (WCA) was determined to verify changes in hydrophilicity for R-HIPS membrane surfaces (Figure 3). The contact angles decreased from 83.8° to 66.2° with an increasing sulfonation degree from 0 to 5%. This reduction was due to the presence of strongly hydrophilic -SO_3H groups, which enhanced membrane surface interaction with water, facilitating the filling of air voids and enhancing overall wetting [35]. In general, porosity, pore size, and hydrophilicity are key factors that determine water permeability and solute rejection properties of membranes [36]. Higher pore size generally allows for increased flux, but it may compromise the membrane ability to reject solutes. On the other hand, a significant challenge in filtration processes is membrane fouling, which occurs when unwanted materials accumulate on the membrane surface or within its pores, due to, among other factors, membrane hydrophobicity. Fouling not only reduces filtering efficiency but also decreases the average membrane life and increases operating costs [37]. Thus, these properties (porosity and hydrophilicity) are crucial for optimizing membrane performance in applications such as water treatment for dye and salt elimination and separation processes, where organic suspended matter is present.

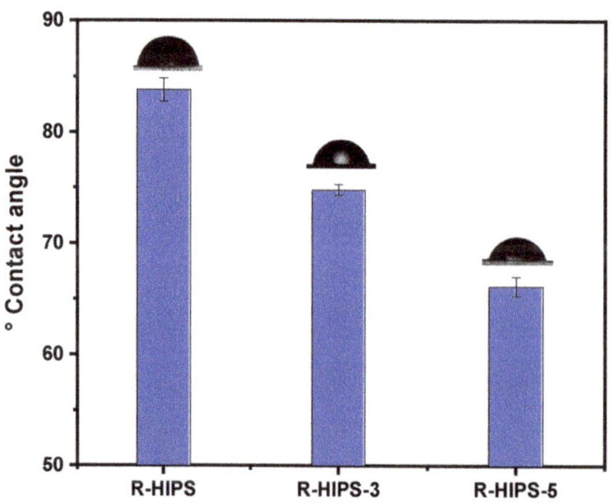

Figure 3. Water contact angles of R-HIPS, R-HIPS-3, and R-HIPS-5 membranes.

Furthermore, HIPS, R-HIPS-3, and R-HIPS-5 membrane swelling tests support the results found by contact angle measurements. HIPS, R-HIPS-3, and R-HIPS-5 showed a higher water uptake as SD increased (see Table 2). After immersion for 7 days in water, R-HIPS and R-HIPS-3 showed water uptake values of 2.2% and 3.6%, respectively, while R-HIPS-5 exhibited a significantly higher uptake of 10.6%, following a similar trend to the contact angle measurements. Moreover, weight losses after immersion were 1.2, 3.2, and 6.0% for R-HIPS, R-HIPS-3, and R-HIPS-5 membranes, respectively (Table 2).

3.3. Membrane Antifouling Properties and Water Permeability

The antifouling performances of R-HIPS, R-HIPS-3, and R-HIPS-5 membranes were evaluated by dynamic permeation experiments using a 1000 ppm BSA solution. The flux recovery ratio (FRR %), irreversible fouling (Rir), recovery capacity (Rr), and the total fouling (R_t) are presented in Figure 4. The flux recovery ratios (FRR) of sulfonated R-HIPS-5 and R-HIPS-3 membranes were greater than that of R-HIPS, indicating that increased hydrophilicity enhanced the antifouling properties of the membranes. The reversible fouling ratios (Rr), which refers to the temporary adsorption of molecules or particles onto a membrane surface, were 8.2%, 9.6%, and 15.6% for R-HIPS, R-HIPS-3, and R-HIPS-5, respectively. In contrast, the irreversible fouling ratio (Rir), which occurs when foulants such as proteins, oils, or particulates become permanently fixed onto the pore surfaces of a membrane [38], was lower in R-HIPS-5 membrane, following the same tendency (R-HIPS > R-HIPS-3 > R-HIPS-5) as the total fouling (R_t), which indicates that increasing sulfonation reduces the fouling in R-HIPS membranes. The notably lower Rr values compared to Rir suggest that particles tend to adhere more readily to the pore surfaces in the non-sulfonated R-HIPS membrane, compared to R-HIPS.

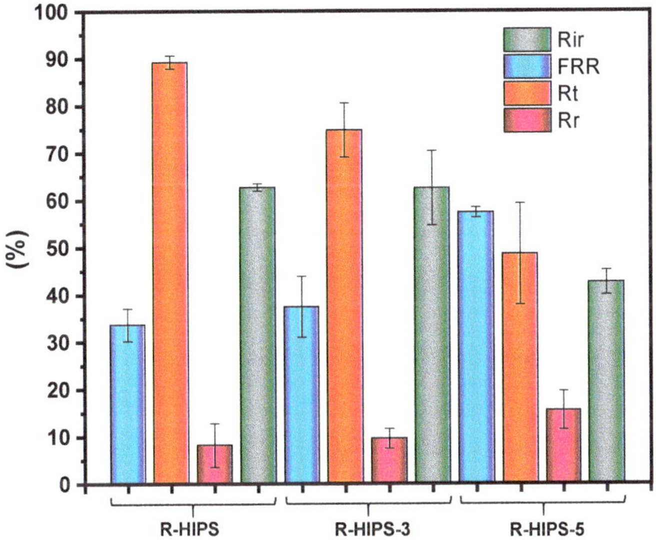

Figure 4. Antifouling performances observed with BSA solutions of R-HIPS, R-HIPS-3, and R-HIPS-5 membranes.

Also, in the dynamic permeation experiments, the R-HIPS membrane exhibited a 6.0 L/m^{2}h water flux, while the sulfonated membranes, R-HIPS-3 and R-HIPS-5, showed reduced fluxes at 4.3 L/m^{2}h and 3.1 L/m^{2}h, respectively. This reduction in water flux is consistent with the membrane porosity results, since a larger degree of sulfonation showed a reduced porosity.

3.4. Separation Performance

To evaluate the efficiency of the recycled HIPS membranes in inorganic and organic separation process, three different solutes were used: magnesium sulphate ($MgSO_4$, a univalent salt), sodium sulphate (Na_2SO_4, a divalent salt), and Reactive Black 5 dye. Figure 5 summarizes the separation performances of R-HIPS and sulfonated R-HIPS membranes, highlighting their effectiveness in filtering these selected solutes. The dye rejection in R-HIPS membranes correlated with a morphology of decreasing pore size, increasing hydrophilicity, and an enhanced negative charge in sulfonated membranes (see Figure 5). R-HIPS membrane exhibited the lowest rejection of Black 5 dye at 5.5%, which can be attributed to the micro-sized pores present on its surface. In contrast, R-HIPS-3 (57% dye rejection) and R-HIPS-5 membranes, which had the smallest pore size (3.3 nm), almost completely blocked the dye, achieving a 94% dye rejection. Due to the fact that R-HIPS sulfonated membranes and the dye have similar charges, the electrostatic repulsion between them can prevent the dye from permeating through the membrane [39]. Additionally, this repulsion helps to minimize the dye adsorption onto the membrane surface, which can contribute to reducing membrane fouling.

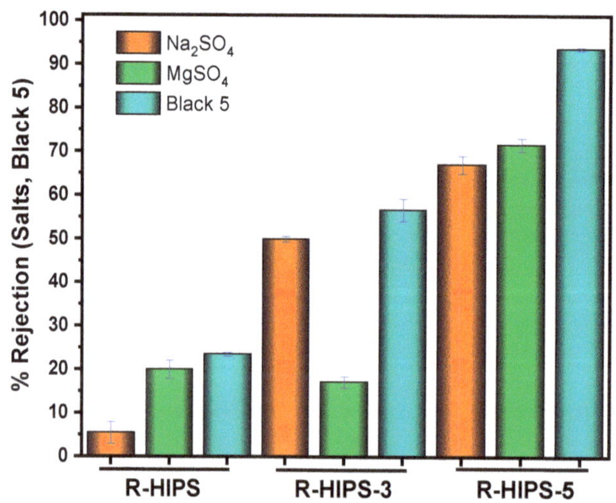

Figure 5. Salts and dye rejection performances of R-HIPS, R-HIPS-3, R-HIPS-5 membranes.

Regarding salts rejection, the R-HIPS membranes showed a tendency where R-HIPS-5 exhibited the highest rejection, followed by R-HIPS-3 and R-HIPS. This order can be attributed to pore size and surface properties. The smaller pore size of the R-HIPS-5 membrane contributed to its enhanced ability to reject divalent and monovalent salts, as smaller pores can physically block larger ions from passing through them. Additionally, the higher negative charge from the sulfonic acid groups enhanced electrostatic repulsion, particularly against positively charged ions, improving rejection efficiency [35].

Moreover, more hydrophilic membranes tend to have better antifouling properties, and they can interact more effectively with water, facilitating selective ion transport while reducing the passage of salts. Thus, the combination of smaller pore size and increased negative charge is crucial for the superior performance of R-HIPS-5 in rejecting salts.

4. Conclusions

It was found that chlorosulfonation in solution is an efficient method to prepare R-HIPS sulfonated material for membrane preparation. Sulfonated recycled HIPS asymmetric membranes were prepared using a phase inversion method, with a coagulation bath of commercial vinegar/isopropanol (15/85% v/v) and 0.12% $NaHCO_3$. This process resulted in the formation of asymmetric membranes, with pore sizes ranging from 4.0 to 3.3 nm and porosities between 75.0% and 63.9%, attributed to a delayed demixing process, particularly for R-HIPS sulfonated membranes. The degree of sulfonation significantly impacted the membrane's morphology, hydrophilicity, antifouling properties, and particle rejection capabilities for substances such as $MgSO_4$, Na_2SO_4, and Black 5 dye. As the sulfonation degree increased, R-HIPS, R-HIPS-3, and R-HIPS-5 membranes exhibited a decrease in pore size and porosity, with a slight reduction in water permeation flux, dropping from 6.0 L/m^2h in the R-HIPS membrane to 3.1 L/m^2h in the R-HIPS-5. Notably, higher sulfonation levels enhanced hydrophilicity, leading to improved antifouling properties and separation performance. In particular, R-HIPS-5 membrane demonstrated promising results, achieving a separation efficiency of 94% for Black 5, 72% for $MgSO_4$, and 67% for Na_2SO_4, with a high flux recovery ratio of 58%. These results are attributed to a smaller pore size and increased sulfonic acid group concentration in R-HIPS-3 and particularly R-HIPS-5 membranes, which enhanced their rejection efficiency.

Given the feasibility of producing flat membranes and their separation properties, these membranes could be useful for the production of spiral-wound modules. Scaling up could enhance their separation efficiency in large-scale wastewater treatment processes for the removal of dyes from water industrial stream and for industrial product purification applications. Recycling materials such as HIPS waste offers an alternative way to reduce plastic waste and reintroduce it into the circular economy using membranes for water purification, increasing its value.

Author Contributions: Conceptualization, M.O.G.-D. and M.A.-V.; Methodology, M.A.Y.-C., R.S.-S., S.D. and E.U.-F.; Validation, M.H.-I., D.P.-C. and E.U.-F.; Formal analysis, M.A.Y.-C. and R.S.-S.; Investigation, M.A.Y.-C., R.S.-S., S.D. and M.H.-I.; Resources, M.O.G.-D. and M.A.-V.; Writing—original draft preparation, M.A.Y.-C.; Writing—review and editing, M.O.G.-D., D.P.-C. and M.A.-V.; Supervision, M.O.G.-D.; Funding acquisition, M.O.G.-D. All authors have read and agreed to the published version of the manuscript.

Funding: This research was funded by Ciencia de Frontera CONAHCYT, grant number CF-2023-I-670.

Data Availability Statement: The original contributions presented in the study are included in the article, further inquiries can be directed to the corresponding author.

Acknowledgments: M. Yam-Cervantes acknowledges the CONAHCYT postdoctoral scholarship 96869. Moreover, the authors thank Jorge Uribe Calderon for the contact angle measurements and Silvia Andrade Canto for the SEM technical assistance.

Conflicts of Interest: The authors declare no conflicts of interest.

References

1. Huang, Z.; Shanmugam, M.; Liu, Z.; Brookfield, A.; Bennett, E.L.; Guan, R.; Herrera, D.E.V.; Lopez-Sanchez, J.A.; Slater, A.G.; McInnes, E.J.L.; et al. Chemical Recycling of Polystyrene to Valuable Chemicals via Selective Acid catalyzed Aerobic Oxidation under Visible Light. *J. Am. Chem. Soc.* **2022**, *144*, 6532–6542. [CrossRef] [PubMed]
2. Secretary of Environmental and Natural Resources of Mexico. Inventario Nacional de Fuentes de Contaminación Plástica, NFCP. Primera Edición. 2023, pp. 1–186. Available online: https://dsiappsdev.semarnat.gob.mx/datos/portal/publicaciones/2023/NFCP_2023.pdf (accessed on 21 September 2024).
3. Halim, A.; Gabriel, A.A.; Ismayati, M.; Nur, P.L.; Azizah, U. Expanded Polystyrene Waste Valorization as a Superhydrophobic Membrane for Oil Spill Remediation. *Waste Biomass Valor.* **2023**, *14*, 2025–2036. [CrossRef]
4. Tarek, S.; Nour, U.; El-Shazly, A.H. Morphology and Thermodynamic Study of a Novel Composite Membrane from Waste Polystyrene/Slag: Experimental Investigation. *ACS Omega* **2024**, *9*, 23512–23522. [CrossRef] [PubMed]
5. Khanzada, N.K.; Al-Juboori, R.A.; Khatri, M.; Ahmed, F.E.; Ibrahim, Y.; Hilal, N. Sustainability in Membrane Technology: Membrane Recycling and Fabrication Using Recycled Waste. *Membranes* **2024**, *4*, 14–52. [CrossRef]
6. Haqi, A.B.; Zulfi, A.; Nur'aini, S.; Sri Hartati, S.; Rafryanto, A.F.; Noviyanto, A.; Ahmad, A.D.; Feriyanto, D.; Khairurrijal, K. High-Efficiency Water Filtration by Electrospun Expanded Polystyrene Waste Nanofibers. *ACS Omega* **2023**, *8*, 23664–23672. [CrossRef]
7. Adamczak, M.; Kamińska, G.; Bohdziewicz, J. Application of Wasteolymers as Basic Material for Ultrafiltration Membranes Preparation. *Water* **2020**, *12*, 179. [CrossRef]

8. De Aquino, F.; Marques, P.; Sguizzato, A.; da Silva, E.; Lopes, M.; Guerra, V. Multifactorial evaluation of an ultrafast process electrospinning of recycled expanded polystyrene to manufacture high efficiency membranes for nanoparticle airfiltration. *J. Environ. Manag.* **2024**, *362*, 121352. [CrossRef]
9. Zhuang, G.-L.; Tseng, H.-H.; Wey, M.-Y. Feasiility of using waste polystyrene as a membrane material for gas separation. *Chem. Eng. Res. Des.* **2016**, *111*, 204–217. [CrossRef]
10. Tran, A.T.K.; Pham, T.T.; Nguyen, Q.H.; Hoang, N.T.T.; Bui, D.T.; Nguyen, M.T.; Nguyen, M.K.; Van der Bruggen, B. From waste disposal to valuable material: Sulfonating polystyrene waste for heavy metal removal. *J. Environ. Chem. Eng.* **2020**, *8*, 104302. [CrossRef]
11. Nguyen, Q.H.; Tran, A.T.K.; Hoang, N.T.T.; Tran, Y.T.H.; Nguyen, P.H.; Pham, T.T.; Nguyen, M.K.; Van der Bruggen, B. Plastic waste as a valuable resource: Strategy to remove heavy metals from wastewater in bench scale application. *Environ. Sci. Pollut. Res.* **2022**, *29*, 42074–42089. [CrossRef]
12. Asgari, H.; Ghavipanjeh, F.; Sabour, M.R.; Emadzadeh, D. Fabrication of pore-filling cation-exchange membrane from waste polystyrene and Spunbond Meltblown Spunbond (SMS) non-woven polypropylene fabric as the substrate. *Sci. Rep.* **2024**, *14*, 6399. [CrossRef] [PubMed]
13. Liu, S.-H.; Lu, C.-C.; Lin, C.-W.; Chang, S.-H. Rapid modification of waste expanded polystyrene with H_2SO_4/trace persulfate in one pot for effective adsorption of fluoroquinolone antibiotic and its regeneration. *Chemosphere* **2021**, *271*, 129529. [CrossRef] [PubMed]
14. Nady, N.; Abdel Rehim, M.H.; Badawy, A.A. Dye removal membrane from electrospun nanofibers of blended polybutylenesuccinate and sulphonated expanded polystyrene waste. *Sci. Rep.* **2023**, *13*, 15455. [CrossRef] [PubMed]
15. Garcia-Ivars, J.; Wang-Xu, X.; Iborra-Clar, M.I. Application of post-consumer recycled high-impact polystyrene in the preparation of phase-inversion membranes for low-pressure membrane processes. *Sep. Purif. Technol.* **2017**, *175*, 340–351. [CrossRef]
16. Sharma, S.K.; Khaudiyal, S.; Garg, N.; Das, S.K.; Kumar, S. Utilization of High Impact Polystyrene (HIPS) Plastic Waste as a Partial Replacement of Fine Aggregates in Lightweight Concrete T2. *IOP Conf. Ser. Earth Environ. Sci.* **2022**, *1086*, 012051. [CrossRef]
17. Ruziwa, D.; Chaukura, N.; Gwenzi, W.; Pumure, I. Removal of Zn^{2+} and Pb^{2+} ions from aqueous solution using sulphonated waste polystyrene. *J. Environ. Chem. Eng.* **2015**, *3*, 2528–2537. [CrossRef]
18. Zulfi, A.; Ahmad Hapidin, D.; Saputra, C.; Sinta Mustika, W.; Miftahul Munir, M.; Khairurrijal, K. The Synthesis of Fiber Membranes from High-Impact Polystyrene (HIPS) Waste using Needleless Electrospinning as Air Filtration Media. *Mater. Today Proc.* **2019**, *13*, 154–159. [CrossRef]
19. Khaled, M.; Noby, H.; Mansor, E.S.; Elshazly, A.H.; Aissa, W.A. Waste High Impact Polystyrene (HIPS) Microfiltration Membranes for Water Treatment: A Thorough Experimental Study. *Environ. Qual. Manag.* **2023**, *32*, 93–102. [CrossRef]
20. Khaled, M.; Noby, H.; Aissa, W.A.; El-Shazly, A.H. Facile Fabrication of Porous and Hydrophilic Polystyrene Membranes Using Recycled Waste. *Mater. Sci. Forum* **2020**, *1008*, 39–46. [CrossRef]
21. Khaled, M.; Noby, H.; Mansor, E.; El-Shazly, A.; Aissa, W. Water Purification using Recycled Polymeric Microfiltration Membranes. *Int. J. Appl. Energy Syst.* **2020**, *2*, 122–126. [CrossRef]
22. Giakoumakis, N.S.; Vos, C.; Janssens, K.; Vekeman, J.; Denayer, M.; De Proft, F.; Marquez, C.; De Vos, D.E. Total revalorization of high impact polystyrene (HIPS): Enhancing styrene recovery and upcycling of the rubber pase. *Green. Chem.* **2024**, *26*, 340–352. [CrossRef]
23. Mazumder, M.A.J.; Raja, P.H.; Isloor, A.M.; Usman, M.; Chowdhury, S.H.; Ali, S.A.; Inamuddin; Al-Ahmed, A. Assessment of sulfonated homo and co-polyimides incorporated polysulfone ultrafiltration blend membranes for effective removal of heavy metals and proteins. *Sci. Rep.* **2020**, *10*, 7049. [CrossRef] [PubMed]
24. Yam-Cervantes, M.A.; Santiago-García, J.L.; Loría-Bastarrachea, M.I.; Duarte-Aranda, S.; Alberto Ruiz-Treviño, F.; Aguilar-Vega, M. Sulfonated polyphenylsulfone asymmetric membranes: Effect of coagulation bath (acetic acid-$NaHCO_3$/isopropanol) on morphology and antifouling properties. *J. Appl. Polym. Sci.* **2016**, *134*, 45502. [CrossRef]
25. Idibie, C.A.; Abdulkareem, S.A.; Pienaar, C.H.V.; Iyuke, S.E.; Vandyk, L. Mechanism and Kinetics of Sulfonation of Polystyrene-Butadiene Rubber with Chlorosulfonic Acid. *Ind. Eng. Chem. Res.* **2010**, *49*, 1600–1604. [CrossRef]
26. Bekri-Abbes, I.; Bayoudh, S.; Baklouti, M.; Papon, E.; LeClercq, D. Converting Waste Polystyrene into Adsorbent: Optimisation of Reaction Parameters and Properties. *Prog. Rubber Plast. Recycl. Technol.* **2006**, *22*, 179–193. [CrossRef]
27. Zhu, J.; Zheng, J.; Zhang, Q.; Zhang, S. Antifouling ultrafiltration membrane fabricated from poly (arylene ether ketone) bearing hydrophilic hydroxyl groups. *J. Appl. Polym. Sci.* **2015**, *133*, 42809. [CrossRef]
28. Ayyaru, S.; Ahn, Y.H. Application of sulfonic acid group functionalized graphene oxide to improve hydrophilicity, permeability, and antifouling of PVDF nanocomposite ultrafiltration membranas. *J. Membr. Sci.* **2017**, *525*, 210–219. [CrossRef]
29. Palanisamy, G.; Thangarusa, A.; Oh, T.H. Effecct of sulfonated Inorganic Additives Incorporated Hydrid composite Polymer Membranes on Enhanced the performance of Microbial Fuel Cells. *Polymers* **2023**, *15*, 1294. [CrossRef]
30. Lacoste, J.; Delor, F.; Pilichowski, J.F.; Singh, R.P.; Prasad, A.V.; Sivara, S. Polybutadiene content and microstructure in high impact polystyrene. *J. Appl. Polym. Sci.* **1996**, *59*, 953–959. [CrossRef]
31. Yang, C.J.; Jablonsky, M.J.; Mays, J.W. NMR and FT-IR studies of sulfonated styrene-based homopolymers and copolymers. *Polymer* **2002**, *43*, 5125–5132. [CrossRef]
32. Feng, Y.; Han, G.; Zhang, L.; Chen, S.B.; Chung, T.S.; Weber, M.; Staudt, C.; Maletzko, C. Rheology and phase inversion behavior of polyphenylenesulfone (PPSU) and sulfonated PPSU for membrane formation. *Polymer* **2016**, *99*, 72–82. [CrossRef]

33. Othman, N.; Harruddin, N.; Idris, A.; Ooi, Z.-Y.; Fatiha, N.; Sulaiman, R.N.R. Fabrication of polypropylene membrane via thermally induced phase separation as a support matrix of tridodecylamine supported liquid membrane for Red 3BS dye removal. *Desalination Water Treat.* **2016**, *57*, 12287–12301. [CrossRef]
34. Shen, L.; Zhang, X.; Tian, L.; Li, Z.; Ding, C.; Yi, M.; Han, C.; Yu, X.; Wang, Y. Constructing substrate of low structural parameter by salt induction for high performance TFC-FO membranes. *J. Membr. Sci.* **2020**, *600*, 117866. [CrossRef]
35. Alsohaimi, I.H.; Alrashidi, A.N.; Hassan, H.M.A.; Chen, Q. Highly efficient ultrafiltration membrane performance of PES@microcrystalline cellulose extracted from waste fruits for the removal of BrO_3—From drinking water simples. *Colloid Interface Sci. Commun.* **2023**, *54*, 100718. [CrossRef]
36. Xu, R.; Wang, J.; Chen, D.D.; Yang, F.; Kang, J.; Xiang, M.; Lib, L.; Sheng, S. Preparation of pH-responsive asymmetric polysulfone ultrafiltration membranes with enhanced anti-fouling properties and performance by incorporating poly(2-ethyl-2-oxazoline) additive. *RSC Adv.* **2018**, *8*, 41270–41279. [CrossRef] [PubMed]
37. Tarun, S.S.; Dakshesh, S.K.; Arthanareeswaran, G. Tuning of polymeric membranes to mitigate fouling and removal of dissolved compounds for wastewater treatment: A Review. *Front. Membr. Sci. Technol.* **2024**, *3*, 1389239. [CrossRef]
38. Huhn-Ibarra, M.J.; Loría-Bastarrachea, M.I.; Duarte-Aranda, S.; Montes-Luna, A.J.; Ortiz-Espinoza, J.; González-Díaz, M.O.; Aguilar-Vega, M. PPSU dual layer hollow fiber mixed matrix membranas with functionalized MWCNT for enhanced antifouling, salt and dye rejection in water treatment. *J. Appl. Polym. Sci.* **2022**, *139*, e53203. [CrossRef]
39. Elias Jigar Sisay, E.J.; Al-Tayawi, A.N.; László, Z.; Kertész, S. Recent Advances in Organic Fouling Control and Mitigation Strategies in Membrane Separation Processes: A Review. *Sustainability* **2023**, *15*, 13389. [CrossRef]

MDPI AG
Grosspeteranlage 5
4052 Basel
Switzerland
Tel.: +41 61 683 77 34

Membranes Editorial Office
E-mail: membranes@mdpi.com
www.mdpi.com/journal/membranes